Jon E. Lewis is the editor of the best-selling *The Mammoth Book of True War Stories Volume 1*, a former *Daily Mirror* "Book of the Week".

Also available

The Mammoth Book of 20th Century Science Fiction
The Mammoth Book of Awesome Comic Fantasy
The Mammoth Book of Best New Erotica 3
The Mammoth Book of Best New Horror 14
The Mammoth Book of Best New Science Fiction 16
The Mammoth Book of British Kings & Queens
The Mammoth Book of Celtic Myths and Legends
The Mammoth Book of Chess
The Mammoth Book of Fighter Pilots
The Mammoth Book of Future Cops
The Mammoth Book of Great Detective Stories
The Mammoth Book of Great Inventions
The Mammoth Book of Hearts of Oak
The Mammoth Book of Heroes
The Mammoth Book of Historical Whodunnits
The Mammoth Book of How It Happened
The Mammoth Book of How It Happened: America
The Mammoth Book of How It Happened: Ancient Egypt
The Mammoth Book of How It Happened: Ancient Rome
The Mammoth Book of How It Happened: Battles
The Mammoth Book of How It Happened In Britain
The Mammoth Book of International Erotica
The Mammoth Book of Jack the Ripper
The Mammoth Book of Jokes
The Mammoth Book of Maneaters
The Mammoth Book of Mountain Disasters
The Mammoth Book of Native Americans
The Mammoth Book of New Terror
The Mammoth Book of Private Eye Stories
The Mammoth Book of Prophecies
The Mammoth Book of Pulp Action
The Mammoth Book of Roaring Twenties Whodunnits
The Mammoth Book of Roman Whodunnits
The Mammoth Book of Roaring Twenties Whodunnits
The Mammoth Book of SAS & Elite Forces
The Mammoth Book of Science Fiction Century Volume 1
The Mammoth Book of Seriously Comic Fantasy
The Mammoth Book of Sex, Drugs & Rock 'n' Roll
The Mammoth Book of Special Forces
The Mammoth Book of Storms & Sea Disasters
The Mammoth Book of Short Erotic Novels
The Mammoth Book of Sword and Honour
The Mammoth Book of The Edge
The Mammoth Book of Travel in Dangerous Places
The Mammoth Book of True War Stories
The Mammoth Book of UFOs
The Mammoth Book of Vampires
The Mammoth Book of War Correspondents
The Mammoth Book of Women Who Kill
The Mammoth Encyclopedia of Extraterrestrial Encounters
The Mammoth Encyclopedia of Modern Crime Fiction
The Mammoth Encyclopedia of Unsolved Mysteries

THE MAMMOTH BOOK OF

True War Stories

Edited by Jon E. Lewis

ROBINSON
London

ROBINSON

First published in Great Britain by Robinson,
an imprint of Constable & Robinson Ltd, 2005

Reprinted by Robinson in 2017

5 7 9 10 8 6

A CIP catalogue record for this book
is available from the British Library.

UK ISBN: 978-1-84529-148-8

Printed and bound in Great Britain by
CPI Group (UK) Ltd., Croydon CR0 4YY

Papers used by Robinson are from well-managed forests
and other responsible sources

Robinson
An imprint of
Little, Brown Book Group
Carmelite House
50 Victoria Embankment
London EC4Y 0DZ

An Hachette UK Company
www.hachette.co.uk

www.littlebrown.co.uk

Contents

Introduction

War is Hell, the general said. Its glory is all moonshine.

Sherman's observation is proven on almost every page of this anthology of war, which runs the chronological arch of combat from the siege of Troy to the Gulf War. In these nearly 4,000 years, there have been no wars fought in spotless uniforms for spotless causes in which people do not die.

War *is* hell. But as the general might have gone on to say, in some men and women war brings out the best. This is the paradox of war: it is our most evil activity but also allows us to demonstrate our greatest virtues – courage, loyalty, compassion, faith, and duty.

Hell and Heroism. War has both these, and every shade of human experience in between. In his *Recollections*, the British veteran of the Napoleonic War Rifleman Harris made the observation: "The field of death and slaughter, the march, the bivouac, and the retreat, are no bad places in which to judge of men". He was right. If you want to know what it is to be human, read on.

But the true war stories (accounts might be a better term) collected here illuminate not only the face of man but the face of war, ranging as they do from battlefield analyses by historians to memoirs of combat by officers on the Western Front to *reportage* of Vietnam by such a master journalist as Nicholas Tomalin. It behoves us to understand war. How can we ask soldiers to fight and die for us if we do not even appreciate what they do and go through? And of course, if you think about it, there is nothing more important in human affairs than war. As the Ancient Chinese military theorist Sun Tzu noted in the *Art of War*, 400 BC: "War is a matter of vital importance . . . a matter of life or death; the road either to survival or ruin."

He too was right.

Tim O'Brien

FNG

Tim O'Brien was drafted into the US Army in 1968 and served in Vietnam as an infantryman.

The summer of 1968, the summer I turned into a soldier, was a good time for talking about war and peace. Eugene McCarthy was bringing quiet thought to the subject. He was winning votes in the primaries. College students were listening to him, and some of us tried to help out. Lyndon Johnson was almost forgotten, no longer forbidding or feared; Robert Kennedy was dead but not quite forgotten; Richard Nixon looked like a loser. With all the tragedy and change that summer, it was fine weather for discussion.

And, with all of this, there was an induction notice tucked into a corner of my billfold.

So with friends and acquaintances and townspeople, I spent the summer in Fred's antiseptic cafe, drinking coffee and mapping out arguments on Fred's napkins. Or I sat in Chic's tavern, drinking beer with kids from the farms. I played some golf and tore up the pool table down at the bowling alley, keeping an eye open for likely-looking high school girls.

Late at night, the town deserted, two or three of us would drive a car around and around the town's lake, talking about the war, very seriously, moving with care from one argument to the next, trying to make it a dialogue and not a debate. We covered all the big questions: justice, tyranny, self-determination, conscience and the state, God and war and love.

College friends came to visit: "Too bad, I hear you're drafted. What will you do?"

I said I didn't know, that I'd let time decide. Maybe something would change, maybe the war would end. Then we'd turn to discuss the matter, talking long, trying out the questions, sleeping late in the mornings.

The summer conversations, spiked with plenty of references to the philosophers and academicians of war, were thoughtful and long and complex and careful. But, in the end, careful and precise argumentation hurt me. It was painful to tread deliberately over all the axioms and assumptions and corollaries when the people on the town's draft board were calling me to duty, smiling so nicely.

"It won't be bad at all," they said. "Stop in and see us when it's over."

So to bring the conversations to a focus and also to try out in real words my secret fears, I argued for running away.

I was persuaded then, and I remain persuaded now, that the war was wrong. And since it was wrong and since people were dying as a result of it, it was evil. Doubts, of course, hedged all this: I had neither the expertise nor the wisdom to synthesize answers; most of the facts were clouded, and there was no certainty as to the kind of government that would follow a North Vietnamese victory or, for that matter, an American victory, and the specifics of the conflict were hidden away – partly in men's minds, partly in the archives of government, and partly in buried, irretrievable history. The war, I thought, was wrongly conceived and poorly justified. But perhaps I was mistaken, and who really knew, anyway?

Piled on top of this was the town, my family, my teachers, a whole history of the prairie. Like magnets, these things pulled in one direction or the other, almost physical forces weighting the problem, so that, in the end, it was less reason and more gravity that was the final influence.

My family was careful that summer. The decision was mine and it was not talked about. The town lay there, spread out in the corn and watching me, the mouths of old women and Country Club men poised in a kind of eternal readiness to find fault. It was not a town, not a Minneapolis or New York, where the son of a father can sometimes escape scrutiny. More, I owed the prairie something. For twenty-one years I'd lived under its laws, accepted its education, eaten its food, wasted and guzzled its water, slept well at night, driven across its highways, dirtied and breathed its air, wallowed in its luxuries. I'd played on its Little League teams. I remembered Plato's *Crito*, when Socrates, facing certain death – execution, not war – had the chance to escape. But he reminded himself that he had seventy years in which he could have left the country, if he were not satisfied or felt the agreements he'd made with it were unfair. He had not chosen Sparta or Crete. And, I reminded myself, I hadn't thought much about Canada until that summer.

The summer passed this way. Gold afternoons on the golf course, a comforting feeling that the matter of war would never touch me, nights in the pool hall or drug store, talking with towns-folk, turning the questions over and over, being a philosopher.

Near the end of the summer the time came to go to the war. The family indulged in a cautious sort of Last Supper together, and afterward my father, who is brave, said it was time to report at the bus depot. I moped down to my bedroom and looked the place over, feeling quite stupid, thinking that my mother would come in there in a day or two and probably cry a little. I trudged back up to the kitchen and put my satchel down. Everyone gathered around, saying so long and good health and write and let us know if you want anything. My father took up the induction papers, checking on times and dates and all the last-minute things, and when I pecked my mother's face and grabbed the satchel for comfort, he told me to put it down, that I wasn't supposed to report until tomorrow.

After laughing about the mistake, after a flush of red color and a flood of ribbing and a wave of relief had come and gone, I took a long drive around the lake, looking again at the place. Sunset Park, with its picnic table and little beach and a brown wood shelter and some families swimming. The Crippled Children's School. Slater Park, more kids. A long string of split-level houses, painted every color.

The war and my person seemed like twins as I went around the town's lake. Twins grafted together and forever together, as if a separation would kill them both.

The thought made me angry.

In the basement of my house I found some scraps of cardboard and paper. With devilish flair, I printed obscene words on them, declaring my intention to have no part of Vietnam. With delightful viciousness, a secret will, I declared the war evil, the draft board evil, the town evil in its lethargic acceptance of it all. For many minutes, making up the signs, making up my mind, I was outside the town. I was outside the law, all my old ties to my loves and family broken by the old crayon in my hand. I imagined strutting up and down the sidewalks outside the depot, the bus waiting and the driver blaring his horn, the *Daily Globe* photographer trying to push me into line with the other draftees, the frantic telephone calls, my head buzzing at the deed.

On the cardboard, my strokes of bright red were big and ferocious looking. The language was clear and certain and burned with a hard, defiant, criminal, blasphemous sound. I tried reading it aloud.

Later in the evening I tore the signs into pieces and put the shreds in the garbage can outside, clanging the gray cover down and trapping the messages inside. I went back into the basement. I slipped the crayons into their box, the same stubs of color I'd used a long time before to chalk in reds and greens on Roy Rogers' cowboy boots.

I'd never been a demonstrator, except in the loose sense. True, I'd taken a stand in the school newspaper on the war, trying to show why it seemed wrong. But, mostly, I'd just listened.

"No war is worth losing your life for," a college acquaintance used to argue. "The issue isn't a moral one. It's a matter of efficiency: what's the most efficient way to stay alive when your nation is at war? That's the issue."

But others argued that no war is worth losing your country for, and when asked about the case when a country fights a wrong war, those people just shrugged.

Most of my college friends found easy paths away from the problem, all to their credit. Deferments for this and that. Letters from doctors or chaplains. It was hard to find people who had to think much about the problem. Counsel came from two main quarters, pacifists and veterans of foreign wars.

But neither camp had much to offer. It wasn't a matter of peace, as the pacifists argued, but rather a matter of when and when not to join others in making war. And it wasn't a matter of listening to an ex-lieutenant colonel talk about serving in a right war, when the question was whether to serve in what seemed a wrong one.

On August 13, I went to the bus depot. A Worthington *Daily Globe* photographer took my picture standing by a rail fence with four other draftees.

Then the bus took us through corn fields, to little towns along the way – Lismore and Rushmore and Adrian – where other recruits came aboard. With some of the tough guys drinking beer and howling in the back seats, brandishing their empty cans and calling one another "scum" and "trainee" and "GI Joe," with all this noise and hearty farewelling, we went to Sioux Falls. We spent the night in a YMCA. I went out alone for a beer, drank it in a corner booth, then I bought a book and read it in my room.

By noon the next day our hands were in the air, even the tough guys. We recited the proper words, some of us loudly and daringly and others in bewilderment. It was a brightly lighted room, wood paneled. A flag gave the place the right colors, there was some smoke in the air. We said the words, and we were soldiers.

I'd never been much of a fighter. I was afraid of bullies. Their ripe muscles made me angry: a frustrated anger. Still, I deferred to no one. Positively lorded myself over inferiors. And on top of that was the matter of conscience and conviction, uncertain and surface-deep but pure nonetheless: I was a confirmed liberal, not a pacifist; but I would have cast my ballot to end the Vietnam war immediately, I would have voted for Eugene McCarthy, hoping he would make peace. I was not soldier material, that was certain.

But I submitted. All the personal history, all the midnight conversations and books and beliefs and learning, were crumpled by abstention, extinguished by forfeiture, for lack of oxygen, by a sort of sleepwalking default. It was no decision, no chain of ideas or reasons, that steered me into the war.

It was an intellectual and physical stand-off, and I did not have the energy to see it to an end. I did not want to be a soldier, not even an observer to war. But neither did I want to upset a peculiar balance between the order I knew, the people I knew, and my own private world. It was not that I valued that order. But I feared its opposite, inevitable chaos, censure, embarrassment, the end of everything that had happened in my life, the end of it all.

And the stand-off is still there. I would wish this book could take the form of a plea for everlasting peace, a plea from one who knows, from one who's been there and come back, an old soldier looking back at a dying war.

That would be good. It would be fine to integrate it all to persuade my younger brother and perhaps some others to say no to wars and other battles.

Or it would be fine to confirm the odd beliefs about war: it's horrible, but it's a crucible of men and events and, in the end, it makes more of a man out of you.

But, still, none of these notions seems right. Men are killed, dead human beings are heavy and awkward to carry, things smell different in Vietnam, soldiers are afraid and often brave, drill sergeants are boors, some men think the war is proper and just and others don't and most don't care. Is that the stuff for a morality lesson, even for a theme?

Do dreams offer lessons? Do nightmares have themes, do we awaken and analyze them and live our lives and advise others as a result? Can the foot soldier teach anything important about war, merely for having been there? I think not. He can tell war stories.

In advanced infantry training, the soldier learns new ways to kill people.

Claymore mines, booby traps, the M-60 machine gun, the M-70 grenade launcher. The old .45-caliber pistol. Drill sergeants give lessons on the M-16 automatic rifle, standard weapon in Vietnam.

On the outside, AIT looks like basic training. Lots of push-ups, lots of shoe-shining and firing ranges and midnight marches. But AIT is not basic training. The difference is inside the new soldier's skull, locked to his brain, the certainty of being in a war, pending doom that comes in with each day's light and stays with him all the day long.

The soldier in advanced infantry training is doomed, and he knows it and thinks about it. War, a real war. The drill sergeant said it when we formed up for our first inspection: every swinging dick in the company was now a foot soldier, a grunt in the United States Army, the infantry, Queen of Battle. Not a cook in the lot, not a clerk or mechanic among us. And in eight weeks, he said, we were all getting on a plane that would fly to a war.

The man who finds himself in AIT is doomed, and he knows it and thinks about it. There are no more hopes of being made into a rear-echelon trooper. The drill sergeant said it when we formed up for our first inspection: every swingin' dick in the company was now a foot soldier, a grunt in the United States Army. Not a cook or typist in the lot. And in eight weeks, he said, we were all getting on a plane bound for Vietnam.

"I don't want you to mope around thinkin' about Germany or London," he told us. "Don't even think about it, 'cause there just ain't no way. You're leg men now, and we don't need no infantry in Piccadilly or Southampton. Besides, Vietnam ain't all that bad. I been over there twice now, and I'm alive and still screwin' everything in sight. You troops pay attention to the trainin' you get here, and every swingin' dick will be back in one piece, believe me. Just pay attention, try to learn something. The Nam, it ain't so bad, not if you got your shit together."

One of the trainees asked him about rumors that said we would be shipped to Frankfort.

"Christ, you'll hear the crap till it makes you puke. Every swingin' dick is going to Nam, every big fat swingin' dick."

During the first month, I learned that FNG meant "fuckin' new guy," and that I would be one until the Combat Center's next shipment arrived. I learned that GIs in the field can be as lazy and careless and stupid as GIs anywhere. They don't wear helmets and armored vests unless an officer insists; they fall asleep on guard, and for the most part, no one really cares; they throw away or bury ammunition if it gets heavy and hot. I learned that REMF means "rear echelon motherfucker"; that a man is getting "Short" after his third or fourth month; that a hand grenade is really a "frag"; that one bullet is all it takes and that "you never hear the shot that gets you"; that no one in Alpha Company knows or cares about the cause or purpose of their war: it is about "dinks and slopes," and the idea is simply to kill them or avoid them. Except that in Alpha you don't kill a man, you "waste" him. You don't get mangled by a mine, you get fucked up. You don't call a man by his first name – he's the Kid or the Water Buffalo, Buddy Wolf or Buddy Barker or Buddy Barney, or if the fellow is bland or disliked, he's just Smith or Jones or Rodríguez. The NCOs who go through a crash two-month program to earn their stripes are called "instant NCOs"; hence the platoon's squad leaders were named Ready Whip, Nestle's Quick, and Shake and Bake. And when two of them – Tom and Arnold – were killed two months later, the tragedy was somehow mitigated and depersonalized by telling ourselves that ol' Ready Whip and Quick got themselves wasted by the slopes. There was Cop – an Irish fellow who wanted to join the police force in Danbury, Connecticut – and Reno and the Wop and the College Joe. You can go through a year in Vietnam and live with a platoon of sixty or seventy people, some going and some coming, and you can leave without knowing more than a dozen complete names, not that it matters.

Mad Mark was the platoon leader, a first lieutenant and a Green Beret. It was hard to tell if the name or the reason for the name came first. The madness in Mad Mark, at any rate, was not a hysterical, crazy, into-the-brink, to-the-fore madness. Rather, he was insanely calm. He never showed fear. He was a professional soldier, an ideal leader of men in the field. It was that kind of madness, the perfect guardian for the Platonic Republic. His attitude and manner seemed perfectly molded in the genre of the CIA or KCB operative.

This is not to say that Mad Mark ever did the work of the assassin. But it was his manner, and he cultivated it. He walked with a lanky, easy, silent, fearless stride. He wore tiger fatigues, not for their camouflage but for their look. He carried a shotgun – a weapon I'd thought was outlawed in international war – and the shotgun itself was a measure of his professionalism, for to use it effectively requires an exact blend of courage and skill and self-confidence. The weapon is neither accurate nor lethal at much over seventy yards. So it shows the skill of the carrier, a man who must work his way close enough to the prey to make a shot, close enough to see the enemy's retina and the tone of his skin. To get that close requires courage and self-confidence. The shotgun is not an automatic weapon. You must hit once, on the first shot, and the hit must kill. Mad Mark once said that after the war and in the absence of other U.S. wars he might try the mercenary's life in Africa.

He did not yearn for battle. But neither was he concerned about the prospect. Throughout the first month, vacationing on the safe beaches, he did precisely what the mission called for: a few patrols, a few ambushes, staying ready to react, watching for signs of a rocket attack on Chu Lai. But he did not take the mission to excess. Mad Mark was not a fanatic. He was not gung-ho, not a man in search of a fight. It was more or less an Aristotelian ethic that Mad Mark practiced: making war is a necessary and natural profession. It is natural, but it is only a profession, not a crusade: "Hunting is a part of that art; and hunting might be practiced – not only against wild animals, but also against human beings who are intended by nature to be ruled by others and refuse to obey that intention – because war of this order is naturally just." And, like Aristotle, Mad Mark believed in and practiced the virtue of moderation, so he did what was necessary in war, necessary for an officer and platoon leader in war, and he did no more or less.

He lounged with us during the hot days, he led a few patrols and ambushes, he flirted with the girls in our caravan, and, with a concern for only the basics of discipline, he allowed us to enjoy the holiday. Lying in the shade with the children, we learned a little Vietnamese, and they learned words like "motherfucker" and "gook" and "dink" and "tit." Like going to school.

It was not a bad war until we sent a night patrol into a village called Tri Binh 4. Mad Mark led it, taking only his shotgun and five other men. They'd been gone for an hour. Then came a burst of fire and a radio call that they'd opened up on some VC smoking and talking by a well. In ten minutes they were out of the village and back with the platoon.

The Kid was ecstatic. "Christ! They were right out there, right in the open, right in the middle of the ville, in a little clearing, just sitting on their asses! Shit, I almost shit! Ten of 'em, just sitting there. Jesus, we gave 'em hell. Damn, we gave it to 'em!" His face was on fire in the night, his teeth were flashing, he was grinning himself out of his skin. He paced back and forth, wanting to burst.

"Jesus," he said. "Show 'em the ear we got! Let's see the ear!"

Someone turned on a flashlight. Mad Mark sat cross-legged and unwrapped a bundle of cloth and dangled a hunk of brown, fresh human ear under the yellow beam of light. Someone giggled. The ear was clean of blood. It dripped with a little water, as if coming out of a bathtub. Part of the upper lobe was gone. A band of skin flopped away from the ear, at the place where the ear had been held to a man's head. It looked alive. It looked like it would move in Mad Mark's hands, as if it might make a squirm for freedom. It seemed to have the texture of a hunk of elastic.

"Christ, Mad Mark just went up and sliced it off the dead dink! No wonder he's Mad Mark, he did it like he was cuttin' sausages or something."

"What are you gonna do with it? Why don't you eat it, Mad Mark?"

"Bullshit, who's gonna eat a goddamn dink. I eat women, not dead dinks."

"We got some money off the gook, too. A whole shitload."

One of the men pulled out a roll of greasy piasters, and the members of the patrol split it up and pocketed it; then they passed the ear around for everyone to look at.

Mad Mark called in gunships. For an hour the helicopters strafed and rocketed Tri Binh 4. The sky and the trees and the hillsides were lighted up by spotlights and tracers and fires. From our position we could smell the smoke coming from Tri Binh 4. We heard cattle and chickens dying. At two in the morning we started to sleep, one man at a time. Tri Binh 4 turned curiously quiet and dark, except for the sound and light of a last few traces of fire. Smoke continued to billow over to our position all night, however, and when I awakened every hour, it was the first thing to sense and to remind me of the ear. In the morning another patrol was sent into the village. The dead VC was still there, stretched out on his back with his eyes closed and his arms folded and his head cocked to one side so that you could not see where the ear was gone. Little fires burned in some of the huts, and dead animals lay about, but there were no people. We searched Tri Binh 4, then burned most of it down.

Keith Douglas

First Action

Keith Douglas was commissioned in 1940, but had to wait until 1942 for his first tank battle – which came at Alamein in November of that year. Douglas was killed by a shell fragment in Normandy in June 1944.

At five o'clock I woke the Colonel, who lay in his opulent sleeping-bag, in his pyjamas, his clothes and suede boots neatly piled beside him; a scent of pomade drifted from him as he sat up. I told him the time and about the snipers, and handed him over to his batman who already hovered behind me with a cup of tea. I went about stirring the sleeping cocoons of men with my foot. On the way I woke John and said, "Have some whisky. I suppose we've sunk pretty low, taking it for breakfast." Unfortunately there was no more chocolate.

By six o'clock the wireless in every tank was switched on, engines were running, and at six-fifteen, through a thick morning mist, the Crusader squadron began to move out in close formation ahead of the regiment. Andrew had relayed rather vague orders to me: but the only thing that seemed clear to me was that there was now no one between us and the enemy. If that were so, it seemed crazy to go swanning off into the mist; but I was fairly certain it was not free from doubt, because I knew there had been a traffic of one or two vehicles passing through our lines in the early morning – and they were soft-skinned vehicles, not tanks. Presently, as I moved slowly forward, keeping one eye on the vague shape of Andrew's tank in the mist to my left, I saw on my right a truck, with its crew dismounted. I reported it to Andrew, and cruised across to investigate it. It was, of course, a British truck, whose driver told us there was a whole unit of soft vehicles ahead of us, and as far as he knew, no enemy in the immediate neighbourhood.

There was no more mention of snipers, and I imagined these would have been part of some kind of patrol who had now returned to their own lines.

Andrew now began to call me impatiently over the air: "Nuts five, Nuts five, you're miles behind. Come on. Come on. Off." Speeding up, we saw the shape of a tank looming ahead of us again, and made for it. As we came nearer, it was recognizable as a German derelict. I had not realized how derelicts can complicate manœuvres in a bad light. We increased speed again; but there seemed to be no one ahead of us. I began to suppose we had passed Andrew in the mist, and realized that we were lost, without any information of our position or objective. In fact, the regiment had made a sharp turn left while we were halted, and if Andrew had mentioned this to me over the air we could have found them easily. As it was, we continued to move vaguely round until the mist cleared. Seeing some Crusaders on our left when it grew clear enough to pick up objects at a distance, we approached them: they belonged to one of the other regiments in the Brigade, and had no idea (although their colonel was in one of the tanks) on which side of them our regiment was moving. How all this came about I am not sure, because I afterwards found Brigade and regimental orders to be very clear, and there was never an occasion in later actions when every member of a tank crew did not know what troops were on his right and left. This was, however, only the third time the regiment had ever seen action as a tank unit, and I was probably not so far behind the others in experience as I felt.

Meanwhile we rushed eagerly towards every Crusader, like a short-sighted little dog who has got lost on the beach. Andrew continued to call up with such messages as: "Nuts three, Nuts three I still can't see you. Conform. Conform. Off." I perceived that two other tanks of the squadron had attached themselves to me and were following me slavishly about, although the other tank of my own troop was nowhere to be seen.

Another Crusader several hundred yards away attracted our attention, and we rushed towards it, floundering over slit trenches and passing through some of our own infantry. As we approached another trench, I was too late to prevent the driver from running over a man in black overalls who was leaning on the parapet. A moment before the tank struck him I realized he was already dead; the first dead man I had ever seen. Looking back, I saw he was a Negro. "Libyan troops," said Evan. He was pointing. There were several of them scattered about, their clothes soaked with dew; some lacking limbs, although no flesh of these was visible, the clothes seeming to have wrapped themselves round the places where arms, legs, or even heads should have been, as though with an instinct for decency. I have noticed this before in photographs of people killed by explosive.

The Crusader which had attracted our attention was newly painted, covered with bedding and kit: tin hats and binoculars hung on the outside of the turret, and a revolver lay on the turret flap. Although it was outwardly undamaged, we saw that it had been abandoned. As my own field-glasses were old and quite useless, I told Evan to get out and bring the ones hanging on the derelict. He was very reluctant. "It might be a booby trap, sir," said he, rolling his eyes at me. This seemed unlikely at that stage of the battle, and I said, not very sympathetically, "Well, have a look first and make sure nothing's attached to them. And if nothing is, get them." Very gingerly, he climbed on to the other turret, and returned with the glasses. While he was getting them, I had at last caught sight of the regiment and we moved across to them, with our two satellites.

We took position on the right, the Crusaders still lying in front of the regiment, and my own tank being near a derelict Italian M13, apparently no more damaged than the tank we had just left, and covered with a camouflage of scrub. Two burnt-out German tanks stood about fifty yards apart some four hundred yards away to our right front. The other Crusaders were spaced out away to the left of us and over into some dead ground. Nothing seemed to be happening at the moment of our arrival.

I had a look through my new field-glasses: they were certainly an improvement on the ones that had been issued to me. I thought I could make out some lorries and men moving about on the far skyline, and reported them. Two or three other tanks confirmed this and said they could see them too. Unfortunately, our R.H.A. Battery, which had been withdrawn for barrage work, was not yet back with us and these vehicles were out of range of our seventy-fives and six-pounders. So we continued to sit there. Evan produced a thriller and found his place in it. Mudie asked me to pass him a biscuit. I took one myself, and cut us a sliver of cheese each from the tin. I took off my greatcoat and draped it over the turret. We seemed to have settled down for the morning and I began to wonder when we should get a chance to brew up.

I was disturbed from a mental journey through the streets of Jerusalem by the shriek and crash of a shell which threw up dark grey smoke and flame near one of the heavy tanks. During the next hour these shells continued to arrive, with the same tearing and shunting noises as I had heard the night before. Among them, however, was a disturbing new kind of explosion, the air-burst, which the 88 mm. gunners often fired for ranging and to make the occupants of open tank turrets uncomfortable. By tinkering with their fuse they produced a sudden thunderclap overhead, which, beyond drawing a straight line of tiny puffs along the sand, hardly showed after the moment of bursting: so that the first time I heard the bang I was unable to find a dust cloud anywhere to account for it.

The flashes from these guns were not visible: they continued firing spasmodically for about two hours, distributing their fire between the Crusaders and heavy tanks. After a time they began to introduce some sort of oil-shell which burst with a much greater volume of flame and of black smoke. One of these set fire to the bedding strapped to the outside of a Sherman, but the crew soon extinguished this and climbed into their tank again. Apart from this short interlude of excitement, shells continued to arrive and to miss, and we to sit there in sulky silence, reading our magazines and books, eating our biscuits and cheese, and indulging in occasional backchat over the air, for the rest of the morning.

About midday, feeling that the futility of war had been adequately demonstrated to me, I arrived back among the supply vehicles, to refuel. We seized the opportunity to have a brew-up, and ate some of our tinned fruit. An infantry sergeant and three or four men had brought in a German prisoner, a boy of about fifteen, who looked very tired but still defiant. He had remained lying in a patch of scrub while our tanks passed him, and after the supply echelon had arrived, dismounted, brewed up, and settled down to wait for us to come back to them, he had started to snipe them. The sergeant, who regarded this as an underhand piece of work, was for executing the boy at once. "Shoot the bugger. That's what I say," he kept repeating. His more humane companions were for giving the prisoner a cup of tea and a cigarette, which he obviously needed. I think he got them in the end.

Having refreshed ourselves and our vehicles, we went back and sat beside the camouflaged derelict again. Presently an infantry patrol, moving like guilty characters in a melodrama, came slinking and crouching up to my tank. A corporal, forgetting his attitude for the time being, leant against the tank, saying: "You see them Jerry derelicts over there, them two?" He indicated the two burnt-out tanks to our right front and added: "They've got a machine-gun in that right-hand one. We can't get up to them. They open up on us and pin us down, see?" "Well, what would you like us to do?" "I should have thought you could run over the buggers with this," he said, patting the tank. "Well, we'll see. I'll have to ask my squadron leader." I indicated his tank, "Will you go over and tell him all about it?" "Very good, sir," said the corporal, suddenly deciding that I was an officer. He departed. His patrol, who had been slinking aimlessly round in circles, waiting for him, tailed on behind him.

Andrew's instructions were of the kind I was beginning to expect from him. "See what you can do about it. See if you can get those chaps out of it. But be very careful. I don't want you to take any risks." I interpreted this to mean: "If you make a mess of it, I wash my hands of you," and opened the proceedings by ordering Evan to spray the area of the derelict with machine-gun fire.

The machine-gun, however, fired a couple of desultory shots, and jammed; Evan cleared and re-cocked it. It jammed again. A furious argument followed, Evan maintaining that the trouble was due to my not passing the belt of ammunition over the six-pounder and helping it out of the box. I pointed out that the belt was free on my side. Our understanding of each other was not helped by the fact that while I was speaking into the i/c microphone, Evan removed his earphones because they hampered his movements. He then shouted to me, disdaining the microphone, words which I could not hear through my heavy earphones. At length the conversation resolved itself into a shouting match. Evan became more and more truculent, and I ordered the driver to begin advancing slowly towards the enemy. This had the effect I wanted. Evan stopped talking, and applied himself feverishly to mending the machine-gun. After about a hundred yards I halted and scrutinized the derelict through my glasses. I could see no movement. I wondered what the crew of the machine-gun felt like, seeing a tank slowly singling them out and advancing on them. Evan was stripping the gun in the bad light and confined space of the turret, skinning his fingers, swearing and perspiring. At this moment Andrew's voice spoke in my ear, saying airily that he was going to refuel: "Nuts five, I'm going back to the N.A.A.F.I. for lemonade and buns. Take charge. Off." So now I was left to my own devices.

Looking down for a moment at a weapon-pit beside us, I saw a Libyan soldier reclining there. He had no equipment nor arms, and lay on his back as though resting, his arms flung out, one knee bent, his eyes open. He was a big man: his face reminded me of Paul Robeson. I thought of Rimbaud's poem: "Le Dormeur du Val" – but the last line:

Il a deux trous rouges au côté droit

was not applicable. There were no signs of violence. As I looked at him, a fly crawled up his cheek and across the dry pupil of his unblinking right eye. I saw that a pocket of dust had collected in the trough of the lower lid. The fact that for two minutes he had been lying so close to me, without my noticing him, was surprising: it was as though he had come there silently and taken up his position since our arrival.

Evan's swearing approached a crescendo. "I'll have to take the bastard out," he said. "It's the remote control's bust. I'll fire it from the trigger." We got the biscuit tin off the back of the tank and mounted the gun on it loose, on the top of the turret. From this eminence, as we advanced again, Evan sprayed earth and air impartially, burning his fingers on the barrel casing, his temper more furious every minute. At length he succeeded in landing a few shots round the derelict tank. A red-faced infantry subaltern ran up behind us, and climbed on to the tank. He put his hands in his pocket and pulled out two grenades, the pins of which he extracted with his teeth. He sat clutching them and said to me: "Very good of you to help us out, old boy," in a voice much fiercer than his words. We were now only about thirty yards from the derelict, and saw the bodies of men under it. They did not move.

"There they are!" cried the infantryman suddenly. A few yards from the left of the tank, two German soldiers were climbing out of a pit, grinning sheepishly as though they had been caught out in a game of hide and seek. In their pit lay a

Spandau machine-gun with its perforated jacket. So much, I thought with relief, for the machine-gun nest. But men now arose all round us. We were in a maze of pits. Evan flung down the Besa machine-gun, cried impatiently, "Lend us your revolver, sir," and snatching it from my hand, dismounted. He rushed up and down calling "Out of it, come on out of it, you bastards," etc. The infantry officer and I joined in this chorus, and rushed from trench to trench; I picked up a rifle from one of the trenches and aimed it threateningly, although I soon discovered that the safety-catch was stuck and it would not fire. The figures of soldiers continued to arise from the earth as though dragons' teeth had been sown there. I tried to get the prisoners into a body by gesticulating with my useless rifle. To hurry a man up, I pointed a rifle at him, but he cowered to the ground, like a puppy being scolded, evidently thinking I was going to shoot him on the spot. I felt very embarrassed, and lowered the rifle: he shot away after his comrades as though at the start of a race. I began to shout: "Raus, raus, raus," with great enthusiasm at the occupants of some trenches further back, who were craning their necks at us in an undecided way. Evan unluckily discouraged them by blazing off at them with a Spandau which he had picked up, and some high explosive began to land near the tank, which was following us about like a tame animal. Evan now found a man shamming dead in the bottom of a pit and was firing at his heels with my revolver, swearing and cursing at him. Another German lay on the ground on his back, occasionally lifting his head and body off the ground as far as the waist, with his arms stretched stiffly above his head and his face expressive of strenuous effort, like a man in a gymnasium. His companions gesticulated towards him and pointed at their heads, so that I thought he had been shot in the head. But when I looked more closely, I could see no wound, and he told me he was ill. Two of them assisted him away.

From the weapon pits, which were crawling with flies, we loaded the back of the tank with Spandaus, rifles, Luger pistols, Dienstglasse, the lightweight German binoculars, British tinned rations and the flat round German tins of chocolate.

As the main body of the prisoners was marched away under an infantry guard, the high explosive began to land closer to us. I did not feel inclined to attack the further position single-handed, so I moved the tank back and tacked it on to the column of prisoners. The mortar stopped firing at us, and some of the infantry climbed on to the tank to ride back. I reported over the air that we had taken some prisoners.

"Nuts five, how many prisoners?" asked what I presumed to be Andrew's voice. "Nuts five wait. Off." I said, counting, "Nuts five about figures four zero. Over." "Bloody good. Most excellent." Apparently it was the Colonel talking. "Now I want you to send these chaps back to our Niner" – he meant the Brigadier – "so that you'll get the credit for this." This was unfortunately more than my conscience would stand. I felt that all the work had been done by Evan and the infantry officer, and said so. This was a bad thing to say to Piccadilly Jim, because it showed him that I did not agree with him about snatching little gobbets of glory for the regiment whenever possible. The infantry were in another Brigade, as Piccadilly Jim knew. Evan said: "You were a bloody fool to say that, sir. You've as good as thrown away an M.C." I said shortly that if I had, it was an undeserved one.

The reaction on me of all this was an overpowering feeling of insignificance. I went over to the infantry officers who were searching the prisoners and said: "You did most of the dirty work, so you'd better take them back to your Brigade." The one who had ridden on my tank replied. "Yes, we had orders to," in such a supercilious way that I almost decided to insist on my right to escort them after all. The man with a bad head was lying groaning on the ground. He clutched his head and waved it from side to side. I think perhaps he had ostitis: the pain made him roll about and kick his legs like a baby.

The turret, after the removal of the Besa, and our leaping in and out of it, was in utter confusion. During our struggles with the machine-gun the bottom of an ammunition box had dropped out, and the belt of it was coiled everywhere. The empty belt fired from the biscuit box mounting had fallen in whorls on top of this. The microphones, spare headphones, gunner's headphones and all their respective flexes were inextricably entwined among the belts. Empty cartridge and shell cases littered the floor. On the surface of this morass of metal reposed the Besa itself, and an inverted tin of Kraft cheese, which had melted in the sunlight. I rescued a microphone and a pair of headphones, and got permission to retire and reorganize. On my way back I was to call at the Colonel's tank. This I duly did, but my ears were singing so loudly that I could scarcely hear his kind words. As soon as the tank moved away from the prisoners, we were again fired on by a mortar, which followed us as we moved back, dropping shells consistently a few yards behind us. We brewed up in dead ground to the enemy behind a ridge; the mortar continued to search this ground with fire, but never got nearer than thirty yards, and that only with one shot.

We examined our trophies, and were shocked to find that the infantry had stolen all our German binoculars while enjoying our hospitality as passengers on the tank. We all bitterly reproached them, and I regretted ever having wished to give them extra credit. We had left, however, a large stack of machine-guns and rifles, which we dumped. Three Luger pistols, which we kept: these are beautiful weapons, though with a mechanism too delicate for use in sandy country. There were a few odds and ends of rations, cutlery, badges, knives, etc., which we shared out, eating most of the extra rations there and then in a terrific repast, with several pints of coffee. At last I decided we ought to rejoin the squadron, and reported we were on our way back.

After we had been back in our position about a quarter of an hour, someone on the right reported twelve enemy tanks advancing. A second report estimated twenty. Soon after this a very hot fire began to fall around us. Two petrol lorries were hit at once and began to blaze. The Germans came towards us out of the setting sun, firing, and supported by anti-tank and high-explosive fire. Some of the Crusaders, Andrew's tank included, began reversing. I moved back myself, but it was obvious that we should soon get moved up with the heavy tanks. I halted. Someone could be heard calling for smoke. Andrew was berating the squadron for bunching up and his own tank meanwhile avoided reversing into a Grant more by luck than judgment. There was a certain amount of incipient panic apparent in some of the messages coming over the air. The enemy fire grew more intense; it seemed incredible that only the two lorries had so far been hit. I crouched in the turret, expecting at any moment the crash which would bring our disintegration, seeing again the torn shell of the tank we had passed the previous evening – it seemed weeks ago. I could not see the enemy tanks any longer, and was not sure after so much reversing and milling around, exactly what the situation was. These were the intensest moments of physical fear, outside of dreams, I have ever experienced.

Control now instructed us: "Open fire on the enemy. Range

one zero zero zero. Give the buggers every round you've got. Over." With, I think, some relief the various squadrons acknowledged. "One O.K. off"; "Two O.K. off"; "Three O.K. off." I ordered Davis to fire. "I can't see a muckin' thing," he protested. "Never mind, you fire at a thousand as fast as I can load." Every gun was now blazing away into the twilight, the regiment somewhat massed together, firing with every available weapon. I crammed shells into the six-pounder as fast as Evan could lay and fire it. Presently the deflector bag was full of shell cases, and Evan, who had now adjusted the Besa, blazed off a whole belt without a stoppage, while I tossed out the empty cases, too hot to touch with a bare hand. The turret was full of fumes and smoke. I coughed and sweated; fear had given place to exhilaration. Twilight increased to near-darkness, and the air all round us gleamed with the different coloured traces of shells and bullets, brilliant graceful curves travelling from us to the enemy and from him towards us. The din was tremendously exciting. I could see a trail of machine-gun bullets from one of our heavy tanks passing a few yards to the left of my tank, on a level with my head. Above us whistled the shells of the seventy-fives. Overhead the trace of enemy shells could be seen mounting to the top of their flight where, as the shell tilted towards us, it disappeared. Red and orange bursts leapt up beside and in front of us.

Darkness ended the action as suddenly as it had begun; the petrol lorries alone blazed like beacons, answered by distant fires in the direction of the enemy. Gradually we found our way into leaguer, creeping past the beacons after the dim shapes of our companions. My first day in action had been eventful enough: I felt as if I had been fighting for months.

I shall remember that day as a whole, separate from the rest of my time in action, because it was my first, and because we were withdrawn at two o'clock that morning for a four days' rest. We arrived about four o'clock and lay down to sleep at once. My last sensations were of complete satisfaction in the luxury of sleep, without a thought for the future or the past. I did not wake until ten o'clock next morning, and when I opened my eyes, still no one was stirring.

T.E. Lawrence

Guerrilla Attack on a Turkish Outpost

"Lawrence of Arabia" was the British Army's liaison officer with the Arab insurgents fighting the Ottoman Empire in the First World War.

We rode all night, and when dawn came were dismounting on the crest of the hills between Batra and Aba el Lissan, with a wonderful view westwards over the green and gold Guweira plain, and beyond it to the ruddy mountains hiding Akaba and the sea. Gasim abu Dumeik, head of the Dhumaniyeh, was waiting anxiously for us, surrounded by his hard-bitten tribesmen, their grey strained faces flecked with the blood of the fighting yesterday. There was a deep greeting for Auda and Nasir. We made hurried plans, and scattered to the work, knowing we could not go forward to Akaba with this battalion in possession of the pass. Unless we dislodged it, our two months' hazard and effort would fail before yielding even first fruits.

Fortunately the poor handling of the enemy gave us an unearned advantage. They slept on, in the valley, while we crowned the hills in a wide circle about them unobserved. We began to snipe them steadily in their positions under the slopes and rock-faces by the water, hoping to provoke them out and up the hill in a charge against us. Meanwhile, Zaal rode away with our horsemen and cut the Maan telegraph and telephone in the plain.

This went on all day. It was terribly hot – hotter than ever before I had felt it in Arabia – and the anxiety and constant moving made it hard for us. Some even of the tough tribesmen broke down under the cruelty of the sun, and crawled or had to be thrown under rocks to recover in their shade. We ran up and down to supply our lack of numbers by mobility, ever looking over the long ranges of hill for a new spot from which to counter this or that Turkish effort. The hill-sides were steep, and exhausted our breath, and the grasses twined like little hands about our ankles as we ran, and plucked us back. The sharp reefs of limestone which cropped out over the ridges tore our feet, and long before evening the more energetic men were leaving a rusty print upon the ground with every stride.

Our rifles grew so hot with sun and shooting that they seared our hands; and we had to be grudging of our rounds, considering every shot, and spending great pains to make it sure. The rocks on which we flung ourselves for aim were burning, so that they scorched our breasts and arms, from which later the skin drew off in ragged sheets. The present smart made us thirst. Yet even water was rare with us; we could not afford men to fetch enough from Batra, and if all could not drink, it was better that none should.

We consoled ourselves with knowledge that the enemy's enclosed valley would be hotter than our open hills; also that they were Turks, men of white meat, little apt for warm weather. So we clung to them, and did not let them move or mass or sortie out against us cheaply. They could do nothing valid in return. We were no targets for their rifles, since we moved with speed, eccentrically. Also we were able to laugh at the little mountain guns which they fired up at us. The shells passed over our heads, to burst behind us in the air; and yet, of course, for all that they could see from their hollow place, fairly amongst us above the hostile summits of the hill.

Just after noon I had a heat-stroke, or so pretended, for I was dead weary of it all, and cared no longer how it went. So I crept into a hollow where there was a trickle of thick water in a muddy cup of the hills, to suck some moisture off its dirt through the filter of my sleeve. Nasir joined me, panting like a winded animal, with his cracked and bleeding lips shrunk apart in his distress; and old Auda appeared, striding powerfully, his eyes bloodshot and staring, his knotty face working with excitement.

He grinned with malice when he saw us lying there, spread out to find coolness under the bank, and croaked at me harshly, "Well, how is it with the Howeitat? All talk and no work?" "By God, indeed," spat I back again, for I was angry with every one and with myself, "they shoot a lot and hit a little." Auda almost pale with rage, and trembling, tore his headcloth off and threw it on the ground beside me. Then he ran back up the hill like a madman, shouting to the men in his dreadful strained and rustling voice.

They came together to him, and after a moment scattered away down hill. I feared things were going wrong, and struggled to where he stood alone on the hill-top, glaring at the enemy; but all he would say to me was, "Get your camel if you want to see the old man's work." Nasir called for his camel and we mounted.

The Arabs passed before us into a little sunken place, which rose to a low crest; and we knew that the hill beyond went down in a facile slope to the main valley of Aba el Lissan, somewhat below the spring. All our four hundred camel men were here tightly collected, just out of sight of the enemy. We rode to their head, and asked the Shimt what it was and where the horsemen had gone.

He pointed over the ridge to the next valley above us, and said, "With Auda there"; and as he spoke yells and shots poured up in a sudden torrent from beyond the crest. We kicked our camels furiously to the edge, to see our fifty horsemen coming down the last slope into the main valley like a run-away, at full gallop, shooting from the saddle. As we watched, two or three went down, but the rest thundered forward at marvellous speed, and the Turkish infantry, huddled together under the cliff ready to cut their desperate way out towards Maan in the first dusk, began to sway in and out, and finally broke before the rush, adding their flight to Auda's charge.

Nasir screamed at me, "Come on," with his bloody mouth; and we plunged our camels madly over the hill, and down towards the head of the fleeing enemy. The slope was not too steep for a camel-gallop, but steep enough to make their pace terrific, and their course uncontrollable; yet the Arabs were able to extend to right and left and to shoot into the Turkish brown. The Turks had been too bound up in the terror of Auda's furious charge against their rear to notice us as we came over the eastward slope: so we also took them by surprise and in the flank; and a charge of ridden camels going nearly thirty miles an hour was irresistible.

The Howeitat were very fierce, for the slaughter of their women on the day before had been a new and horrible side of warfare suddenly revealed to them. So there were only a hundred and sixty prisoners, many of them wounded; and three hundred dead and dying were scattered over the open valleys.

A few of the enemy got away, the gunners on their teams, and some mounted men and officers with their Jazi guides. Mohammed el Dheilan chased them for three miles into Mreigha, hurling insults as he rode, that they might know him and keep out of his way. The feud of Auda and his cousins had never applied to Mohammed, the political-minded, who showed friendship to all men of his tribe when he was alone to do so. Among the fugitives was Dhaif-Allah, who had done us the good turn about the King's Well at Jefer.

Auda came swinging up on foot, his eyes glazed over with the rapture of battle, and the words bubbling with incoherent speed from his mouth. "Work, work, where are words, work, bullets, Abu Tayi" . . . and he held up his shattered field-glasses, his pierced pistol-holster, and his leather sword-scabbard cut to ribbons. He had been the target of a volley which had killed his mare under him, but the six bullets through his clothes had left him scathless.

He told me later, in strict confidence, that thirteen years before he had bought an amulet Koran for one hundred and twenty pounds and had not since been wounded. Indeed, Death had avoided his face, and gone scurvily about killing brothers, sons and followers. The book was a Glasgow reproduction, costing eighteenpence; but Auda's deadliness did not let people laugh at his superstition.

He was wildly pleased with the fight, most of all because he had confounded me and shown what his tribe could do. Mohammed was wroth with us for a pair of fools, calling me worse than Auda, since I had insulted him by words like flung stones to provoke the folly which had nearly killed us all; though it had killed only two of us, one Rueili and one Sherari.

It was, of course, a pity to lose any one of our men, but time was of importance to us, and so imperative was the need of dominating Maan, to shock the little Turkish garrisons between us and the sea into surrender, that I would have willingly lost much more than two. On occasions like this Death justified himself and was cheap.

William Grattan

The Storming of Ciudad Rodrigo

William Grattan served with the Connaught Rangers (88th Foot) during the Peninsular War. The fortress of Ciudad Rodrigo guarded a northern entrance to Spain, and its taking was a necessity if the English under Wellington were to liberate the country from Napoleon. On 19 January 1812, after two weeks of siege, Wellington ordered that the breaches in the fortress walls be stormed.

Early on the morning of the 19th, the 3rd Division (although not for duty that day) received orders to march to the Convent of La Caridad; and as Lord Wellington was not in the habit of giving us unnecessary marches, we concluded that he intended us the honour of forming one of the corps destined to carry the place. On our march we perceived our old friends and companions, the Light Division, debouching from their cantonments, and the joy expressed by our men when they saw them is not to be described; we were long acquainted, and like horses accustomed to the same harness, we pulled well together. At two o'clock in the afternoon we left La Caridad, and, passing to the rear of the first parallel, formed in column about two gun-shots distant from the main breach. The 4th Division still occupied the works, and it was the general opinion that ours (the 3rd) were to be in reserve. The number of Spaniards, Portuguese, and soldiers' wives in the character of sutlers, was immense, and the neighbourhood, which but a few days before was only an empty plain, now presented the appearance of a vast camp. Wretches of the poorest description hovered round us, in hopes of getting a morsel of food, or of plundering some dead or wounded soldier: their cadaverous countenances expressed a living picture of the greatest want; and it required all our precaution to prevent these miscreants from robbing us the instant we turned our backs from our scanty store of baggage or provisions.

Our bivouac, as may be supposed, presented an animated appearance – groups of soldiers cooking in one place; in another, some dozens collected together, listening to accounts brought from the works by some of their companions whom curiosity had led thither; others relating their past battles to any of the young soldiers who had not as yet come hand-to-hand with a Frenchman; others dancing and singing; officers' servants preparing dinner for their masters; and officers themselves, dressed in whatever way best suited their taste or convenience, mixed with the men, without any distinguishing mark of uniform to denote their rank. The only thing uniform to be discovered amongst a group of between four and five thousand was good conduct and confidence in themselves and their general.

It was now five o'clock in the afternoon, and darkness was approaching fast, yet no order had arrived intimating that we were to take a part in the contest about to be decided. We were in this state of suspense when our attention was attracted by the sound of music; we all stood up, and pressed forward to a ridge, a little in our front, and which separated us from the cause of our movement, but it would be impossible for me to convey an adequate idea of our feelings when we beheld the 43rd Regiment, preceded by their band, going to storm the left breach; they were in the highest spirits, but without the slightest appearance of levity in their demeanour – on the contrary, there was a cast of determined severity thrown over their countenances that expressed in legible characters that they knew the sort of service they were about to perform, and had made up their minds to the issue. They had no knapsacks – their firelocks were slung over their shoulders – their shirt-collars were open, and there was an indescribable *something* about them that at one and the same moment impressed the lookers-on with admiration and awe. In passing us, each officer and soldier stepped out of the ranks for an instant, as he recognised a friend, to press his hand – many for the last time; yet, notwithstanding this animating scene, there was no shouting or huzzaing, no boisterous bravadoing, no unbecoming language; in short, every one seemed to be impressed with the seriousness of the affair entrusted to his charge, and any interchange of words was to this effect: "Well, lads, mind what you're about tonight"; or, "We'll meet in the town by and by"; and other little familiar phrases, all expressive of confidence. The regiment at length passed us, and we stood gazing after it as long as the rear platoon continued in sight: the music grew fainter every moment, until at last it died away altogether; they had no drums, and there was a melting sweetness in the sounds that touched the heart.

The first syllable uttered after this scene was, "And are we to be left behind?" The interrogatory was scarcely put, when the word "Stand to your arms!" answered it. The order was promptly obeyed, and a breathless silence prevailed when our commanding officer, in a few words, announced to us that Lord Wellington had directed our division to carry the grand breach. The soldiers listened to the communication with silent earnestness, and immediately began to disencumber themselves of their knapsacks, which were placed in order by companies and a guard set over them. Each man then began to arrange himself for the combat in such manner as his fancy or the moment would admit of – some by lowering their cartridge-boxes, others by turning theirs to the front in order that they might the more conveniently make use of them; others unclasping their stocks or opening their shirt-collars, and others oiling their bayonets; and more taking leave of their wives and children. This last was an affecting sight, but not so much so as might be expected, because the women, from long habit, were accustomed to scenes of danger, and the order for their husbands to march against the enemy was in their eyes tantamount to a victory; and as the soldier seldom returned without plunder of some sort, the painful suspense which his absence caused was made up by the gaiety which his return was certain to be productive of; or if, unfortunately, he happened to fall, his place was sure to be supplied by some one of the company to which he belonged, so that the women of our army had little cause of alarm on this head. The worst that could happen to them was the chance of being in a state of widowhood for a week.

It was by this time half-past six o'clock, the evening was piercingly cold, and the frost was crisp on the grass; there was a keenness in the air that braced our nerves at least as high as *concert pitch*. We stood quietly to our arms, and told our companies off by files, sections, and sub-divisions; the sergeants called over the rolls – not a man was absent.

It appears it was the wish of General Mackinnon to confer a mark of distinction upon the 88th Regiment, and as it was one of the last acts of his life, I shall mention it. He sent for Major

Thompson, who commanded the battalion, and told him it was his wish to have the forlorn hope[1] of the grand breach led on by a subaltern of the 88th Regiment, adding at the same time that, in the event of his surviving, he should be recommended for a company. The Major acknowledged this mark of the General's favour, and left him folding up some letters he had been writing to his friends in England – this was about twenty minutes before the attack of the breaches. Major Thompson, having called his officers together, briefly told them the wishes of their General; he was about to proceed, when Lieutenant William Mackie (*then senior Lieutenant*) immediately stepped forward, and dropping his sword said, "Major Thompson, I am ready for that service." For once in his life poor old Thompson was affected – Mackie was his own townsman, they had fought together for many years, and when he took hold of his hand and pronounced the words, "God bless you, my boy," his eye filled, his lip quivered, and there was a faltering in his voice which was evidently perceptible to himself, for he instantly resumed his former composure, drew himself up, and gave the word, "Gentlemen, fall in," and at this moment Generals Picton[2] and Mackinnon, accompanied by their respective staffs, made their appearance amongst us.

Long harangues are not necessary to British soldiers, and on this occasion but few words were made use of. Picton said something animating to the different regiments as he passed them, and those of my readers who recollect his deliberate and strong utterance will say with me, that his mode of speaking was indeed very impressive. The address to each was nearly the same, but that delivered by him to the 88th was so characteristic of the General, and so applicable to the men he spoke to, that I shall give it word for word; it was this:-

"Rangers of Connaught! it is not my intention to expend any powder this evening. We'll do this business with the cold iron."

I before said the soldiers were silent – so they were, but the man who could be silent after such an address, made in such a way, and in such a place, had better have stayed at home. It may

1 The van of the storming party, the expression being Anglicized from the Dutch *verloren hope*, "lost party." As Grattan relates, there was no shortgage of volunteers for the "forlorn hope" – effectively a suicide mission – at Ciudad Rodrigo or Badajoz.

2 Lieutenant-General Sir Thomas Picton (1758–1815); one of Wellington's most able – and moodiest subordinates – he took command of the 3rd Division in 1810. Killed leading his men at Waterloo.

be asked what did they do? Why, what would they do, or would any one do, but give the loudest hurrah he was able.

The burst of enthusiasm caused by Picton's address to the Connaught Rangers had scarcely ceased, when the signal-gun announced that the attack was to commence. Generals Picton and Mackinnon dismounted from their horses, and placing themselves at the head of the right brigade, the troops rapidly entered the trenches by sections right in front; the storming party under the command of Major Russell Manners of the 74th heading it, while the forlorn hope, commanded by Lieutenant William Mackie of the 88th, and composed of twenty volunteers from the Connaught Rangers, led the van, followed closely by the 45th, 88th, and 74th British, and the 9th and 21st Portuguese; the 77th and 83rd British, belonging to the left brigade, brought up the rear and completed the dispositions.

While these arrangements were effecting opposite the grand breach, the 5th and 94th, belonging to the left brigade of the 3rd Division, were directed to clear the ramparts and Fausse Braye wall, and the 2nd Regiment of Portuguese Caçadores, commanded by an Irish colonel of the name of O'Toole, was to escalade the curtain to the left of the lesser breach, which was attacked by the Light Division under the command of General Robert Craufurd.

It wanted ten minutes to seven o'clock when these dispositions were completed; the moon occasionally, as the clouds which overcast it passed away, shed a faint ray of light upon the battlements of the fortress, and presented to our view the glittering of the enemy's bayonets as their soldiers stood arrayed upon the ramparts and breach, awaiting our attack; yet, nevertheless, their batteries were silent, and might warrant the supposition to an unobservant spectator that the defence would be but feeble.

The two divisions got clear of the covered way at the same moment, and each advanced to the attack of their respective points with the utmost regularity. The obstacles which presented themselves to both were nearly the same, but every difficulty, no matter how great, merged into insignificance when placed in the scale of the prize about to be contested. The soldiers were full of ardour, but altogether devoid of that blustering and bravadoing which is truly unworthy of men at such a moment; and it would be difficult to convey an adequate idea of the enthusiastic bravery which animated the troops. A cloud that had for some time before obscured the moon, which was at its full, disappeared altogether, and the countenances of the soldiers were for the first time, since Picton addressed them, visible – they presented a material change. In place of that joyous animation which his fervid and impressive address called forth, a look of severity, bordering on ferocity, had taken its place; and although ferocity is by no means one of the characteristics of the British soldier, there was, most unquestionably, a savage expression in the faces of the men that I had never before witnessed. Such is the difference between the storm of a breach and the fighting of a pitched battle.

Once clear of the covered way, and fairly on the plain that separated it from the fortress, the enemy had a full view of all that was passing; their batteries, charged to the muzzle with case-shot, opened a murderous fire upon the columns as they advanced, but nothing could shake the intrepid bravery of the troops. The Light Division soon descended the ditch and gained, although not without a serious struggle, the top of the narrow and difficult breach allotted to them; their gallant General, Robert Craufurd, fell at the head of the 43rd, and his second in command, General Vandeleur, was severely wounded, but there were not wanting others to supply their place; yet these losses, trying as they were to the feelings of the soldiers, in no way damped their ardour, and the brave Light Division carried the left breach at the point of the bayonet. Once established upon the ramparts, they made all the dispositions necessary to ensure their own conquest, as also to render every assistance in their power to the 3rd Division in their attack. They cleared the rampart which separated the lesser from the grand breach, and relieved Picton's division from any anxiety it might have as to its safety on its left flank.

The right brigade, consisting of the 45th, 88th, and 74th, forming the van of the 3rd Division, upon reaching the ditch, to its astonishment, found Major Ridge and Colonel Campbell at the head of the 5th and 94th mounting the Fausse Braye wall. These two regiments, after having performed their task of silencing the fire of the French troops upon the ramparts, with a noble emulation resolved to precede their comrades in the attack of the grand breach. Both parties greeted each other with a cheer, only to be understood by those who have been placed in a similar situation; yet the enemy were in no way daunted by the shout raised by our soldiers – they crowded the breach, and defended it with a bravery that would have made any but troops accustomed to conquer, waver. But the "fighting division" were not the men to be easily turned from their purpose; the breach was speedily mounted, yet, nevertheless, a serious affray took place ere it was gained. A considerable mass of infantry crowned its summit, while in the rear and at each side were stationed men, so placed that they could render every assistance to their comrades at the breach without any great risk to themselves; besides this, two guns of heavy calibre, separated from the breach by a ditch of considerable depth and width, enfiladed it, and as soon as the French infantry were forced from the summit, these guns opened their fire on our troops.

The head of the column had scarcely gained the top, when a discharge of grape cleared the ranks of the three leading battalions, and caused a momentary wavering; at the same instant a frightful explosion near the gun to the left of the breach, which shook the bastion to its foundation, completed the disorder. Mackinnon, at the head of his brigade, was blown into the air. His aide-de-camp, Lieutenant Beresford of the 88th, shared the same fate, and every man on the breach at the moment of the explosion perished. This was unavoidable, because those of the advance, being either killed or wounded, were necessarily flung back upon the troops that followed close upon their footsteps, and there was not a sufficient space for the men who were ready to sustain those placed *hors de combat* to rally. For an instant all was confusion; the blaze of light caused by the explosion resembled a huge meteor, and presented to our sight the havoc which the enemy's fire had caused in our ranks; while from afar the astonished Spaniard viewed for an instant, with horror and dismay, the soldiers of the two nations grappling with each other on the top of the rugged breach which trembled beneath their feet, while the fire of the French artillery played upon our columns with irresistible fury, sweeping from the spot the living and the dead. Amongst the latter was Captain Robert Hardyman and Lieutenant Pearse of the 45th, and many more whose names I cannot recollect. Others were so stunned by the shock, or wounded by the stones which were hurled forth by the explosion, that they were insensible to their situation; of this number I was one, for being close to the magazine when it blew up, I was quite overpowered, and I owed my life to the Sergeant-Major of my regiment, Thorp, who saved me from being trampled to death by our soldiers in their advance, ere I could recover strength sufficient to move forward or protect myself.

The French, animated by this accidental success, hastened once more to the breach which they had abandoned, but the leading regiments of Picton's division, which had been disorganised for the moment by the explosion, rallied, and soon regained its summit, when another discharge from the two flank guns swept away the foremost of those battalions.

There was at this time but one officer alive upon the breach (Major Thomson, of the 74th, acting engineer); he called out to those next to him to seize the gun to the left, which had been so fatal to his companions – but this was a desperate service. The gun was completely cut off from the breach by a deep trench, and soldiers, encumbered with their firelocks, could not pass it in sufficient time to anticipate the next discharge – yet to deliberate was certain death. The French cannoniers, five in number, stood to, and served their gun with as much *sang froid* us if on a parade, and the light which their torches threw forth showed to our men the peril they would have to encounter if they dared to attack a gun so defended; but this was of no avail. Men going to storm a breach generally make up their minds that there is no great probability of their ever returning from it to tell their adventures to their friends; and whether they die at the bottom or top of it, or at the muzzle, or upon the breech of a cannon, is to them pretty nearly the same!

The first who reached the top, after the last discharge, were three of the 88th. Sergeant Pat Brazil – the brave Brazil of the Grenadier company, who saved his captain's life at Busaco – called out to his two companions, Swan and Kelly, to unscrew their bayonets and follow him; the three men passed the trench in a moment, and engaged the French cannoniers hand to hand; a terrific but short combat was the consequence. Swan was the first, and was met by the two gunners on the right of the gun, but, no way daunted, he engaged them, and plunged his bayonet into the breast of one; he was about to repeat the blow upon the other, but before he could disentangle the weapon from his bleeding adversary, the second Frenchman closed upon him, and by a *coup de sabre* severed his left arm from his body a little above the elbow; he fell from the shock, and was on the eve of being massacred, when Kelly, after having scrambled under the gun, rushed onward to succour his comrade. He bayoneted two

Frenchmen on the spot, and at this instant Brazil came up; three of the five gunners lay lifeless, while Swan, resting against an ammunition chest, was bleeding to death. It was now equal numbers, two against two, but Brazil in his over-anxiety to engage was near losing his life at the onset; in making a lunge at the man next to him, his foot slipped upon the bloody platform, and he fell forward against his antagonist, but as both rolled under the gun, Brazil felt the socket of his bayonet strike hard against the buttons of the Frenchman's coat. The remaining gunner, in attempting to escape under the carriage from Kelly, was killed by some soldiers of the 5th, who just now reached the top of the breach, and seeing the serious dispute at the gun, pressed forward to the assistance of the three men of the Connaught Rangers.

While this was taking place on the left, the head of the column remounted the breach, and regardless of the cries of their wounded companions, whom they indiscriminately trampled to death, pressed forward in one irregular but heroic mass, and putting every man to death who opposed their progress, forced the enemy from the ramparts at the bayonet's point. Yet the garrison still rallied, and defended the several streets with the most unflinching bravery; nor was it until the musketry of the Light Division was heard in the direction of the Plaza Mayor, that they gave up the contest, but from this moment all regular resistance ceased, and they fled in disorder to the Citadel. There were, nevertheless, several minor combats in the streets, and in many instances the inhabitants fired from the windows, but whether their efforts were directed against us or the French is a point that I do not feel myself competent to decide; be this as it may, many lives were lost on both sides by this circumstance, for the Spaniards, firing without much attention to regularity, killed or wounded indiscriminately all who came within their range.

During a contest of such a nature, kept up in the night, as may be supposed, much was of necessity left to the guidance of the subordinate officers, if not to the soldiers themselves. Each affray in the streets was conducted in the best manner the moment would admit of, and decided more by personal valour than discipline, and in some instances officers as well as privates had to combat with the imperial troops. In one of these encounters Lieutenant George Faris, of the 88th, by an accident so likely to occur in an affair of this kind, separated a little too far from a dozen or so of his regiment, and found himself opposed to a French soldier who, apparently, was similarly placed. It was a curious coincidence, and it would seem as if each felt that he individually was the representative of the country to which he belonged; and had the fate of the two nations hung upon the issue of the combat I am about to describe, it could not have been more heroically contested. The Frenchman fired at and wounded Faris in the thigh, and made a desperate push with his bayonet at his body, but Faris parried the thrust, and the bayonet only lodged in his leg. He saw at a glance the peril of his situation, and that nothing short of a miracle could save him; the odds against him were too great, and if he continued a scientific fight he must inevitably be vanquished. He sprang forward, and, seizing hold of the Frenchman by the collar, a struggle of a most nervous kind took place; in their mutual efforts to gain an advantage they lost their caps, and as they were men of nearly equal strength, it was doubtful what the issue would be. They were so entangled with each other their weapons were of no avail, but Faris at length disengaged himself from the grasp which held him, and he was able to use his sabre; he pushed the Frenchman from him, and ere he could recover himself he laid his head open nearly to the chin. His sword-blade, a heavy, soft, ill-made Portuguese one, was doubled up with the force of the blow, and retained some pieces of the skull and clotted hair! At this moment I reached the spot with about twenty men, composed of different regiments, all being by this time mixed *pell mell* with each other. I ran up to Faris – he was nearly exhausted, but he was safe. The French grenadier lay upon the pavement, while Faris, though tottering from fatigue, held his sword firmly in his grasp, and it was crimson to the hilt. The appearance of the two combatants was frightful! – one lying dead on the ground, the other faint from agitation and loss of blood; but the soldiers loudly applauded him, and the feeling uppermost with them was, that our man had the best of it! It was a shocking sight, but it would be rather a hazardous experiment to begin moralising at such a moment and in such a place.

Those of the garrison who escaped death were made prisoners, and the necessary guards being placed, and everything secured, the troops not selected for duty commenced a very diligent search for those articles which they most fancied, and which they considered themselves entitled to by "right of conquest." I believe on a service such as the present, there is a sort of tacit acknowledgment of this "right"; but be this as it may, a good deal of property most indubitably changed owners on the night of the 19th of January 1812. The conduct of the soldiers, too, within the last hour, had undergone a complete change; before, it was all order and regularity, now it was nothing but licentiousness and confusion – subordination was at an end; plunder and blood was the order of the day, and many an officer on this night was compelled to show that he carried a sabre.

The doors of the houses in a large Spanish town are remarkable for their strength, and resemble those of a prison more than anything else; their locks are of huge dimensions, and it is a most difficult task to force them. The mode adopted by the men of my regiment (the 88th) in this dilemma was as effective as it was novel; the muzzles of a couple of muskets were applied to each side of the keyhole, while a third soldier, fulfilling the functions of an officer, deliberately gave the word, "make ready" – "present" – "fire!" and in an instant the ponderous lock gave way before the combined operations of the three individuals, and doors that rarely opened to the knock of a stranger in Rodrigo, now flew off their hinges to receive the Rangers of Connaught.

The chapels and chandlers' houses were the first captured, in both of which was found a most essential ingredient in the shape of large wax candles; these the soldiers lighted, and commenced their perambulations in search of plunder, and the glare of light which they threw across the faces of the men, as they carried them through the streets, displayed their countenances, which were of that cast that might well terrify the unfortunate inhabitants. Many of the soldiers with their faces scorched by the explosion of the magazine at the grand breach; others with their lips blackened from biting off the ends of their cartridges, more covered with blood, and all looking ferocious, presented a combination sufficient to appeal the stoutest heart.

Scenes of the greatest outrage now took place, and it was pitiable to see groups of the inhabitants half naked in the streets – the females clinging to the officers for protection – while their respective houses were undergoing the strictest scrutiny. Some of the soldiers turned to the wine and spirit houses, where, having drunk sufficiently, they again sallied out in quest of more plunder; others got so intoxicated that they lay in a helpless state in different parts of the town, and lost what they had previously gained, either by the hands of any passing Spaniard, who could venture unobserved to stoop down, or by those of their own companions, who in their wandering surveys happened to recognise a comrade lying with half a dozen silk gowns, or some such thing, wrapped about him. Others wished to attack the different stores, and as there is something marvellously attractive in the very name of a brandy one, it is not to be wondered at that many of our heroes turned not only their thoughts, but their steps also, in the direction in which these houses lay; and from the unsparing hand with which they supplied themselves, it might be imagined they intended to change their habits of life and turn spirit-venders, and that too in the wholesale line!

It was astonishing to see with what rapidity and accuracy these fellows traversed the different parts of the town, and found out the shops and storehouses. A stranger would have supposed they were natives of the place, and it was not until the following morning that I discovered the cause of what was to me before incomprehensible.

In all military movements in a country which an army is not thoroughly acquainted with (and why not in a large town?), there are no more useful appendages than good guides. Lord Wellington was most particular on this point, and had attached to his army a corps of this description. I suppose it was this knowledge of tactics which suggested to the soldiers the necessity of so wise a precaution; accordingly, every group of individuals was preceded by a Spaniard, who, upon learning the species of plunder wished for by his employers, instantly conducted them to the most favourable ground for their operations. By this means the houses were unfurnished with less confusion than can be supposed; and had it not been for the state of intoxication that some of the young soldiers – mere tyros in the art of sacking a town – had indulged themselves in, it is inconceivable with what facility the city of Ciudad Rodrigo would have been eased of its superfluities. And the *conducteur* himself was not always an idle spectator. Many of these fellows realised something considerable from their more wealthy neighbours, and being also right well paid by the soldiers, who were liberal enough, they found themselves in the morning in far better circumstances than they had been the preceding night, so that all things considered, there were about as many cheerful faces as sad ones. But although the inhabitants were, by this sort of transfer, put more on an equality with each other, the town itself was greatly impoverished. Many things of value were destroyed, but in the hurry so natural to the occasion, many also escaped; besides, our men were as yet young hands in the arcana of plundering a town in that *au fait* manner with which a French army would have done a business of the sort: but they most unquestionably made up for their want of tact by the great inclination they showed to profit by any occasion that offered itself for their improvement.

By some mistake, a large spirit store situated in the Plaza Mayor took fire, and the flames spreading with incredible fury, despite the exertions of the troops, the building was totally destroyed; but in this instance, like many others which we are obliged to struggle against through life, there was something that neutralised the disappointment which the loss of so much brandy occasioned the soldiers: the light which shone forth from the building was of material service to them, inasmuch as it tended to facilitate their movements in their excursions for plunder; the heat also was far from disagreeable, for the night was piercingly cold, yet, nevertheless, the soldiers exerted themselves to the utmost to put a stop to this calamity. General Picton was to be seen in the midst of them, encouraging them by his example and presence to make still greater efforts; but all would not do, and floor after floor fell in, until at last it was nothing but a burning heap of ruins.

Some houses were altogether saved from plunder by the interference of the officers, for in several instances the women ran out into the streets, and seizing hold of three or four of us, would force us away to their houses, and by this stroke of political hospitality saved their property. A good supper was then provided, and while all outside was noise and pillage, affairs within went on agreeably enough. These instances were, however, but few.

Winston Churchill

The 21st Lancers at Omdurman

Winston Churchill was a subaltern in the 21st Lancers at the Battle of Omdurman, 2 September 1898, which ended the Mahdi revolt in Sudan.

LONG before the dawn we were astir, and by five o'clock the 21st Lancers were drawn up mounted outside the zeriba. My squadron-leader Major Finn, an Australian by birth, had promised me some days before that he would give me "a show" when the time came. I was afraid that he would count my mission to Lord Kitchener the day before as quittance; but I was now called out from my troop to advance with a patrol and reconnoitre the ridge between the rocky peak of Jebel Surgham and the river. Other patrols from our squadron and from the Egyptian cavalry were also sent hurrying forward in the darkness. I took six men and a corporal. We trotted fast over the plain and soon began to breast the unknown slopes of the ridge. There is nothing like the dawn. The quarter of an hour before the curtain is lifted upon an unknowable situation is an intense experience of war. Was the ridge held by the enemy or not? Were we riding through the gloom into thousands of ferocious savages? Every step might be deadly; yet there was no time for overmuch precaution. The regiment was coming on behind us, and dawn was breaking. It was already half light as we climbed the slope. What should we find at the summit? For cool, tense excitement I commend such moments.

Now we are near the top of the ridge. I make one man follow a hundred yards behind, so that whatever happens, he may tell the tale. There is no sound but our own clatter. We have reached the crest line. We rein in our horses. Every minute the horizon extends; we can already see 200 yards. Now we can see perhaps a quarter of a mile. All is quiet; no life but our own breathes among the rocks and sand hummocks of the ridge. No ambuscade, no occupation in force! The farther plain is bare below us: we can now see more than half a mile.

So they have all decamped! Just what we said! All bolted off to Kordofan; no battle! But wait! The dawn is growing fast. Veil after veil is lifted from the landscape. What is this shimmering in the distant plain? Nay – it is lighter now – what are these dark markings beneath the shimmer? *They are there!* These enormous black smears are thousands of men; the shimmering is the glinting of their weapons. It is now daylight. I slip off my horse; I write in my field service notebook "The Dervish army is still in position a mile and a half south-west of Jebel Surgham." I send this message by the corporal direct as ordered to the Commander-in-Chief. I mark it XXX. In the words of the drill book "with all despatch" or as one would say "Hell for leather".

A glorious sunrise is taking place behind us; but we are admiring something else. It is already light enough to use field-glasses. The dark masses are changing their values. They are already becoming lighter than the plain; they are fawn-coloured. Now they are a kind of white, while the plain is dun. In front of us is a vast array four or five miles long. It fills the horizon till it is blocked out on our right by the serrated silhouette of Surgham Peak. This is an hour to live. We mount again, and suddenly new impressions strike the eye and mind. These masses are not stationary. They are advancing, and they are advancing fast. A tide is coming in. But what is this sound which we hear: a deadened roar coming up to us in waves? They are cheering for God, his Prophet and his holy Khalifa. They think they are going to win. We shall see about that presently. Still I must admit that we check our horses and hang upon the crest of the ridge for a few moments before advancing down its slopes.

But now it is broad morning and the slanting sun adds brilliant colour to the scene. The masses have defined themselves into swarms of men, in ordered ranks bright with glittering weapons, and above them dance a multitude of gorgeous flags. We see for ourselves what the Crusaders saw. We must see more of it. I trot briskly forward to somewhere near the sand-hills where the 21st Lancers had halted the day before. Here we are scarcely 400 yards away from the great masses. We halt again and I make four troopers fire upon them, while the other two hold their horses. The enemy come on like the sea. A crackle of musketry breaks out on our front and to our left. Dust spurts rise among the sandhills. This is no place for Christians. We scamper off; and luckily no man nor horse is hurt. We climb back on to the ridge, and almost at this moment there returns the corporal on a panting horse. He comes direct from Kitchener with an order signed by the Chief of Staff. "Remain as long as possible, and report how the masses of attack are moving." Talk of Fun! Where will you beat this! On horseback, at daybreak, within shot of an advancing army, seeing everything, and corresponding direct with Headquarters.

So we remained on the ridge for nearly half an hour and I watched close up a scene which few have witnessed. All the masses except one passed for a time out of our view beyond the peak of Surgham on our right. But one, a division of certainly 6,000 men moved directly over the shoulder of the ridge. Already they were climbing its forward slopes. From where we sat on our horses we could see both sides. There was our army ranked and massed by the river. There were the gunboats lying expectant in the stream. There were all the batteries ready to open. And meanwhile on the other side, this large oblong gay-coloured crowd in fairly good order climbed swiftly up to the crest of exposure. We were about 2,500 yards from our own batteries, but little more than 200 from the approaching target. I called these Dervishes "The White Flags". They reminded me of the armies in the Bayeux tapestries, because of their rows of white and yellow standards held upright. Meanwhile the Dervish centre far out in the plain had come within range, and one after another the British and Egyptian batteries opened upon it. My eyes were riveted by a nearer scene. At the top of the hill "The White Flags" paused to rearrange their ranks and drew out a broad and solid parade along the crest. Then the cannonade turned upon them. Two or three batteries and all the gunboats, at least thirty guns, opened an intense fire. Their shells shrieked towards us and burst in scores over the heads and among the masses of the White Flagmen. We were so close, as we sat spellbound on our horses, that we almost shared their perils. I saw the full blast of Death strike this human wall. Down went their standards by dozens and their men by hundreds. Wide gaps and shapeless heaps appeared in their array. One saw them jumping and tumbling under the shrapnel bursts; but none turned back. Line after line they all streamed over the shoulder and advanced towards our zeriba, opening a heavy rifle fire which wreathed them in smoke.

Hitherto no one had taken any notice of us; but I now saw Baggara horsemen in twos and threes riding across the plain on our left towards the ridge. One of these patrols of three men came within pistol range. They were dark, cowled figures, like monks on horseback – ugly, sinister brutes with long spears. I fired a few shots at them from the saddle, and they sheered off. I did not see why we should not stop out on this ridge during the assault. I thought we could edge back towards the Nile and so watch both sides while keeping out of harm's way. But now arrived a positive order from Major Finn saying "Come back at once into the zeriba as the infantry are about to open fire." We should in fact have been safer on the ridge, for we only just got into the infantry lines before the rifle-storm began . . .

As soon as the fire began to slacken and it was said on all sides that the attack had been repulsed, a General arrived with his staff at a gallop with instant orders to mount and advance. In two minutes the four squadrons were mounted and trotting out of the zeriba in a southerly direction. We ascended again the slopes of Jebel Surgham which had played its part in the first stages of the action, and from its ridges soon saw before us the whole plain of Omdurman with the vast mud city, its minarets and domes, spread before us six or seven miles away. After various halts and reconnoitrings we found ourselves walking forward in what is called "column of troops". There are four troops in a squadron and four squadrons in a regiment. Each of these troops now followed the other. I commanded the second troop from the rear, comprising between twenty and twenty-five Lancers.

Everyone expected that we were going to make a charge. That was the one idea that had been in all minds since we had started from Cairo. Of course there would be a charge. In those days, before the Boer War, British cavalry had been taught little else. Here was clearly the occasion for a charge. But against what body of enemy, over what ground, in which direction or with what purpose, were matters hidden from the rank and file. We continued to pace forward over the hard sand, peering into the mirage-twisted plain in a high state of suppressed excitement. Presently I noticed, 300 yards away on our flank and parallel to the line on which we were advancing, a long row of blue-black objects, two or three yards apart. I thought there were about a hundred and fifty. Then I became sure that these were men – enemy men – squatting on the ground. Almost at the same moment the trumpet sounded "Trot", and the whole long column of cavalry began to jingle and clatter across the front of these crouching figures. We were in the lull of the battle and there was perfect silence. Forthwith from every blue-black blob came a white puff of smoke, and a loud volley of musketry broke the odd stillness. Such a target at such a distance could scarcely be missed, and all along the column here and there horses bounded and a few men fell.

The intentions of our Colonel had no doubt been to move round the flank of the body of Dervishes he had now located, and who, concealed in a fold of the ground behind their riflemen, were invisible to us, and then to attack them from a more advantageous quarter; but once the fire was opened and losses began to grow, he must have judged it inexpedient to prolong his procession across the open plain. The trumpet sounded "Right wheel into line", and all the sixteen troops swung round towards the blue-black riflemen. Almost immediately the regiment broke into a gallop, and the 21st Lancers were committed to their first charge in war!

I propose to describe exactly what happened to me: what I saw and what I felt. The troop I commanded was, when we wheeled into line, the second from the right of the regiment. I was riding a handy, sure-footed, grey Arab polo pony. Before we wheeled and began to gallop, the officers had been marching with drawn swords. On account of my shoulder I had always decided that if I were involved in hand-to-hand fighting, I must use a pistol and not a sword. I had purchased in London a Mauser automatic pistol, then the newest and latest design. I had practised carefully with this during our march and journey up the river. This then was the weapon with which I determined to fight. I had first of all to return my sword into its scabbard, which is not the easiest thing to do at a gallop. I had then to draw my pistol from its wooden holster and bring it to full cock. This dual operation took an appreciable time, and until it was finished, apart from a few glances to my left to see what effect the fire was producing, I did not look up at the general scene.

Then I saw immediately before me, and now only half the length of a polo ground away, the row of crouching blue figures firing frantically, wreathed in white smoke. On my right and left my neighbouring troop leaders made a good line. Immediately behind was a long dancing row of lances couched for the charge. We were going at a fast but steady gallop. There was too much trampling and rifle fire to hear any bullets. After this glance to the right and left and at my troop, I looked again towards the enemy. The scene appeared to be suddenly transformed. The blue-black men were still firing, but behind them there now came into view a depression like a shallow sunken road. This was crowded and crammed with men rising up from the ground where they had hidden. Bright flags appeared as if by magic, and I saw arriving from nowhere Emirs on horseback among and around the mass of the enemy. The Dervishes appeared to be ten or twelve deep at the thickest, a great grey mass gleaming with steel, filling the dry watercourse. In the same twinkling of an eye I saw also that our right overlapped their left, that my troop would just strike the edge of their array, and that the troop on my right would charge into air. My subaltern comrade on the right, Wormald of the 7th Hussars, could see the situation too; and we both increased our speed to the very fastest gallop and curved inwards like the horns of the moon. One really had not time to be frightened or to think of anything else but these particular necessary actions which I have described. They completely occupied mind and senses.

The collision was now very near. I saw immediately before me, not ten yards away, the two blue men who lay in my path. They were perhaps a couple of yards apart. I rode at the interval between them. They both fired. I passed through the smoke conscious that I was unhurt. The trooper immediately behind me was killed at this place and at this moment, whether by these shots or not I do not know. I checked my pony as the ground began to fall away beneath his feet. The clever animal dropped like a cat four or five feet down on the sandy bed of the watercourse, and in this sandy bed I found myself surrounded by what seemed to be dozens of men. They were not thickly packed enough at this point for me to experience any actual collision with them. Whereas Grenfell's troop next but one on my left was brought to a complete standstill and suffered very heavy losses, we seemed to push our way through as one has sometimes seen mounted policemen break up a crowd. In less time than it takes to relate, my pony had scrambled up the other side of the ditch. I looked round.

Once again I was on the hard, crisp desert, my horse at a trot. I had the impression of scattered Dervishes running to and fro in all directions. Straight before me a man threw himself on the ground. The reader must remember that I had been trained as a cavalry soldier to believe that if ever cavalry broke into a mass of infantry, the latter would be at their mercy. My first idea therefore was that the man was terrified. But simultaneously I saw the gleam of his curved sword as he drew it back for a ham-stringing cut. I had room and time enough to turn my pony out of his reach, and leaning over on the off side I fired two shots into him at about three yards. As I straightened myself in the saddle, I saw before me another figure with uplifted sword. I raised my pistol and fired. So close were we that the pistol itself actually struck him. Man and sword disappeared below and behind me. On my left, ten yards away, was an Arab horseman in a bright-coloured tunic and steel helmet, with chain-mail hangings. I fired at him. He turned aside. I pulled my horse into a walk and looked around again . . . There was a mass of Dervishes about forty or fifty yards away on my left. They were huddling and clumping themselves together, rallying for mutual protection. They seemed wild with excitement, dancing about on their feet, shaking their spears up and down. The whole scene seemed to flicker. I have an impression, but it is too fleeting to define, of brown-clad Lancers mixed up here and there with this surging mob. The scattered individuals in my immediate neighbourhood made no attempt to molest me. Where was my troop? Where were the other troops of the squadron? Within a hundred yards of me I could not see a single officer or man. I looked back at the Dervish mass. I saw two or three riflemen crouching and aiming their rifles at me from the fringe of it. Then for the first time that morning I experienced a sudden sensation of fear. I felt myself absolutely alone. I thought these riflemen would hit me and the rest devour me like wolves. What a fool I was to loiter like this in the midst of the enemy! I crouched over the saddle, spurred my horse into a gallop and drew clear of the *mêlée*. Two or three hundred yards away I found my troop all ready faced about and partly formed up.

The other three troops of the squadron were reforming close by. Suddenly in the midst of the troop up sprang a Dervish. How he got there I do not know. He must have leaped out of some scrub or hole. All the troopers turned upon him thrusting with their lances: but he darted to and fro causing for the moment a frantic commotion. Wounded several times, he staggered towards me raising his spear. I shot him at less than a yard. He fell on the sand, and lay there dead. How easy to kill a man! But I did not worry about it. I found I had fired the whole magazine of my Mauser pistol, so I put in a new clip of ten cartridges before thinking of anything else.

I was still prepossessed with the idea that we had inflicted great slaughter on the enemy and had scarcely suffered at all ourselves. Three or four men were missing from my troop. Six men and nine or ten horses were bleeding from spear thrusts or sword cuts. We all expected to be ordered immediately to charge back again. The men were ready, though they all looked serious. Several asked to be allowed to throw away their lances and draw their swords. I asked my second sergeant if he had enjoyed himself. His answer was "Well, I don't exactly say I enjoyed it, Sir; but I think I'll get more used to it next time." At this the whole troop laughed.

Lloyd Lewis

Bloody Shiloh

At Shiloh the losses to the Union and Confederacy were 10,000 men apiece. Of the battles of the American Civil war, only Gettysburg and Antietam surpassed such bloodiness.

THE dawn came up on Sunday, April 6, to shine red on the peach blossoms that were flowering in Tennessee. Among the fluttering petals, buglers in blue uniforms stood up and their horns wailed "The-devil-is-loose, the-devil-is-loose." The routine reveille snarled through the tents and the Army of the Tennessee awakened to remember that they were soldiers face to face with another day of camp life. There had been a little scare on Friday evening when some gray cavalry had galloped up with a few cannon to annoy the outposts, but that meant nothing more than bluff. Sherman had pursued the enemy for five miles with his brigade, only to find no respectable force menacing him. On Saturday afternoon Colonel Jesse J. Appler of the Fifty-third Ohio, holding the most advanced position, had sent Sherman word that a large force of the foe was approaching, and the red-haired commander, bulging with confidence, had answered, "Take your damned regiment back to Ohio. There is no enemy nearer than Corinth." That afternoon he had wired Grant, "I do not apprehend anything like an attack upon our position."

Ever since his arrival at Pittsburgh Landing, Sherman had been listening to wild-eyed pickets rushing in with tales of massed armies "out there," and always he had found on investigation merely a few squads of Southern cavalrymen scampering away. He had had enough of these camp rumors in Kentucky and would not make the same mistake again. In Sherman's tent, his new aide-de-camp, Lieutenant John T. Taylor, asked why he didn't march out to fight the "Rebs" over in Corinth. Sherman replied, "Never mind, young man, you'll have all the fighting you want before this war is over."

Now the Confederates, looking at the red dawn, exclaimed, "The sun of Austerlitz!" – so filled were they with Napoleonic mottoes. No bugles blew; the whole Southern army stepped quietly into battle array. "Tonight we will water our horses in the Tennessee," said Johnston, his large mustachios flaring. At half-past five the brigades, spread wide, came marching through the dew, straight down the ribs of the giant fan, aiming at the Landing in the handle. Men in the ranks carried their muskets at right-shoulder shift; the skirmishers ahead bore their guns like quail-hunters.

Johnston's battle scheme was to strike the Union right, then let the whole Southern line, as it came up, roll down the length of the Union front – a method that would begin with Sherman, proceed to Prentiss, then engage Hurlbut and W. H. L. Wallace on the left.

"What a beautiful morning this is!" said boys of the Eighty-first Ohio as they washed their faces in front of their tents, stuffed their shirt tails inside their trousers, and stretched themselves. The birds and insects sang with that especial loudness which they seemed to possess on Sundays. Breakfast was cooking. Shots popped among the trees, far away. "Those pickets again," everybody said. The Eighty-first, well to the rear – they were in W. H. L. Wallace's division – did not know that the shooting came from skirmishers whom Prentiss had sent out to reconnoiter. Prentiss, the volunteer officer, was warier than his neighbor Sherman, the trained soldier.

In Sherman's lines, so much nearer the sound of this first clash between the opposing skirmishers, there was deadly calm. One man, that timorous leader, Colonel Appler of the Fifty-third Ohio, took alarm and had his drums sound the long roll. He had cried "Wolf" so often that his men, grumbling, took their own time about falling into formation. Suddenly a private of the Twenty-fifth Missouri, one of Prentiss's outposts, stumbled out of the thicket, holding a wound and calling, "Get into line, the Rebels are coming!" Appler sent a courier to Sherman, who sent back word, "You must be badly scared over there." Neighboring regiments, accustomed to Appler's chronic uneasiness, went on with their breakfasts.

An officer of the Fifty-third who had gone into the bushes half dressed came scrambling back howling, "Colonel, the

Rebels are crossing the field!" Appler hurried two companies out to see and one of their captains rushed back with the news, "The Rebels are out there thicker than fleas on a dog's back!" At that moment the quail-hunting skirmishers of the Confederate advance stalked into view within musket shot of Appler's right flank. "Look, Colonel!" an officer shouted. A Union skirmisher dashed in yelling, "Get ready, the Johnnies are here thicker than Spanish needles in a fence corner!"

"This is no place for us," wailed Appler, and ordering battalions right to meet the Southern threat, shook in his shoes. His men, who had never held a battalion drill, were confused and milled about pathetically. Cooks left their camp kettles and ran. The sick, one third of the regiment, were carried to the rear. Sherman, one orderly behind him, rode up and trained his glasses on a part of the field that was as yet clear. The quail-hunters raised their rifles. "Sherman will be shot!" cried the Fifty-third. "General, look to your right!"

The general looked, threw up his hand, snapping, "My God, we're attacked!" As he said it the Confederates fired and his orderly fell dead, the first mortality at Shiloh. "Colonel Appler, hold your position! I'll support you!" shouted Sherman, and he spurred away for reinforcements. Appler received the encouraging news, walked over to a tree, and lay down behind it, his face like ashes. His men, forming in a wavering line, began to shoot at the Confederates, whose main line, guns flashing in the sun, came out of the woods. "Retreat! Save yourselves!" bawled Appler, and jumping up from the shelter of the tree, he bounded away to the rear and so out of the Civil War. The Fifty-third wavered. Some boys followed their colonel; the rest began to shoot at the enemy.

An incessant humming was going on among the tree tops. The boys said it sounded like a swarm of bees. Then the leaden swarm drew lower and lower until men began to fall down under its stings. It was all new and puzzling. When a man fell wounded his friends dropped their guns and helped him to the rear, staring at the blood with horror and curiosity. There were no stretchers, no hospital attendants at hand, no first-aid kits. Men bled to death because no comrade knew how to stanch the flow with a twisted handkerchief. One boy of the Fifty-third was hit a glancing blow in the shin and sat down, rubbing the place and squalling loudly. It hurt, bad!

Private A. C. Voris of the Seventeenth Illinois, which stood close by, left his regiment and came over to help the leaderless

Fifty-third. He had served at Fort Donelson and was therefore a veteran among these apprentice killers. Walking calmly among them, Voris taught the trembling youngsters how to use their guns. He would aim, fire, reload, and talk. "I've met the elephant before and the way to do is to keep cool and aim low." His rifle would go "Crack!" then his voice would resume: "It's just like shooting squirrels, only these squirrels have guns, that's all." The Fifty-third began to do better. Soon Voris, seeing his own regiment moving off, called "Good-by!" and left, but the Ohio boys never forgot him, even if a little later they all ran away. After their flight they re-formed, promoted Captain Jones to the colonelcy, and marched back into the fight in scattered units.

Recruits like those of the Fifty-third were scampering away from all parts of the field before nine o'clock, and soon a number, estimated by Grant to be 8,000 were hiding under the bluffs by the river screeching in terror. Grant, who had hurried down from Savannah at the first sound of guns, wasted no time trying to re-form the fugitives. "Excluding these troops who fled, panic-stricken before they had fired a shot, there was not a time," he said, "when during the day we had more than 25,000 men in line." This 25,000, however, learned the business of battle quickly. Considering their lack of training it would not have been surprising if they had all run; so said British military critics when they studied the battle years later. The average Federal stood his ground, shooting at enemies sometimes not more than thirty feet away. When the Confederates derisively shouted "Bull Run!" the Union boys gave them back "Donelson!" in a jeering bellow.

To join them came a thin trickle of soldiers who, after fleeing, regained self-command on the river bank. Surgeon Horace Wardner, Twelfth Illinois, was working among the wounded on the wharf when he heard a large splash and looked up to see a demoralized horseman trying to swim the river on horseback. Some fifty yards from shore the animal wheeled, unseating its rider, and headed back. Frantically the cavalryman caught the passing tail and was towed to land. The ducking had cooled his blood and, gathering up weapons, he mounted and rode toward the battle.

So stoutly did Sherman hold the Union right that Johnston failed in his scheme for rolling up the Federal line like a sheet of paper. With his face and red beard black with powder, Sherman dashed up and down the field, re-forming regiments as fast as they crumbled, plugging leaks in the human dike, drawing back his force, step by step, and succeeding, somehow, in keeping the stormy tide of Southerners from breaking through. Confederate batteries were shelling his force heavily and volleys of musket balls and buckshot swept the ground. One buckshot penetrated his palm, but without taking his eyes off the enemy he wrapped a handkerchief about it and thrust his hand into his breast. Another ball tore his shoulder strap, scratching the skin. Captain William Reuben Rowley, aide-de-camp to Grant, arriving to ask how the battle was going, found Sherman standing with his uninjured hand resting on a tree, his eyes watching his skirmishers.

"Tell Grant," he said, "if he has any men to spare I can use them; if not, I will do the best I can. We are holding them pretty well just now – pretty well – but it's hot as hell."

Four horses had died between Sherman's knees. At the death of the first, Lieutenant Taylor dismounted and handed his reins to the general. Swinging into the saddle, Sherman said, "Well, my boy, didn't I promise you all the fighting you could do?" Albert D. Richardson, collecting descriptions of Sherman from his men after the battle, said that at this point in the encounter:

> All around him were excited orderlies and officers, but though his face was besmeared with powder and blood, battle seemed to have cooled his usually hot nerves.

Other soldiers said that during the battle Sherman hadn't waved his arms when he talked, nor talked so much, as in the past. His lips were shut tight, his eyelids narrowed to a slit. He let his cigars go out more often than in peace times. He didn't puff smoke as furiously as in camp. John Day of Battery A in the Chicago Light Artillery saw Sherman halt his spurring progress over the field by the guns, again and again. Brass missionaries, the cannoneers called their pieces, having vowed to "convert the Rebels or send 'em to Kingdom Come." Day remembered that during the fight "Sherman had trouble keeping his cigar lit and he used up all his matches and most of the men's."

Thomas Kilby Smith, officer of the Fifty-fourth Ohio, and a family friend of the Ewings, watched Sherman with worshipful eyes and wrote home, "Sherman's cheek never blanched." For the second time in his life, Sherman had found something to make him forget himself, completely, utterly. He had caught that sharp rapture of absorption as a youth painting pictures on canvas in South Carolina. Now he had found it again – the strange joy of profound selflessness. Here in the storm and thunder of Shiloh, the artist found his art. His nerves, so close to the thin skin, congealed into ice. Sometimes he held his horse motionless for a period, studying the enemy, the dead and wounded piled high before him. He did not notice them, yet they were the same boys for whose safety he had worried himself in Louisville to the brink of lunacy.

Soldiers around him thought he saw and foresaw everything. When his right wing fell back, he grinned, saying, "I was looking for that," and loosed a battery that halted the charging Confederates in stricken postures. When his chief of staff, Major Dan Sanger, pointed out Southern cavalry charging the battery, Sherman produced two companies of infantry that had been held for this emergency. They shot riders from saddles while Sherman went on with his cannonade.

For all its absorption, his mind – perhaps his subconscious mind – was photographing hideous pictures, sharp negatives, and storing them away. Later on they would become vivid positives:

> our wounded mingled with rebels, charred and blackened by the burning tents and grass, crawling about begging for some one to end their misery . . . the bones of living men crushed beneath the cannon wheels coming left about . . . 10,000 men lying in a field not more than a mile by half a mile.

The field of which he spoke was a cocklebur meadow in the front. Across it Beauregard had sent his Irish-born dare-devil General Patrick R. Cleburne, to lose one third of the brigade in the fury of Sherman's fire. Probably all that saved the life of Cleburne was an accident; his horse stumbled at the start of the charge, sinking the general in the mud and separating him from his command. When the day was done, many observers said that a man could have walked all over the cocklebur meadow using bodies for stepping-stones.

Novitiates though the Northern boys might be at the profession of war, most of them were trained squirrel-shooters who, once they had mastered the complexities of the newfangled muskets, did lavish execution at point-blank range. After the first flurry, nothing could terrify them, not even the Rebel yell that had first been heard at Fort Donelson. This incoherent battle cry was distinguished by a peculiar shrillness from the deeper shouts of the Federals.

By ten o'clock the Northerners had steadied enough to begin counter-charges, the Twentieth Illinois, for instance, fighting back and forth through its camps a half-dozen times. At this hour Grant, making the rounds, had ridden quietly up to Sherman, upon whom the full fury of Southern determination continued to fall. Grant said that he had anticipated Sherman's need of cartridges, and that he was satisfied the enemy could be held. He said that he was needed more elsewhere and galloped away. It was their first meeting under fire, and in the smoke they gauged each other. Later Grant said, "In thus moving along the line, I never deemed it important to stay long with Sherman."

It was at this hour of 10 a.m. that the battle settled into what most of those participants who survived the war would describe as the fiercest they ever saw. Regiments mixed, blue and gray, in the hit-trip-smother. Men carried away confused memories – awful sheets of flame . . . the endless zip-zip of musket balls, canister . . . the shudder of grapeshot . . . dirt, gravel, twigs, pieces of bark, flying in their faces . . . splinters like knives ripping open bodies . . . men tearing paper cartridges, ramming them down musket barrels, capping the guns, firing, and as likely as not forgetting to remove the ramrods, not missing them, in fact, until they saw them quivering like arrows in the throats of enemies fifty feet away. Sense and hearing were stunned by the crash of exploding powder and the death shrieks of boys. Fountains of warm wet blood sprayed on the faces and hands of the living, brains spattered on coat sleeves. Men moved convulsively, wondering whether this moment – now – would be their last. When the Fifty-fifth Illinois retreated into a blind ravine, Confederates slaughtered them from the gully edge. "It was like shooting into a flock of sheep," said Major Whitfield of the Ninth Missippi, and years later he was still saying, "I never saw such cruel work during the war."

Prentiss, who had been forced slowly backward, finally anchored his regiments in a sunken road and by a concentrated fire was achieving a carnage hitherto unimagined by any one of the youths involved. The Hornet's Nest, the Southerners called this sector as they worked in it for six hours, trampling their own dead and wounded. Federals noted how the charging lines would wave like standing grain when a volley cut through them. Others said the lines when hit wobbled like a loose rope shaken at one end. At times the graycoats simply bent their heads as to a sleet storm. For beginners the Southerners were as brave as the Federals, and vice versa – farm boys all, learning a new trade.

A young private of the Fourteenth Illinois came up to Lieutenant Colonel Cam, fumbling at his entrails, which were trying to escape through a great slit in his abdomen, made by a passing shell. The slippery intestines kept working through his fingers. "Oh, Colonel, what shall I do?" he pleaded. Cam laid him gently behind a tree, wiped tears off his own cheeks, then walked back into the killing. Johnson, an officer of the same regiment, spurred his horse after an elderly Confederate officer, shot him through the body, reached out and seized his victim by the hair. To his horror the whole scalp came off as the Southerner slipped dead from the saddle. A roar of laughter arose above the battle clash, and Johnson saw that he held a wig.

Private Robert Oliver of the Fifty-fifth Illinois saw Private James Goodwin walk off the field resembling Mephistopheles in a play: "He looked like he had been dipped in a barrel of blood." Goodwin carried seven bullet holes in his skin. The Union Sergeant Lacey saw George F. Farwell, a company bugler, sitting against a tree reading a letter. Lacey shook him and found that he was dead, his sightless eyes still fast upon his wife's handwriting. Colonel (afterwards General) Joseph Wheeler of the Southern force said, "The Yankee bullets were so thick I imagined if I held up a bushel basket it would fill in a minute." A boy of the Fifty-third Ohio, joining another outfit, was wounded and sent to the rear, but was soon back saying, "Captain, give me a gun, this damned fight ain't got any rear." Units were surrounded at times without knowing it and were rescued only by the equality of their enemy's ignorance.

Lieutenant James H. Wilson of Grant's staff caught a youth starting for the rear, shook him, and called him a coward. The soldier protested indignantly. "I've only lost confidence in my colonel," he said. Private Sam Durkee of Waterhouse's Battery, close to Sherman often during the day, felt a blow on the seat of his trousers as he bent over his cannon, and looking around, he saw the heels of his lieutenant's horse flirting past. Durkee yelled above the cannonade, "Why did you let your horse kick me?" "I didn't," screamed the officer. Sam felt his posterior with his hand and found blood. "Oh, I'm wounded!" he screeched. Ed Russell, thumbing the vent of a cannon nearby, went down with a solid shot through his abdomen, lived twenty minutes, and shook hands with every man in the battery before he died. Men with lung wounds lay heaving, every breath hissing through holes in their chests.

The most ghastly killing of all took place when Albert Sidney Johnston assailed the Peach Orchard, a knob left of the Union center. Hurlbut, defending it, placed his men on their stomachs in a double row to shoot Johnston's men like rabbits "a-settin'." Before such a blast the Confederate boys at length withered and refused to try again. Johnston rode along the front. In one hand he carried a small tin cup that he had picked up in the sack of a Union camp and forgotten to drop. He touched bayonets with it, crying, "Men, they are stubborn; we must use the bayonet!" The Southern boys admitted that Johnston was magnificent and that his horse Fire-eater was beautiful, but they did not want to go into that sheet-flame death again. Suddenly Johnston swung his horse toward the foe and shouted, "Come, I will lead you!" Boys felt hot blood in their veins once more and, rushing past him, took the Peach Orchard, although they left comrades in rows behind them.

Fire-eater was hit four times. Johnston's clothes were pierced, one ball ripping the sole of his boot. He flapped the sole, laughing "They didn't trip me that time!" Then he reeled. Searching hands could at first find no wound, but at length came upon a boot full of blood. Johnston was dead at two-thirty in the afternoon. A tourniquet might have saved him from the thigh wound that drained his life.

The capture of the Peach Orchard was not decisive. Hurlbut fell back to another strong position, and his men fired so rapidly as to shave down saplings and thickets as if with gardeners' shears. Between the lines thirsty men from both forces drank side by side. Wounded soldiers died while drinking, staining the water red. The Bloody Pool, it was called long afterwards.

In the retirement from the orchard, two wounded gunners tried to move their cannon with one horse. Mud stalled them. They decided to give up. Just then a stray bullet obligingly struck the horse at the base of the tail and with an astonished snort and lurch the animal took the gun off to safety.

Battered slowly, steadily backward, Grant did not lose confidence. At 3 p.m. he calmly began to assemble cannon on high ground near the Landing, parking the guns wheel to wheel and collecting enough ammunition for a final burst of flame, which at close quarters was expected to destroy any possible number of assailants.

At 4 p.m. the crisis of the battle came. In nine hours of fighting the Confederates had captured 23 cannon, and pushed the Union line back a mile or more. At the beginning Sherman had prevented them from turning the Union flank, yet they had seized three out of five Northern division camps, shooting some Federals in night clothes among the tent ropes. Some of the attackers who reached the camps so suddenly owed their success to their blue uniforms, Union batteries having let them advance unmolested. These rows of tents had helped save the Northern battle line from complete breakage, for the Confederates halted to loot the camps. It had been almost twelve hours since many of them had eaten, and they forgot the battle in their hunger for the half-cooked breakfasts standing on Union fires. They rifled haversacks, drank whisky, and read the letters of Federal privates that fluttered on tent floors.

The Confederates had been at fault, too, in attacking in long lines on so rough and broken a front. The battle had promptly split up into many individual struggles, with cooperation between generals impossible. Bragg, smashing fiercely at Prentiss, finally surrounded him, but could find no brother Confederate to push in through the open spaces on right and left to divide the Union line into three sections. By the time Prentiss had surrendered to save the lives of his remaining 2,000 men, Grant had patched up a solid front line again, Sherman and McClelland had fallen back into a more solid array, and the Federals waited for the next assault.

It never came. Beauregard, succeeding to the Confederate command on Johnston's death, saw that his men had had enough. Many had left their posts to go over and stare at the Union prisoners. Organization was broken, officers were separated from men, losses had been frightful. Furthermore, a dull, heavy, and monotonous pum-pum had begun to sound from the river. Two Union gunboats, escorting Buell's army in its advance up the stream, had begun to throw shells into the Confederate lines.

Shortly after six o'clock that morning, Grant had sent word for Buell's advance guard, under General "Bull" Nelson of Kentucky, to make haste. A steamer with rush orders had gone on to tell Buell at Savannah to bring up his whole force. To his men Buell seemed negligent as he listened to the distant guns. Boys of the Fifty-first Indiana Volunteers said he was "seemingly unconcerned – a condition of mind and heart almost universally attributed to him by the men of his command." Their regimental historian described the scene:

> Colonel Streight stormed around at a great rate and Captain Will Searce became so impatient that he cried like a child and railed out against Buell, characterizing him as a rebel. Looking up he saw Buell not forty feet away. He had certainly heard the remark but took no notice. We paced up and down the bank like caged animals.

Although there was no convincing proof of the not uncommon charge that Buell's loyalty was doubtful, the man had too much of his friend McClelland's jealousies and prima donna's outlook ever to fit into the Western way of war. Twenty-two years after Shiloh, Sherman wrote James B. Fry, one of Buell's officers:

> General Grant believes, and we all do, that you [Buell's army] were derelict in coming by the short line . . . so deliberately and slowly as to show a purpose, while Sidney Johnston moved around by the longer line and made his concentration and attack on us before you arrived, and long after you should have been there to help us on the *first day*.

Buell arrived at the Landing in mid-afternoon, in advance of his men, and concluded from the sight of 8,000 fugitives at the wharf that the Army of the Tennessee had been defeated. He later insisted that Grant, whom he soon met, gave him a similar impression. But Sherman, who conversed with Grant at almost the same time, declared that the latter had talked quite differently, saying that at Donelson he had noticed that there came a time "when either side was ready to give way if the other showed a bold front." He had decided to be the bold one, and had won. Now, he said, the enemy had shot its bolt and with Buell's force available by morning, victory was sure.

Near dusk, Sherman, meeting Buell and Fry, told them that the Army of the Tennessee had 18,000 men in line, that Lew Wallace's 6,000 "had just come in and that I had orders from General Grant in person to attack at daylight the next morning." He was glad Buell had come, but thought victory certain even without him. Buell regarded this as a poor way to welcome him, "the savior of the day."

The battle dwindled as twilight spread. Grant and Sherman had narrow escapes at almost the same moment. A shell, missing Grant, tore the whole head, except for a strip of chin, from a captain beside him, ripped a cantle from a saddle behind him, and bowled on to clip both legs from one of Nelson's men as he came up from the river bank.

Sherman was swinging into his saddle when his horse pranced sufficiently to tangle around his neck the reins held by Major Hammond. As he bowed while Hammond raised the reins, a cannon ball cut the straps two inches below the major's hand and tore the crown and back rim of Sherman's hat.

Up from the Landing poured Nelson's men, stepping over piles of wounded on the wharf. The 8,000 fugitives had already trampled these bloody victims, sailors had dragged heavy cables across them, and they were now so caked with mud and dried blood that they were as black as Negroes.

The fresh legions cursed the cowards at the Landing. They thought them as terrorized as sheep who have been visited by killer dogs. In answer to these taunts, the deserters answered, "*You'll* catch it; *you'll* see. They'll cut you to pieces!" Nelson wanted to fire upon them. Colonel Jacob Ammen, the Virginia-born leader of a Union brigade, found his way blocked by a clergyman who exhorted the refugees, "Rally for God and country! Oh, rally round the flag! Oh, rally!" Always a pious Episcopalian, Ammen forgot himself this once and burst out, "Shut-up, you God-damned old fool! Get out of the way!"

The first of Nelson's men to reach the top were Rousseau's brigade – the Kentuckians who had originally disliked Sherman at Muldraugh's Hill. Now when they saw him with his hat in tatters, black powder on his red beard, his hand bandaged, they put their hats on their bayonets and cheered for Old Sherman. He pretended not to notice, but he remembered it always. It was the first really good word he had had since the beginning of the war.

While the Union officers rearranged their battered forces, Bragg had been moaning, "My God! My God!" because Beauregard would not order the one final charge that, Bragg was sure, would bring complete victory. But several days later he admitted to his wife that "our force was disorganized, demoralized and exhausted and hungry." Some of his men he described as

> too lazy to hunt the enemy's camps for provisions. They were mostly out of ammunition and though millions of cartridges were around them, not one officer in ten supplied his men. . . . Our failure is entirely due to a want of discipline and a want of officers. Universal suffrage, furloughs and whisky have ruined us.

It was just such a letter as Sherman would have written had he been in Bragg's shoes.

That night Bragg and Beauregard slept in Sherman's vacated tent. Nearby, the captive Prentiss slept among Confederates he had known before the war. He twitted his hosts about the defeat awaiting them on the morrow. "Do you hear that?" he would say, awakening them in the night, when the boom of the United States Navy cannon came from the river. Colonel Nathan Bedford Forrest, lately a slave-trader, now an unmilitary but surpassingly warlike cavalry leader in the Southern army, walked through the bivouacs of his men confiding to brother officers, "If the enemy attack us in the morning they will whip us like hell."

Grant was riding through the Union camps with substantially the same message, hunting out his commanders in the chaos to tell them that he was going to attack at daylight. To General Rusling, he said quietly, "Whichever side takes the initiative in the morning will make the other retire, and Beauregard will be mighty smart if he attacks before I do."

Across the torn field, men slept with the roar of the gunboats and the screams of the wounded ripping the air. Hospitals had broken down. Surgeons, swamped with work, did what they could, slicing and sawing in desperate haste. Flies had been blackening wounds all day. A mixture of whisky and chloroform was the only antiseptic, and when it was poured on mangled flesh it brought out maggots "on a canter," as the sufferers grimly said. Rain fell, bringing misery to the tentless warriors and relief to the burning lips of the sufferers.

When the lightning flashed the wet and weary Confederates saw sickening sights all around them – naked, bloating flesh, ghastly white faces – and they heard the moaning refrains, "Water! Water!" in the storm. A. H. Mecklin, a Bible student who had joined a Mississippi regiment, thought he heard wild hogs in the bushes. "Through the dark I heard the sound of hogs quarreling over their carnival feasts." He admitted, however, that the sound was not unmistakable.

As the night grew gray with morning, Lieutenant William George Stevenson, of Beauregard's staff, rode the field searching for his chief. Stevenson had seen a cannon ball take off the head of an earlier mount and was sick of everything. His new horse balked at a little ravine. He said afterward:

> He hesitated and I glanced down to detect the cause. The rain had washed leaves out of the narrow channel down the gully some six inches wide, leaving the hard clay exposed. Down this pathway ran a band of blood nearly an inch thick, filling the channel. Striking my rowels into the horse to escape the horrible sight, he plunged his foot into the stream of blood and threw the already thickening mass in ropy folds up on the dead leaves on the bank.

Through both battered and bleeding armies ran the folk saying, "Nobody ever wins who starts a battle on a Sunday."

Monday saw sharp, bitter fighting but victory for the North was certain. General Lew Wallace with 5,000 men arrived and took their places in the line. Wallace had started early on Sunday morning to march the five miles to Shiloh Meeting-house, but had wandered around the country all day within sound of the battle without being able to find it. Whether the mistake was his or that of Grant's aides was a matter of dispute for years to come. Beauregard, calling the roll at dawn, found only half of his original 40,000 men at hand – and these were disorganized. Nevertheless the Confederates fought stoutly for eight hours more.

At 3 p.m. on Monday Grant, gathering up fragments of regiments led them in one last charge that broke Confederate resistance, and Shiloh was won. That evening the cold, drizzling rain resumed, gradually turning to sleet and hail that bruised the butchered Southern boys who lay in the young spring grass or who had been piled like bags of grain into open wagons for the jolting trip to Corinth. Against orders, Confederate privates crowded into the tent of General Breckinridge and stood there packed and wretched while the water ran in under the tent flap. In defeat they had lost their awe of great men.

The hail pelted Union wounded too, as they lay shrieking on the field of victory; it knocked from the trees the last few peach blossoms that the bullets had spared.

Robert Southey

Nelson at the Nile

Fought off Aboukir Bay in 1798, the Battle of the Nile established the fame of the victorious Horatio Nelson RN, who was promptly raised to the peerage as Baron Nelson of the Nile. Seven years later, at Trafalgar, Nelson delivered another – and decisive – coup to Napoleonic naval pretensions. Southey was Nelson's biographer.

THE first news of the enemy's armament was that it had surprised Malta. Nelson formed a plan for attacking it while at anchor at Gozo; but on the 22nd of June intelligence reached him that the French had left that island on the 16th, the day after their arrival. It was clear that their destination was eastward – he thought for Egypt – and for Egypt, therefore, he made all sail. Had the frigates been with him he could scarcely have failed to gain information of the enemy: for want of them, he only spoke three vessels on the way; two came from Alexandria, one from the Archipelago; and neither of them had seen anything of the French. He arrived off Alexandria on the 28th, and the enemy were not there, neither was there any account of them; but the governor was endeavouring to put the city in a state of defence, having received advice from Leghorn that the French expedition was intended against Egypt, after it had taken Malta. Nelson then shaped his course to the northward for Caramania, and steered from thence along the southern side of Candia, carrying a press of sail both night and day, with a contrary wind. It would have been his delight, he said, to have tried Buonaparte on a wind. It would have been the delight of Europe too, and the blessing of the world, if that fleet had been overtaken with its general on board. But of the myriads and millions of human beings who would have been preserved by that day's victory, there is not one to whom such essential benefit would have resulted as to Buonaparte himself. It would have spared him his defeat at Acre – his only disgrace; for to have been defeated by Nelson upon the seas would not have been disgraceful: it would have spared him all his after enormities. Hitherto his career had been glorious; the baneful principles of his heart had never yet passed his lips: history would have represented him as a soldier of fortune, who had faithfully served the cause in which he engaged; and whose career had been distinguished by a series of successes unexampled in modern times. A romantic obscurity would have hung over the expedition to Egypt, and he would have escaped the perpetration of those crimes which have incarnadined his soul with a deeper dye than that of the purple for which he committed them – those acts of perfidy, midnight murder, usurpation, and remorseless tyranny, which have consigned his name to universal execration, now and for ever.

Conceiving that when an officer is not successful in his plans it is absolutely necessary that he should explain the motives upon which they were founded, Nelson wrote at this time an account and vindication of his conduct for having carried the fleet to Egypt. The objection which he anticipated was, that he ought not to have made so long a voyage without more certain information. "My answer," said he, "is ready – who was I to get it from? The Governments of Naples and Sicily either knew not, or chose to keep me in ignorance. Was I to wait patiently until I heard certain accounts? If Egypt were their object, before I could hear of them they would have been in India. To do nothing was disgraceful; therefore I made use of my understanding. I am before your lordships' judgment; and if, under all circumstances, it is decided that I am wrong, I ought, for the sake of our country, to be superseded; for at this moment, when I know the French are not in Alexandria, I hold the same opinion as off Cape Passaro – that, under all circumstances, I was right in steering for Alexandria: and by that opinion I must stand or fall." Captain Ball, to whom he showed this paper, told him he should recommend a friend never to begin a defence of his conduct before he was accused of error: he might give the fullest reasons for what he had done, expressed in such terms as would evince that he had acted from the strongest conviction of being right; and of course he must expect that the public would view it in the same light. Captain Ball judged rightly of the public, whose first impulses, though from want of sufficient information they must frequently be erroneous, are generally founded upon just feelings. But the public are easily misled, and there are always persons ready to mislead them. Nelson had not yet attained that fame which compels envy to be silent; and when it was known in England that he had returned after an unsuccessful pursuit, it was said that he deserved impeachment; and Earl St. Vincent was severely censured for having sent so young an officer upon so important a service.

Baffled in his pursuit, he returned to Sicily. The Neapolitan ministry had determined to give his squadron no assistance, being resolved to do nothing which could possibly endanger their peace with the French Directory. By means, however, of Lady Hamilton's influence at court, he procured secret orders to the Sicilian governors; and, under those orders, obtained everything which he wanted at Syracuse – a timely supply, without which, he always said, he could not have recommenced his pursuit with any hope of success. "It is an old saying," said he in his letter, "that 'the devil's children have the devil's luck.' I cannot to this moment learn, beyond vague conjecture, where the French fleet are gone to; and having gone a round of six hundred leagues at this season of the year, with an expedition incredible, here I am, as ignorant of the situation of the enemy as I was twenty-seven days ago. Every moment I have to regret the frigates having left me; had one-half of them been with me, I could not have wanted information. Should the French be so strongly secured in port that I cannot get at them, I shall immediately shift my flag into some other ship, and send the *Vanguard* to Naples to be refitted, for hardly any person but myself would have continued on service so long in such a wretched state." Vexed, however, and disappointed as he was, Nelson, with the true spirit of a hero, was still full of hope. "Thanks to your exertions," said he, writing to Sir W. and Lady Hamilton, "we have victualled and watered; and surely watering at the fountain of Arethusa, we must have victory. We shall sail with the first breeze; and be assured I will return either crowned with laurel or covered with cypress." Earl St. Vincent he assured, that if the French were above water he would find them out – he still held his opinion that they were bound for Egypt; "but," said he to the First Lord of the Admiralty, "be they bound to the Antipodes, your lordship may rely that I will not lose a moment in bringing them to action."

On the 25th of July he sailed from Syracuse for the Morea. Anxious beyond measure, and irritated that the enemy should so long have eluded him, the tediousness of the nights made him impatient; and the officer of the watch was repeatedly called on to let him know the hour, and convince him, who measured time by his own eagerness, that it was not yet daybreak. The Squadron made the Gulf of Coron on the 28th. Trowbridge entered the port, and returned with intelligence that the French had been seen about four weeks before steering to the South-east, from Candia. Nelson then determined immediately to return to Alexandria; and the British fleet accordingly, with every sail set, stood once more for the coast of Egypt. On the 1st of August they came in sight of Alexandria; and at four in the afternoon, Captain Hood, in the *Zealous*, made the signal for the French fleet. For many preceding days Nelson had hardly taken either sleep or food; he now ordered his dinner to be served, while preparations were making for battle; and when his officers rose from the table, and went to their separate stations, he said to them: "Before this time tomorrow I shall have gained a peerage, or Westminster Abbey."

The French, steering direct for Candia, had made an angular passage for Alexandria; whereas Nelson, in pursuit of them, made straight for that place, and thus materially shortened the distance. The comparative smallness of his force made it necessary to sail in close order, and it covered a less space than it would have done if the frigates had been with him: the weather also was constantly hazy. These circumstances prevented the English from falling in with the enemy on the way to Egypt, and during the return to Syracuse there was still less probability of discovering them.

Why Buonaparte, having effected his landing, should not have suffered the fleet to return, has never yet been explained. This much is certain, that it was detained by his command; though, with his accustomed falsehood, he accused Admiral Brueys, after that officer's death, of having lingered on the coast, contrary to orders. The French fleet arrived at Alexandria on the 1st of July; and Brueys, not being able to enter the port, which time and neglect had ruined, moored his ships in Aboukir Bay, in a strong and compact line of battle; the headmost vessel, according to his own account, being as close as possible to a shoal on the North-west, and the rest of the fleet forming a kind of a curve along the line of deep water, so as not to be turned by any means in the South-west. By Buonaparte's desire he had offered a reward of ten thousand livres to any pilot of the country who would carry the squadron in; but none could be found who would venture to take charge of a single vessel drawing more than twenty feet. He had therefore made the best of his situation, and chosen the strongest position which he could possibly take in an open road. The commissary of the fleet said they were moored in such a manner as to bid defiance to a force more than double their own. This presumption could not then be thought unreasonable. Admiral Barrington, when moored in a similar manner off St. Lucia, in the year 1778, beat off the Comte d'Estaign in three several attacks, though his force was inferior by almost one-third to that which assailed it. Here, the advantage of numbers, both in ships, guns, and men, was in favour of the French. They had thirteen ships of the line and four frigates, carrying 1,196 guns and 11,230 men. The English had the same number of ships of the line, and one 50-gun ship, carrying 1,012 guns and 8,068 men. The English ships were all 74s: the French had three 80-gun ships, and one three-decker of 120.

During the whole pursuit, it had been Nelson's practice, whenever circumstances would permit, to have his captains on board the *Vanguard*, and explain to them his own ideas of the different and best modes of attack, and such plans as he proposed to execute on falling in with the enemy, whatever their situation might be. There is no possible position, it is said, which he did not take into calculation. His officers were thus fully acquainted with his principles of tactics: and such was his confidence in their abilities, that the only thing determined upon, in case they should find the French at anchor, was for the ships to form as most convenient for their mutual support, and to anchor by the stern. "First gain the victory," he said, "and then make the best use of it you can." The moment he perceived the position of the French, that intuitive genius with which Nelson was endowed displayed itself; and it instantly struck him, that where there was room for an enemy's ship to swing, there was room for one of ours to anchor. The plan which he intended to pursue, therefore, was to keep entirely on the outer side of the French line, and station his ships, as far as he was able, one on the outer bow, and another on the outer quarter, of each of the enemy's. This plan of doubling on the enemy's ships was projected by Lord Hood, when he designed to attack the

French fleet at their anchorage in Gourjean Road. Lord Hood found it impossible to make the attempt; but the thought was not lost upon Nelson, who acknowledged himself on this occasion indebted for it to his old and excellent commander. Captain Berry, when he comprehended the scope of the design, exclaimed with transport, "If we succeed, what will the world say?" "There is no *if* in the case," replied the admiral: "that we shall succeed is certain: who may live to tell the story is a very different question."

As the squadron advanced, the enemy opened a steady fire from the starboard side of their whole line, full into the bows of our van ships. It was received in silence: the men on board of every ship were employed aloft in furling sails, and below in tending the braces, and making ready for anchoring. A miserable sight for the French; who with all their skill, and all their courage, and all their advantages of numbers and situation, were upon that element, on which, when the hour of trial comes, a Frenchman has no hope. Admiral Brueys was a brave and able man; yet the indelible character of his country broke out in one of his letters, wherein he delivered it as his private opinion that the English had missed him, because, not being superior in force, they did not think it prudent to try their strength with him. The moment was now come in which he was to be undeceived.

A French brig was instructed to decoy the English, by manœuvring so as to tempt them toward a shoal lying off the island of Bequieres; but Nelson either knew the danger, or suspected some deceit, and the lure was unsuccessful. Captain Foley led the way in the *Goliath*, outsailing the *Zealous*, which for some minutes disputed this post of honour with him. He had long conceived, that if the enemy were moored in line of battle in with the land, the best plan of attack would be to lead between them and the shore, because the French guns on that side were not likely to be manned, nor even ready for action. Intending, therefore to fix himself on the inner bow of the *Guerrier*, he kept as near the edge of the bank as the depth of water would admit; but his anchor hung, and having opened his fire, he drifted to the second ship, the *Conquérant*, before it was clear; then anchored by the stern, inside of her, and in ten minutes shot away her masts. Hood, in the *Zealous*, perceiving this, took the station which the *Goliath* intended to have occupied, and he totally disabled the *Guerrier* in twelve minutes. The third ship which doubled the enemy's van was the *Orion*, Sir J. Saumarez; she passed to windward of the *Zealous*, and opened her larboard guns as long as they bore on the *Guerrier*; then passing inside the *Goliath*, sunk a frigate which annoyed her, hauled round toward the French line; and anchoring inside, between the fifth and sixth ships from the *Guerrier*, took her station on the larboard bow of the *Franklin*, and the quarter of the *Peuple Souverain*, receiving and returning the fire of both. The sun was now nearly down. The *Audacious*, Captain Gould, pouring a heavy fire into the *Guerrier* and the *Conquérant*, fixed herself on the larboard bow of the latter; and when that ship struck, passed on to the *Peuple Souverain*. The *Theseus*, Captain Miller, followed, brought down the *Guerrier*'s remaining main and mizen masts, then anchored inside of the *Spartiate*, the third in the French line.

While these advanced ships doubled the French line, the *Vanguard* was the first that anchored on the outer side of the enemy, within half-pistol-shot of their third ship, the *Spartiate*. Nelson had six colours flying in different parts of his rigging, lest they should be shot away – that they should be struck, no British admiral considers as a possibility. He veered half a cable, and instantly opened a tremendous fire; under cover of which the other four ships of his division, the *Minotaur*, *Bellerophon*, *Defence*, and *Majestic*, sailed on ahead of the admiral. In a few minutes every man stationed at the first six guns in the fore part of the *Vanguard*'s deck was killed or wounded – these guns were three times cleared. Captain Louis, in the *Minotaur*, anchored next ahead, and took off the fire of the *Aquilon*, the fourth in the enemy's line. The *Bellerophon*, Captain Darby, passed ahead, and dropped her stern anchor on the starboard bow of the *Orient*, seventh in the line, Brueys' own ship of 120 guns, whose difference of force was in proportion of more than seven to three, and whose weight of ball, from the lower deck alone, exceeded that from the whole broadside of the *Bellerophon*. Captain Peyton, in the *Defence*, took his station ahead of the *Minotaur*, and engaged the *Franklin*, the sixth in line, by which judicious movement the British line remained unbroken. The *Majestic*, Captain Westcott, got entangled with the main rigging of one of the French ships astern of the *Orient*, and suffered dreadfully from that three-decker's fire: but she swung clear, and closely engaging the *Heureux*, the ninth ship on the starboard bow, received also the fire of the *Tonnant*, which was the eighth in the line. The other four ships of the British squadron, having been detached previously at the discovery of the French, were at a considerable distance when the action began. It commenced at half after six; about seven, night closed, and there was no other light than that from the fire of the contending fleets.

Trowbridge, in the *Culloden*, the foremost of the remaining ships, was two leagues astern. He came on sounding, as the others had done; as he advanced, the increasing darkness increased the difficulty of the navigation; and suddenly, after having found eleven fathoms' water, before the lead could be hove again, he was fast aground; nor could all his own exertions, joined to those of the *Leander* and the *Mutiné* brig, which came to his assistance, get him off in time to bear a part in the action. His ship, however, served as a beacon to the *Alexander* and *Swiftsure*, which would else, from the course which they were holding, have gone considerably farther on the reef, and must inevitably have been lost. These ships entered the bay, and took their stations, in the darkness, in a manner still spoken of with admiration by all who remember it. Captain Hallowell, in the *Swiftsure*, as he was bearing down, fell in with what seemed to be a strange sail; Nelson had directed his ships to hoist four lights horizontally at the mizen-peak, as soon as it became dark; and this vessel had no such distinction. Hallowell, however, with great judgment, ordered his men not to fire; if she was an enemy, he said, she was in too disabled a state to escape; but, from her sails being loose, and the way in which her head was, it was probable she might be an English ship. It was the *Bellerophon*, overpowered by the huge *Orient*; her lights had gone overboard, nearly two hundred of her crew were killed or wounded, all her masts and cables had been shot away, and she was drifting out of the line, toward the lee side of the bay. Her station, at this important time, was occupied by the *Swiftsure*, which opened a steady fire on the quarter of the *Franklin*, and the bows of the French admiral. At the same instant, Captain Ball, with the *Alexander*, passed under his stern, and anchored within side on his larboard quarter, raking him, and keeping up a severe fire of musketry upon his decks. The last ship which arrived to complete the destruction of the enemy was the *Leander*. Captain Thompson, finding that nothing could be done that night to get off the *Culloden*, advanced with the intention of anchoring athwart hawse of the *Orient*. The *Frank-*

lin was so near her ahead, that there was not room for him to pass clear of the two; he therefore took his station athwart hawse of the latter, in such a position as to rake both.

The two first ships of the French line had been dismasted within a quarter of an hour after the commencement of the action; and the others had in that time suffered so severely that victory was already certain. The third, fourth, and fifth were taken possession of at half-past eight. Meantime, Nelson received a severe wound on the head from a piece of langridge shot. Captain Berry caught him in his arms as he was falling. The great effusion of blood occasioned apprehension that the wound was mortal; Nelson himself thought so; a large flap of the skin of the forehead, cut from the bone, had fallen over one eye, and the other being blind, he was in total darkness. When he was carried down, the surgeon – in the midst of a scene scarcely to be conceived by those who have never seen a cock-pit in time of action, and the heroism which is displayed amid its horrors – with a natural and pardonable eagerness, quitted the poor fellow then under his hands, that he might instantly attend the admiral. "No!" said Nelson, "I will take my turn with my brave fellows." Nor would he suffer his own wound to be examined till every man who had been previously wounded was properly attended to. Fully believing that the wound was mortal, and that he was about to die, as he had ever desired, in battle and in victory, he called the chaplain, and desired him to deliver what he supposed to be his dying remembrance to Lady Nelson; he then sent for Captain Louis on board from the *Minotaur*, that he might thank him personally for the great assistance which he had rendered to the *Vanguard*; and, ever mindful of those who deserved to be his friends, appointed Captain Hardy from the brig to the command of his own ship, Captain Berry having to go home with the news of the victory. When the surgeon came in due time to examine his wound (for it was in vain to entreat him to let it be examined sooner), the most anxious silence prevailed; and the joy of the wounded men, and of the whole crew, when they heard that the hurt was merely superficial, gave Nelson deeper pleasure than the unexpected assurance that his life was in no danger. The surgeon requested, and, as far as he could, ordered him to remain quiet, but Nelson could not rest. He called for his secretary, Mr. Campbell, to write the despatches. Campbell had himself been wounded, and so affected at the blind and suffering state of the admiral, that he was unable to write. The chaplain was then sent for; but before he came, Nelson, with his characteristic eagerness, took the pen, and contrived to trace a few words, marking his devout sense of the success which had already been obtained. He was now left alone; when suddenly a cry was heard on the deck, that the *Orient* was on fire. In the confusion he found his way up, unassisted and unnoticed; and, to the astonishment of every one, appeared on the quarter-deck, where he immediately gave the order that boats should be sent to the relief of the enemy.

It was soon after nine that the fire on board the *Orient* broke out. Bruey was dead: he had received three wounds, yet would not leave his post: a fourth cut him almost in two. He desired not to be carried below, but to be left to die upon deck. The flames soon mastered his ship. Her sides had just been painted, and the oil-jars and paint-buckets were lying on the poop. By the prodigious light of this conflagration, the situation of the two fleets could now be perceived, the colours of both being clearly distinguishable. About ten o'clock the ship blew up. This tremendous explosion was followed by a silence not less awful: the firing immediately ceased on both sides; and the first sound which broke the silence was the dash of her shattered masts and yards falling into the water from the vast height to which they had been exploded. It is upon record that a battle between two armies was once broken off by an earthquake – such an event would be felt like a miracle; but no incident in war, produced by human means, has ever equalled the sublimity of this co-instantaneous pause, and all its circumstances.

About seventy of the *Orient*'s crew were saved by the English boats. Among the many hundreds who perished were the commodore, Casa-Bianca, and his son, a brave boy only ten years old. They were seen floating on the wreck of a mast when the ship blew up. She had money on board to the amount of £600,000 sterling. A port fire from her fell into the main-royal of the *Alexander*; the fire which it occasioned was speedily extinguished. Captain Ball had provided, as far as human foresight could provide, against any such danger. All the shrouds and sails of his ship which were not absolutely necessary for its immediate management were thoroughly wetted, and so rolled up, that they were as hard and as little inflammable as so many solid cylinders.

The firing recommenced with the ships to leeward of the centre, and continued till about three. At daybreak the *Guil-*

laume Tell and the *Généreux*, the two rear ships of the enemy, were the only French ships of the line which had their colours flying: they cut their cables in the forenoon, not having been engaged, and stood out to sea, and two frigates with them. The *Zealous* pursued; but as there was no other ship in a condition to support Captain Hood, he was recalled. It was generally believed by the officers, that if Nelson had not been wounded, not one of these ships could have escaped: the four certainly could not, if the *Culloden* had got into action: and if the frigates belonging to the squadron had been present, not one of the enemy's fleet would have left Aboukir Bay. These four vessels, however, were all that escaped; and the victory was the most complete and glorious in the annals of naval history. "Victory," said Nelson, "is not a name strong enough for such a scene"; he called it a conquest. Of thirteen sail of the line, nine were taken and two burnt: of the four frigates, one burnt, another sunk. The British loss in killed and wounded amounted to eight hundred and ninety-five. Westcott was the only captain who fell. Three thousand one hundred and five of the French, including the wounded, were sent on shore by cartel: and five thousand two hundred and twenty-five perished.

As soon as the conquest was completed, Nelson sent orders through the fleet, to return thanksgiving in every ship for the victory with which Almighty God had blessed His Majesty's arms. The French at Rosetta, who with miserable fear beheld the engagement, were at a loss to understand the stillness of the fleet during the performance of this solemn duty; but it seemed to affect many of the prisoners, officers as well as men: and graceless and godless as the officers were, some of them remarked, that it was no wonder such order was preserved in the British Navy, when the minds of our men could be impressed with such sentiments after so great a victory, and at a moment of such confusion.

Guy Gibson

Enemy Coast Ahead

Guy Gibson VC led the "dambusters" raid on the Moehne and Eder reservoirs. Here he recounts a "bread and butter" mission for RAF Bomber Command – a night raid on Berlin.

"BERLIN to-night, chaps," says the Squadron-Commander. "Maximum effort; get cracking."

On the airfield a few minutes later there is a roar of engines as Lancaster after Lancaster takes off to be given its air test. These air tests take about half an hour each, and are very thoroughly done. Everything is tested – wireless, guns, navigational instruments, bomb doors. Sometimes even a few bombs are dropped for practice to make sure that the bomb-aimer will be on his target to-night.

Then comes lunch. A short, absent-minded meal taken in the minimum of time. Not much is said, most minds are preoccupied, often the Squadron-Commander does not have time for lunch at all. The 'phone bell is ringing incessantly.

After this comes briefing. Here all the information is transmitted to the crews who are going to take part in to-night's big effort. The room is packed, many of the boys, in their roll-necked pullovers, are standing crowding up against the back of the room. In the corner there may be one or two war correspondents and perhaps a visiting army officer.

The Squadron-Commander comes in, followed by the navigation officer, and the crews get up and stop talking. The babble of conversation dies away and he begins his briefing. "O.K., chaps; sit down. Berlin to-night, aiming point X. This is the centre of a cluster of factories making Daimler-Benz engines. You can see it quite clearly here on the map." He points to a position somewhere in Berlin. "To-night a total of 700 bombers are going. They are all four-engined types, so if you see anything twin-engined you can shoot at it. The bomb-load will be one 4,000-pounder and sixteen cans of incendiaries, so the total load will be about 2,000 tons. It ought to be a pretty good prang. The met. man[1] says the weather will be clear all the way, which is pretty phenomenal. Let's hope he's right. The Pathfinders are going to attack from zero hour minus one to zero hour plus 35; it is going to be a quick, concentrated attack. Your bombing height will be 21,000 feet. Don't get out of this height band or you will run into other aircraft. As it is, we are very lucky not to be the bottom squadron; they will probably see a few bombs whistling past them on the way down. The route is the usual one marked on the board here. The Pathfinder procedure will be detailed by the navigation officer." He cocks his head over to the corner of the room and calls, "Nav."

The navigation officer, a big round man with the D.F.C. and bar, gets up and begins talking. "Zero hour is 18.45 hours. At zero hour minus three and a third minutes the Pathfinders will sky-mark the lane to the target with red flares which will change to green after 120 seconds. At this time, too, they will mark a point on the ground exactly fifteen miles short of the aiming point. With the ground speed of 240 miles an hour this should give three and three-quarter minutes to go to the target. The timing has got to be done in seconds. If anybody is late, he will probably get a packet, so pilots must keep their air-speeds dead right. The target-indicating marker will go down at exactly zero minus one, and should be right on the factory roof. The sky above will also be marked by green flares in case the T.I.s[2] are obscured by fog or smoke. I will see all navigators after the briefing to give them the tracks and distances."

He turns round to the Squadron-Commander, who gets up again and gives his final orders. "Now don't forget, chaps," he begins; "once you have reached the preliminary target indicator you turn on to a course of 135 degrees magnetic and hold it for four minutes. You are to take no evasive action, but to keep straight on past the target. Once you have dropped your bombs you may weave about slightly and gain speed by going down in a gentle dive. The Pathfinders will drop a cluster of green and red flares thirty miles beyond Berlin, and you are to concentrate on these, and return home in a gaggle. Now don't forget, no

1 Meteorological expert.
2 Target-indicators.

straggling. We've had pretty low losses so far and we don't want any to-night; and don't forget to twist your tails a bit so that you can see those fighters, which come up from below. I think that's about all. Don't forget your landing discipline when you come to base. I will see you down in the crew room before take-off. O.K."

The boys go out noisily. Some are on their first trip and look a bit worried. The veterans look as if they are just going to a tea-party, but inside they feel differently. After the briefing a war correspondent comes up and asks the Squadron-Commander a few questions.

"Why all this concentration?" he asks. "What is the exact idea?" The Squadron-Commander is a busy man, but he gives him the whole answer. How there are so many guns in Germany, all depending on short-wave electricity for their prediction, so that if one aircraft were to go over every five minutes, each gun would have that aircraft all to itself. Similarly with the night-fighters. But if all the aircraft go over more or less simultaneously then the guns cannot pick out and fire at any one aircraft nor can the night-fighters be vectored on to any one aircraft. With the result that losses are kept down. Moreover, the bombing takes a more concentrated form when all aircraft bomb together.

"How about collisions?" the war correspondent asks.

"There won't be any," says the Squadron-Leader, "provided they keep straight, and if the Pathfinders are on time. Sometimes this doesn't happen. One night at Stuttgart the Pathfinders were fifteen minutes late and there were some 400 bombers circling the target waiting for them; eighteen didn't come back. Some of those were collisions, I think."

The time after the briefing is not very pleasant. No one knows what to do. Some sit in the Mess, listening to the radio, and wishing they were far away from all this. A few play billiards. But most of them just sit in chairs picking up papers and throwing them down, staring into space and waiting for the clock on the wall to show the time when they must go down to get on their flying-clothes.

The time passes slowly, minutes seem like hours, but it is a busy time for the Squadron-Commander and his Flight-Commanders. First Group telephones to confirm that there is the full number of aircraft on from the squadron. Then the maintenance officer to say that C Charlie has blown an engine, shall he put on the reserve? Yes, put on reserve.

A call from the armament officer – a cookie[3] has dropped off Z Zebra.

"Is everyone all right?"

"Yes, everyone's all right."

"Well, put it on again, then."

The oxygen has leaked from G George – get on to the maintenance flight to have new oxygen bottles put in. And so it goes on, the 'phone ringing the whole time. He does not have time to think, and presently everyone is in the crew rooms dressing for the big raid, putting on their multiple underwear and electrically-heated suits before going out to the aircraft.

All the boys are chattering happily, but this is only to cover up their true feelings. But they all know that they will be quite all right once they get into their aircraft.

"Prang it good, boys," says an Australian who isn't coming to-night; one of his crew is sick.

Then comes the take-off. A thrilling sight to the layman. Exactly at the right time they taxi out, led by the Squadron-Commander in his own aircraft with a gaudy design painted on the nose. They come out one after another, like a long string of ducks, and line up on the runway waiting to take off. There is a cheery wave of goodbyes from the well-wishers on the first flare. Then the pilot slams his window shut and pushes open the throttles. The ground underneath the well-wishers shivers and shakes, sending a funny feeling up their spine, and the Lancasters lumber off one after another down the mile-long flare-path. And off they go into the dusk.

Over to a farm labourer sitting on his tractor in a field . . . He has just done his ploughing and is about to go home. He is looking forward to his evening meal. Looking up, he can see hundreds of specks in the sky, black specks, all getting smaller and smaller as they climb higher and higher into the night air. He turns to his tractor and says, "They be going out again to-night. I 'ope they give 'em bastards hell. May they all come back again, God bless 'em. Good boys they be." Then he begins to trudge home.

Over to a girl typist about to get out of a bus in the nearby city. She hears the roar of the aircraft and says to her companion, "Oh, there they go again. I do hope they will come back early; otherwise they will wake me up . . ."

3 A bomb in R.A.F. slang.

Over to one of our aircraft flying high. . . . They have just reached their operational height. The engines are throttled back to cruising revolutions. "Hullo, navigator. Skipper calling. What time must I set course for rendezvous point?"

The navigator gets a quick fix. "We are about sixty miles away. If you circle here for five minutes, then set course at 240 miles an hour, you will be there dead on time."

"O.K.," says the skipper. "You all right, rear-gunner?"

"Yes," comes the voice from the back.

In five minutes' time he sets course and the blunt nose of the Lancaster points towards the east. At that moment nearly all the bombers have done the same thing and, with navigation lights on at their various heights, they all converge on to the rendezvous spot at exactly zero minus two hours. They reach it more or less together, then all navigation lights go out simultaneously and they straighten up on their course for Berlin. The captain yells to his crew to check that all lights are out on board. The bomb-aimer fuses the bombs, the gunners cock their guns and they are on their way.

To describe this big bomber force flying out in this formation is not easy. But imagine a glass brick two miles across, twenty miles long and 8,000 feet thick, filled with hundreds of Lancasters, and move it slowly towards the Dutch coast, and there you have a concentrated wave on its way. The Dutch coast looms up incredibly soon, rather too soon . . .

It is now five o'clock. At this hour in Germany operational messages have come in from Gruppen and Staffeln of night-fighters scattered throughout German territory. Messerschmitt, Focke-Wulf and other types of fighters are fully loaded with fuel and ammunition, ready for take-off from the operational bases. Aircraft and personnel are ready, mechanics, engineers, armourers are on duty on many airfields ready to supply suddenly arriving aircraft with fresh fuel and more ammunition. Everything has been done to ensure the quickest possible employment of the night-fighter arm.

At this hour it is quiet at the German searchlight and flak batteries. Ammunition stocks have been made up again since the last raid. The enormous power-plants of the searchlights need only be switched on by the young Luftwaffe helpers to convert the electric current, enough to supply a medium-sized town, into shimmering light and send it up into the night sky. The sentries on the large 8.8-cm. guns pace up and down and watch the approaching night. It will soon be pitch dark, as the sky is covered with heavy rain-clouds, and the crescent moon will not rise until later. Even then its light will scarcely pierce the dark clouds. The British prefer nights such as this.

1740 hours. A message comes into the centre near Berlin from the Channel coast. An alarm bell rings. Strong British bomber units are crossing the Dutch coast. A telephone call warns the air-defence forces of the Continent. The night-fighter units in Holland have already taken off and are on the look-out for the enemy on his eastern course, to attach themselves to his units, and while the first night engagements between the German night-fighters and the British bombers are setting the stage for the great night battle, the ground crews of countless other Geschwader in the region of Central Germany are putting the final touches to the aircraft as they stand ready to take off.

Behind the great glass map stand female signals auxiliaries wearing head-phones and laryngophones, with a thick stick of charcoal in their right hands with which they draw in the position of the enemy units. From the control room only their shadows moving behind the glass plate can be seen. Ceaselessly the strokes and arrows on the great map give place to new markings.

Every officer and man takes up his position. Each knows exactly what he has to do, and all work together without friction.

The glass map shows that the enemy is advancing along several different directions, but it is clear that the main force is continuing eastwards. The enemy bombers have crossed the frontier of Western Germany. Suddenly they swing round towards the south-east. A few weaker formations are flying southwards up the Rhine. Cascades are dropped over two West German towns; it may be that the main attack is to be directed against these towns, but it may also be that this is a feint movement designed to lure the German night-fighters into the wrong areas. The enemy hopes that a wrong German order will gain him valuable minutes to get his main attacking force into the prescribed target area, where he would then find weaker German night-fighter forces.

The control officer, who is fully acquainted with the many different problems and questions, the possibilities of attack and defence, makes his decision after conscientiously checking the situation and a brief talk with the O.C. The British force is still on its way towards Central Germany. The main force of the bombers has made another turn and is again flying east. The last message reads: "Front of enemy formation in Dora-Heinrich area, course east."

1830 hours. At this moment fighter unit X, whose aircraft are ready at the end of the runway with their engines roaring, receives the order "Unit X – village take-off by visual beacon Y."

A few minutes later the aircraft are racing over the ground, climbing rapidly, flying towards the flashing light of visual beacon Y. In Berlin the Underground is still running, and traffic goes on as usual. Then the population gets its first warning; the Deutschlandsender goes off the air. The bright lights at the marshalling yard are switched off. The great city sinks into darkness.

The enemy has meanwhile flown past to the north of the first large central German town. In a bare hour he may be over Berlin. At a height of 6,500 m.[4] the four-engine bombers are roaring on their way eastwards.

1845 hours. A message in the head-phones: the enemy has already lost seven aircraft before reaching Osnabrück.

Other night-fighter units are ready to take off to protect the capital. The meteorologist is describing the weather situation. Cloudless sky over South Germany, where night-fighters can land after the battle.

Meanwhile the night-fighter units, which have assembled in certain areas, are guided closer to the enemy. The German fighters have already made contact everywhere with the enemy bomber formations. Now the sirens are sounded in Berlin.

Important decisions are taken relating to the activity of the searchlight batteries, taking into consideration the weather situation. Orders are issued to the batteries of the Berlin flak division.

1916 hours. The enemy is 100 kilometres from Berlin. A large number of night-fighters are accompanying the British bombers.

The O.C. sits next to the 1A (Intelligence) officer. In order to clear up a question quickly he asks to speak to the O.C. in another Luftgau; command priority call to X town. In a matter of seconds a female signals auxiliary has made the desired telephone connection.

4 21,000 ft.

On the great glass map the arrows draw closer and closer to Berlin. The positions of the night-fighter units are exactly shown.

1941 hours. Is the enemy going straight for Berlin? At 1943 hours fire is opened by a heavy flak battery in the west. It is still impossible to say whether the mass of the enemy bombers will not again make a sharp turn short of Berlin and perhaps attack Leipzig.

Above the inner part of the town the enemy drops streams and cascades of flares. Strong forces are reported over various suburbs. A hail of H.E. shells from the heavy flak rushes up to the heights of the approaching bombers.

In spite of the difficulties of the weather the night-fighters hunt out the enemy. In the brilliant beams of the searchlights the British aircraft are clearly recognizable. The enemy drops his bombs on the city's industrial areas and then tries to get away as quickly as possible. At top speed other German night-fighters chase after him to shoot up as many of his forces as possible.

Over to the leading Pathfinder aircraft.

"How far are we from the target, Nav.?"

"About twenty-five miles."

"O.K. Stand by to drop preliminary target marker."

"Standing by."

A voice from the mid-upper turret. "Flak coming up port behind, skipper."

"O.K."

The guns are just beginning to open up down below. Ahead lies Berlin, still and silent. Berlin seems to be lying down there like a gigantic mouse, frightened to move, petrified. Suddenly it is galvanized into life; hundreds of gun-flashes come up from its roofs, its parks and its railway flak.

"Don't weave, for Christ's sake, skipper; only another minute." This from the navigator.

Again the captain's voice: "O.K."

He is not saying much. Both hands are on the wheel, his eyes are darting everywhere, looking for trouble and hoping not to find it. His aircraft seems huge, it appears to be the only one in the sky, every gun down below seems to be aiming at him, the gun-flashes are vicious, short and cruel.

Down below, to the Germans, he is the first of many hundreds of small spots on cathode-ray tubes. The civilians have long since gone to their shelters, but those of the A.R.P., police and fire-watching services are beginning to hear the loud, angry roar of the invading force.

"Coming up now, skipper. Steady – coming up – coming up – now! O.K. T.I. gone."

A few seconds later it bursts and cascades on to the ground; a mass of green bells, shining brightly, for all the world like a lit-up merry-go-round, an unmistakable spot of light . . .

Over to one of the main-force aircraft.

"There she is, skip; straight ahead." This from the bomb-aimer.

"Fine; the Pathfinders are dead on time."

The navigator looks at his watch and makes a note to that effect. The bomb-aimer starts his stop-watch. Three minutes and twenty seconds to go. On all sides other bomb-aimers are doing the same, beginning their straight fifteen-mile run through a curtain of steel. Flak is coming up all round, leaving black balloons which float by at an alarming speed. Searchlights are weaving, trying to pick up a straggler. The bomb-aimer begins to count.

"Three minutes to go, skipper."

Like a fleet of battleships the force sails in. Above are hundreds of fighter flares, lighting up the long lane of bombers like day-light. Now and then Junkers 88s and Me.110s come darting in and out like black moths trying to deliver their attack. The sky is full of tracer bullets, some going up, some going down. Others hose-pipe out horizontally as one of our rear-gunners gets in a good squirt.

Two minutes to go.

More flares have gone down. It seems even lighter than day. Searchlights, usually so bright themselves, can hardly pierce the dazzling glow of flares up above. Now the tracers are coming up in all colours as combats take place left, right and centre. On all sides bombers are blowing up, as they get direct hits – great, slow flashes in the sky, leaving a vast trail of black smoke as they disintegrate earthwards. Someone bales out.

One minute to go – bomb-doors open.

The bomb-aimer is still counting.

"Fifty seconds."

"Forty seconds."

There is flak all round now. The leading wave of bombers has not been broken up, a few have been shot down, but the rest have held their course. But the short time they held that course seemed like a lifetime.

There comes the bomb-aimer's voice again. "Red T.I.s straight ahead."

"Good show; there's the sky marker, too."

"Thirty seconds."

Still dead level. Someone in front has already started a fire. Great sticks of incendiaries are beginning to criss-cross across the target-indicating marker. These sticks are a mile long, but from this height they look about the length of a match-stick.

"Twenty seconds."

"Steady – hold it" – and then the bomb-aimer shouts: "Bombs gone." There is a note of relief in his voice.

The Lancaster leaps forward, relieved of its burden, diving, slithering. But it keeps straight on over the burning city. Throttles are slammed wide open, the engines are in fine pitch; they make a noise as of an aircraft in pain.

A volcano is raging down below, great sticks of incendiaries are slapping across the point where the target-markers had first gone in. Now black smoke is beginning to rise, but as these target-markers burst and drop slowly into the flaming mass, the later bomb-aimers have a good chance of aiming at the middle. Cookies are exploding one after another with their slow, red flashes, photo-flashes are bursting at all heights as each aircraft takes its photographs. This is a galaxy of light, a living nightmare.

As the last wave of bombers roar over, the fires started by the first are beginning to take hold. Against their vivid light can be seen the bottom squadrons, flying steadily on, over the battered city.

The flak is beginning to die, the searchlights have gone out. Once again the ground defences have been beaten.

A few leaflets drift down through the bluish glare, only to be burnt in the flames of the burning houses.

Soon the area is one mass of flames and the last bomber has dropped its bombs. At last the rendezvous is reached and the surviving bombers turn for home.

That is how it is done, by young men with guts, by science and by skill. The Germans do everything in their power to stop it, but in vain. There are too many variations; feint attacks can be made, or the bombers can attack in waves. They can come in at hourly intervals; they might come over one a night when the German fighters cannot get up. And on every raid new devices are carried, made by scientists, to help defeat the German defences.

This was the beginning, the end of three years' hard experiment. The real answer had been found, and the bomber could at last hit hard. It could choose tactical or strategical targets. Both were allergic to bombs.

Hans Rudel

Stukas Dive-Bomb the Soviet Fleet

Rudel flew 2,530 missions for the Luftwaffe during the Second World War, destroying 500 tanks, one battleship, one cruiser, one destroyer, 800 military vehicles and 70 landing craft. To mark Rudel's unique achievement, Hitler introduced a new award, the Golden Oakleaves Swords & Diamonds, which the Führer presented to the flier on 1 January 1945. Rudel was the award's first and only recipient. He survived the war (minus a leg, amputated in February 1945 after he was brought down by Russian flak) and died in 1982, aged 66. Below, Rudel describes his famous attack on the Soviet fleet at Kronstadt in September 1941.

Brilliant blue sky, without a rack of cloud. The same even over the sea. We are already attacked by Russian fighters above the narrow coastal strip; but they cannot deflect us from our objective, there is no question of that. We are flying at 9,000 feet; the flak is deadly. About ten miles ahead we see Kronstadt; it seems an infinite distance away. With this intensity of flak one stands a good chance of being hit at any moment. The waiting makes the time long. Dourly, Steen and I keep on our course. We tell ourselves that Ivan is not firing at single aircraft; he is merely putting up a flak barrage at a certain altitude. The others are all over the shop, not only in the squadrons and the flights, but even in the pairs. They think that by varying height and zigzagging they can make the A.A. gunners' task more difficult. There go the two blue-nosed staff aircraft sweeping through all the formations, even the separate flights. Now one of them loses her bomb. A wild helter-skelter in the sky over Kronstadt; the danger of ramming is great. We are still a few miles from our objective; at an angle ahead of me I can already make out the *Marat* berthed in the harbour. The guns boom, the shells scream up at us, bursting in flashes of livid colours; the flak forms small fleecy clouds that frolic around us. If it was not in such deadly earnest one might use the phrase: an aerial carnival. I look down on the *Marat*. Behind her lies the cruiser *Kirov*. Or is it the *Maxim Gorki?* These ships have not yet joined in the general bombardment. But it was the same the last time. They do not open up on us until we are diving to the attack. Never has our flight through the defence seemed so slow or so uncomfortable. Will Steen use his diving brakes today or in the face of this opposition will he go in for once "without"? There he goes. He has already used his brakes. I follow suit, throwing a final glance into his cockpit. His grim face wears an expression of concentration. Now we are in a dive, close beside each other. Our diving angle must be between 70 and 80 degrees. I have already picked up the *Marat* in my sights. We race down towards her; slowly she grows to a gigantic size. All their A.A. guns are now directed at us. Now nothing matters but our target, our objective; if we achieve our task it will save our brothers in arms on the ground much bloodshed. But what is happening? Steen's aircraft suddenly leaves mine far behind. He is travelling much faster. Has he after all again retracted his diving brakes in order to get down more quickly? So I do the same. I race after his aircraft going all out. I am right on his tail, travelling much too fast and unable to check my speed. Straight ahead of me I see the horrified face of W.O. Lehmann, Steen's rear-gunner. He expects every second that I shall cut off his tail unit with my propeller and ram him.

I increase my diving angle with all the strength I have got – it must surely be 90 degrees – sit tight as if I were sitting on a powder-keg. Shall I graze Steen's aircraft which is right on me or shall I get safely past and down? I streak past him within a hair's breadth. Is this an omen of success? The ship is centered plumb in the middle of my sights. My Ju 87 keeps perfectly steady as I dive; she does not swerve an inch. I have the feeling that to miss is now impossible. Then I see the *Marat* large as life in front of me. Sailors are running across the deck, carrying ammunition. Now I press the bomb release switch on my stick and pull with all my strength. Can I still manage to pull out? I doubt it, for I am diving without brakes and the height at which I have released my bomb is not more than 900 feet. The skipper has said when briefing us that the two thousand pounder must not be dropped from lower than 3,000 feet as the fragmentation effect of this bomb reaches 3,000 feet and to drop it at a lower altitude is to endanger one's aircraft. But now I have forgotten that! – I am intent on hitting the *Marat*. I tug at my stick, without feeling, merely exerting all my strength. My acceleration is too great. I see nothing, my sight is blurred in a momentary blackout, a new experience for me. But if it can be managed at all I must pull out. My head has not yet cleared when I hear Scharnovski's voice:

"She is blowing up, sir!"

Now I look out. We are skimming the water at a level of ten or twelve feet and I bank round a little. Yonder lies the *Marat* below a cloud of smoke rising up to 1,200 feet; apparently the magazine has exploded.

"Congratulations, sir."

Scharnovski is the first. Now there is a babel of congratulations from all the other aircraft over the radio. From all sides I catch the words: "Good show!" Hold on, surely I recognize the Wing Commander's voice? I am conscious of a pleasant glow of exhilaration such as one feels after a successful athletic feat. Then I fancy that I am looking into the eyes of thousands of grateful infantrymen. Back at low level in the direction of the coast.

"Two Russian fighters, sir," reports Scharnovski.

"Where are they?"

"Chasing us, sir. – They are circling round the fleet in their own flak. – Cripes! They will both be shot down together by their own flak."

This expletive and, above all, the excitement in Scharnovski's voice are something quite new to me. This has never happened before. We fly on a level with the concrete blocks on which A.A. guns have also been posted. We could almost knock the Russian crews off them with our wings. They are still firing at our comrades who are now attacking the other ships. Then for a moment there is nothing visible through the pall of smoke rising from the *Marat*. The din down below on the surface of the water must be terrific, for it is not until now that a few flak crews spot my aircraft as it roars close past them. Then they swivel their guns and fire after me; all have had their attention diverted by the main formation flying off high above them. So the luck is with me, an isolated aircraft. The whole neighbourhood is full of A.A. guns; the air is peppered with shrapnel. But it is a comfort to know that this weight of iron is not meant exclusively for me! I am now crossing the coast line. The narrow strip is very unpleasant. It would be impossible to gain height because I could not climb fast enough to reach a safe altitude. So I stay down. Past machine guns and flak. Panic-stricken Russians hurl themselves flat on the ground. Then again Scharnovski shouts:

"A *Rata* coming up behind us!"

I look round and see a Russian fighter about 300 yards astern.

"Let him have it, Scharnovski!"

Scharnovski does not utter a sound. Ivan is blazing away at a range of only a few inches. I take wild evasive action.

"Are you mad, Scharnovski? Fire! I'll have you put under arrest." I yell at him!

Scharnovski does not fire. Now he says deliberately:

"I am holding fire, sir, because I can see a German ME coming up behind and if I open up on the *Rata* I may damage the Messerschmitt." That closes the subject, as far as Scharnovski is concerned; but I am sweating with the suspense. The tracers are going wider on either side of me. I weave like mad.

"You can turn round now, sir. The ME has shot down the *Rata*." I bank round slightly and look back. It is as Scharnovski says; there she lies down below. Now a ME passes groggily.

"Scharnovski, it will be a pleasure to confirm our fighter's claim to have shot that one down." He does not reply. He is rather hurt that I was not content to trust his judgment before. I know him; he will sit there and sulk until we land. How many operational flights have we made together when he has not opened his lips the whole time we have been in the air.

After landing, all the crews are paraded in front of the squadron tent. We are told by Flt./Lt. Steen that the Wing Commander has already rung up to congratulate the 3rd squadron on its achievement. He had personally witnessed the very impressive explosion. Steen is instructed to report the name of the officer who was the first to dive and drop the successful two thousand pounder in order that he may be recommended for the Knight's Cross of the Iron Cross.

With a side-glance in my direction he says:

"Forgive me for telling the Kommodore that I am so proud of the whole squadron that I would prefer it if our success is attributed to the squadron as a whole."

In the tent he wrings my hand. "You no longer need a battleship for a special mention in despatches," he says with a boyish laugh.

The Wing Commander rings up. "It is sinking day for the 3rd. You are to take off immediately for another attack on the *Kirov* berthed behind the *Marat*. Good hunting!" The photographs taken by our latest aircraft show that the *Marat* has split in two. This can be seen on the picture taken after the tremendous cloud of smoke from the explosion had begun to dissipate. The telephone rings again:

"I say, Steen, did you see my bomb? I didn't and neither did Pekrun."

"It fell into the sea, sir, a few minutes before the attack."

We youngsters in the tent are hard put to it to keep a straight face. A short crackling on the receiver and that is all. We are not the ones to blame our Wing Commander, who is old enough to be our father, if presumably out of nervousness he pressed the bomb release switch prematurely. He deserves all praise for flying with us himself on such a difficult mission. There is a big difference between the ages of fifty and twenty five. In dive-bomber flying this is particularly true.

Out we go again on a further sortie to attack the *Kirov*. Steen had a slight accident taxying back after landing from the first sortie: one wheel ran into a large crater, his aircraft pancaked and damaged the propeller. The 7th flight provides us with a substitute aircraft, the flights are already on dispersal and we taxi off from our squadron base airfield. Flt./ Lt. Steen again hits an obstacle and this aircraft is also unserviceable. There is no replacement available from the flights; they are of course already on dispersal. No one else on the staff is flying except myself. He therefore gets out of his aircraft and climbs onto my wingplane.

"I know you are going to be mad at me for taking your aircraft, but as I am in command I must fly with the squadron. I will take Scharnovski with me for this one sortie."

Vexed and disgruntled I walk over to where our aircraft are overhauled and devote myself for a time to my job as engineer officer. The squadron returns at the end of an hour and a half. No. 1, the green-nosed staff aircraft – mine – is missing. I assume the skipper has made a forced landing somewhere within our lines.

As soon as my colleagues have all come in I ask what has happened to the skipper. No one will give me a straight answer until one of them says:

"Steen dived onto the *Kirov*. He was caught by a direct hit at 5,000 or 6,000 feet. The flak smashed his rudder and his aircraft was out of control. I saw him try to steer straight at the cruiser by using the ailerons, but he missed her and nose-dived into the sea. The explosion of his two thousand pounder seriously damaged the *Kirov*."

The loss of our skipper and my faithful Cpl. Scharnovski is a heavy blow to the whole squadron and makes a tragic climax to our otherwise successful day. That fine lad Scharnovski gone! Steen gone! Both in their way were paragons and they can never be fully replaced. They are lucky to have died at a time when they could still hold the conviction that the end of all this misery would bring freedom to Germany and to Europe.

Mark E. Berent

Night Mission Over Nam

Berent served with the 497th Tactical Fighter Squadron of the USAAF during the Vietnam War.

It's cool this evening, thank God. The night is beautiful, moody, an easy rain falling. Thunder rumbles comfortably in the distance. Just the right texture to erase the oppressive heat memories of a few hours ago. Strange how the Thai monsoon heat sucks the energy from your mind and body by day, only to restore it by the cool night rain.

I am pleased by the tranquil sights and sounds outside the BOQ room door. Distant ramp lights, glare softened by the rain, glisten the leaves and flowers. The straight-down, light rain splashes gently, nicely on the walkways, on the roads, the roofs. Inside the room I put some slow California swing on the recorder (*You gotta go where you wanta go. . . .*) and warm some soup on the hot-plate. Warm music, warm smell . . . I am in a different world. (*Do what you wanta, wanta do . . .*) I've left the door open – I like the sound of the rain out there.

A few hours later, slightly after midnight, I am sitting in the cockpit of my airplane. It is a jet fighter, a Phantom, and it's a good airplane. We don't actually get into the thing – we put it on. I am attached to my craft by two hoses, three wires, lap belt, shoulder harness and two calf garters to keep my legs from flailing about in a highspeed bailout. The gear I wear – gun, G-suit, survival vest, parachute harness – is bulky, uncomfortable, and means life or death.

I start the engines, check the myriad systems – electronic, radar, engine, fire control, navigation – all systems; receive certain information from the control tower, and am ready to taxi. With hand signals we are cleared out of the revetment and down the ramp to the arming area.

I have closed the canopy to keep the rain out, and switch the heavy windscreen blower on and off to hold visibility. I can only keep its hot air on for seconds at a time while on the ground, to prevent cracking the heavy screen. The arming crew, wearing bright colours to indicate their duties, swarm under the plane: electrical continuity – checked; weapons – armed; pins – pulled. Last all-round look-see by the chief – a salute, a thumbs-up, we are cleared. God, the rapport between pilot and ground crew – their last sign, thumbs-up – they are with me. You see them quivering, straining bodies posed forward as they watch *their* airplane take off and leave them.

And we are ready, my craft and I. Throttles forward and outboard, gauges OK, afterburners ignite, nose-wheel steering, rudder effective, line speed, rotation speed – we are off, leaving behind only a ripping, tearing, gut noise as we split into the low black overcast, afterburner glow not even visible anymore.

Steadily we climb, turning a few degrees, easing stick forward some, trimming, climbing, then suddenly – on top! On top where the moonlight is so damn marvellously bright and the undercast appears a gently rolling snow-covered field. It's just so clear and good up here, I could fly forever. This is part of what flying is all about. I surge and strain against my harness, taking a few seconds to stretch and enjoy this privileged sight.

I've already set course to rendezvous with a tanker, to take on more fuel for my work tonight. We meet after a long cut-off turn, and I nestle under him as he flies his long, delicate boom toward my innards. A slight thump/bump, and I'm receiving. No words – all light signals. Can't even thank the boomer. We cruise silently together for several minutes. Suddenly he snatches it back, a clean break, and I'm cleared, off and away.

Now I turn east and very soon cross the fence far below. Those tanker guys will take you to hell, then come in and pull you right out again with their flying fuel trucks. Hairy work. They're grand guys.

Soon I make radio contact with another craft, a big one, a gunship, painted black and flying very low. Like the proverbial spectre, he wheels and turns just above the guns, the limestone outcropping, called karst, and the mountains – probing, searching with infra-red eyes for supply trucks headed south. He has many engines and more guns. His scanner gets something in his scope, and the pilot goes into a steep bank – right over the target. His guns flick and flash, scream and moan, long amber tongues lick the ground, the trail, the trucks. I am there to keep enemy guns off him and to help him kill trucks. Funny – he can see the trucks but not the guns till they're on him. I cannot see the trucks but pick the guns up as soon as the first rounds flash out of the muzzles.

Inside my cockpit all the lights are off or down to a dim glow, showing the instruments I need. The headset in my helmet tells me in a crackling, sometimes joking voice the information I must have: how high and how close the nearest karst, target elevation, altimeter setting, safe bail-out area, guns, what the other pilot sees on the trails, where he will be when I roll in.

Then, in the blackest of black, he lets out an air-burning flare to float down and illuminate the sharp rising ground. At least then I can mentally photograph the target area. Or he might throw out a big log, a flare marker, that will fall to the ground and give off a steady glow. From that point he will tell me where to strike: 50 metres east, or 100 metres south, or, if there are two logs; hit between the two.

I push the power up now, recheck the weapons settings, gun switches, gunsight setting, airspeed, altitude – roll in! Peering, straining, leaning way forward in the harness, trying so hard to pick up the area where I know the target to be – it's so dark down there.

Sometimes when I drop, pass after pass, great fire balls will roll and boil upward and a large, rather rectangular fire will let us know we've hit another supply truck. Then we will probe with firepower all around that truck to find if there are more. Often we will touch off several, their fires outlining the trail or truck park. There are no villages or hooches for miles around; the locals have been gone for years. They silently stole away the first day those big trucks started plunging down the trails from up north. But there are gun pits down there – pits, holes, reveted sites, guns in caves, guns on the karst, guns on the hills, in the jungles, big ones, little ones.

Many times garden-hose streams of cherry balls will arc and curve up, seeming to float so slowly toward me. Those from the smaller-calibre, rapid-fire quads; and then the big stuff opens up, clip after clip of 37 mm and 57 mm follow the garden hose, which is trying to pinpoint me like a searchlight. Good fire discipline – no one shoots except on command.

But my lights are out, and I'm moving, jinking. The master fire controller down there tries to find me by sound. His rising shells burst harmlessly around me. The heavier stuff in clips of five and seven rounds goes off way behind.

Tonight we are lucky – no "golden BB". The golden BB is that one stray shell that gets you. Not always so lucky. One night we had four down in Death Valley – that's just south of Mu Gia Pass. Only got two people out the next day, and that cost a Sandy (A-1) pilot. "And if the big guns don't get you, the black karst will," goes the song. It is black, karsty country down there.

Soon I have no more ammunition. We, the gunship and I, gravely thank each other, and I pull up to thirty or so thousand feet, turn my navigation lights back on, and start across the Lao border to my home base. In spite of an air-conditioning system working hard enough to cool a five-room house, I'm sweating. I'm tired. My neck is sore. In fact, I'm sore all over. All those roll-ins and diving pull-outs, jinking, craning your head, looking, always looking around, in the cockpit, outside, behind, left, right, up, down. But I am headed home, my aircraft is light and more responsive.

Too quickly I am in the thick, puffy thunder clouds and rain of the southwest monsoon. Wild, the psychedelic green, wiry, and twisty St Elmo's fire flows liquid and surrealistic on the canopy a few inches away. I am used to it – fascinating. It's comforting, actually, sitting snugged up in the cockpit, harness and lap belt tight, seat lowered, facing a panel of red-glowing instruments, plane buffeting slightly from the storm. Moving without conscious thought, I place the stick and rudder pedals and throttles in this or that position – not so much mechanically moving things, rather just willing the craft to do what I see should be done by what the instruments tell me.

I'm used to flying night missions now. We "night owls" do feel rather élite, I suppose. We speak of the day pilots in somewhat condescending tones. We have a black pilot who says, "Well, day pilots are OK, I guess, but I wouldn't want my daughter to *marry* one." We have all kinds: quiet guys, jokey guys (the Jewish pilot with the fierce black bristly moustache who asks, "What is a nice Jewish boy like me doing over here, killing Buddhists to make the world safe for Christianity?"), noisy guys, scared guys, whatever. But all of them do their job. I mean night after night they go out and get hammered and hosed, and yet keep right at it. And all that effort, sacrifice, blood going down the tubes. Well, these thoughts aren't going to get me home. This is not time to be thinking about anything but what I'm doing right now.

I call up some people on the ground who are sitting in darkened, black-out rooms, staring at phosphorescent screens that are their eyes to the night sky. Radar energy reflecting from me shows them where I am. I flick a switch at their command and trigger an extra burst of energy at them so they have positive identification. By radio they direct me, crisply, clearly, to a point in space and time that another man in another darkened room by a runway watches anxiously. His eyes follow a little electronic bug crawling down a radar screen between two converging lines. His voice tells me how the bug is doing, or how it should be doing. In a flat, precise voice the radar controller keeps up a constant patter – "Turn left two degrees . . . approaching glide path . . . prepare to start descent in four miles."

Inside the cockpit I move a few levers and feel the heavy landing gear thud into place and then counteract the nose rise as the flaps grind down. I try to follow his machine-like instructions quite accurately, as I am very near the ground now. More voice, more commands, then a glimmer of approach lights, and suddenly the wet runway is beneath me. I slip over the end, engines whistling a down note as I retard the throttles, and I'm on the ground at last.

If the runway is heavy with rain, I lower a hook to snatch a cable laid across the runway that connects to a friction device on each side. The deceleration throws me violently into my harness as I stop in less than 900 ft from nearly 175 m.p.h. And this is a gut-good feeling.

Then the slow taxi back, the easing of tension, the good feeling. Crew chiefs with lighted wands in their hands direct me where to park; they chock the wheels and signal me with a throat-cutting motion to shut down the engines. Six or seven people gather around the airplane as the engines coast off, and I unstrap and climb down, soaking wet with sweat.

"You OK? How did it go? See anything, get anything?" They want to know these things and they have a right to know. Then they ask, "How's the airplane?" That concern always last. We confer briefly on this or that device or instrument that needs looking after. And then I tell them what I saw, what I did. They nod, grouped around, swear softly, spit once or twice. They are tough, and it pleases them to hear results.

The crew van arrives, I enter and ride through the rain – smoking a cigarette and becoming thoughtful. It's dark in there, and I need this silent time to myself before going back to the world. We arrive and, with my equipment jangling and thumping about me, I enter the squadron locker room, where there is always easy joking among those who have just come down. Those that are suiting up are quiet, serious, going over the mission brief in their minds, for once on a night strike they cannot look at maps or notes or weapon settings.

They glance at me and ask how the weather is at The Pass. Did I see any thunderstorms over the Dog's Head? They want to ask about the guns up tonight, but know I'll say how it was without their questioning. Saw some light ZPU (automatic weapons fire) at The Pass, saw someone getting hosed at Ban Karai, nothing from across the border. Nobody down, quiet night. Now all they have to worry about is thrashing through a couple of hundred miles of lousy weather, letting down on instruments and radar into the black karst country and finding their targets. Each pilot has his own thoughts on that.

Me, I'll start warming up once the lethargy of finally being back from a mission drains from me. Funny how the mind/body combination works. You are all "hypoed" just after you land, then comes a slump, then you're back up again but not as high as you were when you first landed. By now I'm ready for some hot coffee or a drink (sometimes too many), or maybe just letter writing. A lot of what you want to do depends on how the mission went.

I debrief and prepare to leave the squadron, But before I do, I look at the next day's schedule. Is it an escort? Am I leading? Where are we going? What are we carrying? My mind unrolls pictures of mosaics and gun-camera film of the area. Already I'm mechanically preparing for the next mission.

And so it goes – for a year. And I like it. But every so often, especially during your first few months, a little wisp of thought floats up from way deep in your mind when you see the schedule. "Ah, no, not tonight," you say to yourself. "Tonight I'm sick – or could be sick. Just really not up to par, you know. Maybe, maybe I shouldn't go." There's a feeling – the premonition that tonight is the night I don't come back. But you go anyhow and pretty soon you don't think about it much anymore. You just don't give a fat damn. After a while, when you've been there and see what you see, you just want to go fight! To strike back, destroy. And then sometimes you're pensive – every sense savouring each and every sight and sound and smell. Enjoying the camaraderie, the feeling of doing something. Have to watch that camaraderie thing though – don't get too close. You might lose somebody one night and that can mess up your mind. It happens, and when it does, you get all black and karsty inside your head.

I leave the squadron and walk back through the ever-present rain that's running in little rivulets down and off my poncho. The rain glistens off trees and grass and bushes, and a ripping, tearing sound upsets the balances as another black Phantom rises to pierce the clouds.

Philip Neame

2 Para at Goose Green

The battle fought by the 2nd Battalion of the Parachute Regiment at Goose Green on 28/9 May 1982 was arguably the hardest of the Falklands War, and saw the death of commanding officer, Lieutenant Colonel H. Jones.

In the evening the O Group was called and we were told that Goose Green was held by four or five hundred men who weren't up to much and that their defences were facing seawards and southwards rather than in our direction and all one had to do was to knock hard at the front door, or the back door in our case, and we'd just sort of walk in. I had spent that day looking at the map and there was a very narrow strip of land we had to advance down. I was concerned, to say the least, as this meant there would be little room for manoeuvre and therefore little scope for bold and imaginative tactics – just a straight slog. Also, as they knew we were coming they would have obviously deployed north. I remember saying to Nobby Clark, my Sergeant-Major, that this was either going to be a cakewalk because they would just give up as everyone predicted, or a very bloody do. Not much in between. I came back to my HQ after the O Group with red blobs representing enemy positions smeared all over my map. I tossed it down in front of Nobby and said that I thought it was going to be the second of the two options.

There was nothing very startling or original about the battalion plan – A and B forward and D in reserve (it's totally against the training of the military mind to do anything other than follow the alphabet). So we started off down there, following the rest of the battalion, and it really should have been very easy because all we really had to do was follow everyone else. The only trouble was Battalion HQ stopped off half-way down to the start line and we moved through them expecting guides from the recce patrol to be on the track to show us exactly where to go. But there were no guides and there was a complete mass of tracks leading off in every direction. With so many tracks around we got hopelessly lost and overshot the track in question. The last thing I wanted to do was to end up ahead of A and B Companies and get caught up in their crossfire. So we trod a very careful path back to a known start point, found the track and sat down to wait for the battle to start.

Fortunately A and B Companies were still ahead of us, but what I hadn't taken into account was that we had got ahead of H.'s Tactical HQ. He came stomping down the track, found us there, and took this as a most immense personal affront that his reserve company was actually closer to the battle than he was. Suitably chastened we just sat where we were and watched him go stomping further down the track only to find himself caught in crossfire further down. By this time both A and B Companies had put in an attack of sorts. H. came stomping back and, having been shot at, identified one position where he thought the fire came from and directed me to go and destroy it. My only difficulty was that I couldn't really see where this position was and he didn't really know exactly where it was on the map. So we called up a fire mission from the ship that was offshore, hoping it was one of the pre-targeted objectives. They gave us about two rounds and then the gun jammed on the ship, so that was a great start! We were already underway, so it just became an advance to contact and hope for the best.

By then we were ahead of the other two companies. We dimly saw a position on the skyline ahead of us which offered no opposition at all. So we just went straight into a frontal assault which was the first time I'd been in action in my life. It all seemed to be going well, when suddenly two machine guns opened up on us from the right. Up until then I had thought, if this is war, it's all dead easy. But now we were suddenly really caught flat-footed. There was already one platoon clearing the position in front of us, the platoon on my right was completely pinned down by the two machine guns, and the difficulty was getting any troops available to manoeuvre around and actually assault this position. My only other force available was my third platoon on my left, and any direction they were likely to attack from would mean assaulting straight in towards the direction that I thought B Company was. After a certain amount of flat-footedness, sucking of teeth and wondering what the hell to do,

I saw that Chris Waddington had already started bringing his platoon across so that they could assault. I was still concerned that they'd be shooting up B Company in the process, but there was no option.

By this time H. was yelling to find out what the hell was holding us up. So I told Chris to go in and assault and in he went. I got a few expletives from John Crosland about the number of rounds that were coming his way and I answered with expletives about the number of rounds that were coming my way and we just got on with it. This assault led to four casualties. One of those killed was Corporal Bingley who was very brave. He'd gone to ground not really knowing quite where these machine guns were and found himself virtually overlooking the position. He and Grayling just went in and did an immediate assault, and the two of them took the five-strong position out between them. But Bingley was killed in the process and Grayling slightly injured. It was that sort of immediate get up and go and flare that really got us out of a very sticky situation.

The real problems started because we found ourselves scattered to the four winds. We'd taken out these two machine-gun positions and another platoon position on top of the hill in a single company assault. But in the process people had been going everywhere and it was very featureless ground. Trying to regroup everyone was almost impossible. Much as the School of Infantry would have decried it, I felt the only way to get them together was to put up some light myself to tell everyone where I was. It was also telling the enemy where we were, but I had to take the chance on that. It worked and we got everyone together, less two unaccounted for. There was a long delay while we tried to find out where they were; they weren't found until daylight. Corporal Cork had been shot and Fletcher had been bending over applying a field dressing when he was shot himself and so they had both been killed.

To reorganize took us about an hour and a half and it was quite clear that this was a problem besetting the other companies, which was really why H. had fed us, the reserve, in so soon. The whole encounter had been a little chaotic but of course at this stage we had no perspective of the normal. We just accepted this as the norm – feeling it was not totally different from the average exercise! Life became reasonably simple for us for the next hour or so, we just trogged down behind everyone else. At about daylight we ended up on this little knoll about 1,000 metres short of what became known as the Gorse Line, with A Company at this time fairly heavily engaged around Darwin Hill and B Company brought to a stop on the Gorse Line itself, overlooking Boca House. Then everything began to bog down and I started to move my company up closer to the other two lead companies to get under the lee of the hill and out of sight of the enemy. I was told in very certain terms by H. that he didn't want me getting any closer. So we amused ourselves by taking the odd pot shot at some stray Argentinians who we could see about 1,000 yards away. I had to put a stop to that otherwise we wouldn't have had the ammunition when we needed it.

Life then began to get a bit uncomfortable. There was a minefield either side of the track ahead of us and we were on a very exposed knoll. The enemy artillery started sending in fire periodically. The first rounds were some way away but the next came closer. It suddenly dawned on me that whatever else was going on, they still had an observation post that could see this far back up the peninsula. That observation post was busy getting the enemy artillery zeroed in on us. "Orders notwithstanding", as they say, I decided to push on into the lee of the hill as soon as possible rather than stay around and cop the whole lot – just in time, because as we moved off the hill a fire mission landed smack on it right where we'd been sitting!

We suffered our first daylight casualty, Mechan. At night in the confusion one couldn't see immediately what had happened but with Mechan everyone saw it happen and it obviously had some effect on people's confidence.

We moved round into the lee of this hill and then closer towards the west coast. From there it became obvious to me that there was scope for exploiting the position that we were in. A Company were well bogged in and fighting a fairly fierce battle stage by stage. B Company were apparently in a position where they really couldn't move forward at all. I felt I could move down to the right of everyone else along the shoreline and possibly turn the enemy's position; it seemed worth having a look at least. I put this idea to H. and he was clearly of the frame of mind where he didn't want his reserve committed at this stage and I suspected he felt that things were very much on a tightrope where he was, so he gave me pretty short shrift.

By then we'd been on the go for eight hours and it seemed obvious to me that we weren't going anywhere for at least half an hour so I decided the most sensible thing to do was to get a brew on because it looked like it was going to be a very long day. Stopping in mid-battle and having a brew was met with complete amazement by my blokes. It is not in the book of rules but there seemed nothing better to do. My porridge had just come to the boil when the news came over that H. had been shot. News travelled fast and it wasn't something that could be kept quiet for long, especially as soon after that the battalion Second-in-Command, Chris Keeble, came along giving orders. He gave me orders straight away to move up and join John Crosland to see what we could do to help him. John was at this stage temporarily in charge of the battalion. Well, I was buggered if I was going to waste my porridge so this vagabond army got on the move with everyone trying to take the odd sip of their brew as they went and I was trying to get down the odd spoonful of hot porridge.

We got up to the Gorse Line where John was, crested the hill, and could see the enemy 1,200 metres away. I was convinced that we were out of small-arms range and was bowling along quite confidently when I suddenly felt this "thing" whip past my leg and looked at my signaller who'd just had his ammunition pouch shot away! We gingerly reversed a few crucial yards behind this slope, back to relatively safe ground. I couldn't see where John was exactly and by the appearance of things he was fairly far forward himself, and I figured that if I was going to go off and find him, all that was going to happen was that I was going to get shot, which didn't seem an attractive idea. So I had another look at the shoreline which did seem to offer quite a lot of promise for an approach to Boca House. At Boca House the enemy had their own heavy machine guns which were simply out-ranging our stuff. I thought that even if we couldn't get to a position where we could could assault Boca House, at least we might be able to get our machine guns in range and start causing some damage.

I got on to the radio to let Chris know what I was doing. I went down with a section along the shoreline, and got within about 500 yards of the position before it became clear that we were going to be fairly exposed. So I got all the company down, less one platoon and Pete Adams who I left up on the hilltop to liaise between myself and John Crosland and Chris Keeble. At that stage half my radios had packed up and really it could only be done by me relaying to Pete and him passing on the message.

Our six machine guns were in range so I lined them up on this spur just down by the beach.

Then it all happened. John Crosland started blasting away at Boca House with his Milan and with our machine guns in position we set up a rather good duo, with John blasting the sangars and us chopping off the rather stunned survivors who were staggering to other sangars. This seemed to have a very salutary effect on the Argentinians. They gathered very quickly that they didn't have much of a future going for them and after John had got three or four Milans off, all of a sudden white flags started appearing all over the place. Sitting where I was, looking at the position through field glasses, it was quite clear that these buggers were absolutely knackered and they just wanted no more of it. I had all my six machine guns and half the company ready to move on to their position. The Milans and the guns were also set up to cover us. Then nothing seemed to happen for twenty minutes or so and I was getting more and more impatient, feeling the longer I stayed down on this beach, sooner or later someone was going to spot us and start to direct some shit our way. So I got this wheeze to Chris on the radio that if he didn't give us permission to advance straight away we would get cut off by the tide, which was coming in. I don't think that it would have cut us off but it seemed a useful excuse! We got permission to move.

I decided that it was a moment of commitment when someone had to expose himself first and it looked like this time officers would earn their pay. I was about to start forward when Corporal Harley went dashing ahead of me, saying, "This isn't your job, Sir, you're too valuable. This is toms' work." So he was really the first guy to take the chance about the surrender. I always had rather a soft spot for him after that, especially having such faith in my judgement! But it was quite clear that it was a completely genuine surrender and they had totally lost interest. We advanced up to the position and one of the platoons, in their eagerness to be first into the position, blundered straight into a minefield rather than follow my directive to stay on the beach. One of them tripped a mine which turned him head over heels. I think Argentinian mines were much like the rest of the Argentinians – not too effective – so he picked himself up, shrugged and carried on going! When we got to Boca House we found a considerable scene of carnage – I suppose thirty or forty casualties and probably eighteen to twenty dead. In the distance one could see the fitter ones who had hightailed at the last moment and were literally fleeing across the airfield.

After ten minutes Chris Keeble came up on the radio and congratulated us on securing Boca House and told me to head straight for Goose Green. With some smugness I told him that we were already on our way. As we were about half the way there we saw what looked like a deserted HQ so I sent one of the platoons off that way and headed with the remaining two down towards Goose Green. At this stage a combination of ack-ack fire overhead and some mines diverted us into a shallow valley which led to the schoolhouse. So we went down towards it and got to the stage where we were almost surrounded by minefields. They were not well laid and were partially visible but the lead platoon at this stage was getting just a little bit nervous. We eventually got into a little hollow ground just short of the schoolhouse where we could actually start forming up ready for an assault. The way to Goose Green lay up the track and there was no way we could move without being exposed to the schoolhouse. In addition there was another position on the skyline with a flag flying a bit further up the track. This became known as the Flagpole Position. It was quite clear what had to be done: first the schoolhouse with fire support from our present position and then the Flagpole Position with support from the school.

I suppose the real difficulty at this stage was that we were really a little bit off balance as the platoon which I had detached to check the enemy HQ had by this time come under fire from the Flagpole Position and was unable to join us. The nature of the game from Boca House had been attack and exploitation and almost hot pursuit. Now suddenly we were not exactly in the face of fierce opposition but were clearly in a potentially very dangerous situation. I left one platoon to try and neutralize the Flagpole Position and with the remaining platoon I got ready to assault the schoolhouse. Then things really began to happen in a fierce way. We got small-arms fire down on us from the Flagpole Position and also from the schoolhouse. More alarming still was that we began to get extremely accurate and heavy artillery fire down on us. I suppose the only saving grace was that the ground was so soft the rounds were landing relatively close to you but not having any really serious effect. However, life began to get rather unpleasant. We were also at this stage very much on our own, the rest of the battalion separated by a forward slope behind us which was being raked with ack-ack fire.

We were then joined by one of C Company's platoons which gave us the added momentum we needed for the school. Just as we were about to assault the school, I got the news that Jim Barry, the other platoon commander, had been shot when he had gone up to take a surrender under a white flag. He and half the section had been shot down. It was such a tragic waste of life. After a little deliberation as to where my priorities were, I left Pete Adams to command the assault on the school and I went back to join 12 Platoon to find that Sergeant Meredith by this stage had got the situation firmly under control. His platoon was busily knocking shit out of the Flagpole Position with 66 rocket launchers and machine guns. We didn't know who had been killed or injured with Jim Barry, but certainly some of the injured were trying to get back. There were one or two very brave people there – Shevill who was very badly shot managed to pull himself back about 200 yards, finding his own cover, refusing help from others who would have had to expose themselves, and a couple of others who performed extraordinarily well for just private soldiers in organizing themselves and getting their injured companions back under covering fire from Meredith and his crew. Meredith, of course, held it all together, and made sure the platoon continued to work together – a really solid number, hard as nails and with the ability to think. He never appeared fussed which is what I think really helped at this time, at least for his blokes. Private Carter was the other guy who really came through. He'd been one of the blokes up with Jim Barry and was perhaps the first to recover from the shock and get the four of them still alive to start reacting. He'd only just joined the company and for a young inexperienced soldier he showed incredible resilience and presence of mind and initiative. Carter and Meredith, between them, probably saved the lives of the other three involved in the incident.

The assault went in on the school with no problems and we made many schoolchildren happy by burning down the schoolhouse. It was all they could talk about when we finally entered the village. However, we couldn't stay in the area as we were coming under very heavy direct fire from Goose Green itself, and we had no way of neutralizing it because there were civilians in the village. For the same reason, we couldn't actually occupy the Flagpole Position, although Meredith's crew had knocked seven kinds of shit out of it, having set off an ammunition dump. This continued to give an excellent firework display for the rest of the day.

Chris Waddington and his platoon had joined us by this stage. We were all tightly grouped on the track leading to Goose Green. We couldn't move off to the left without coming under fire from Goose Green nor to the right because of the minefield. So we were sitting there, having been told that there was going to be a Harrier strike onto the enemy gun position. Over came this aircraft which wasn't a Harrier at all but an enemy Skyhawk. I saw this cannon-fire zipping towards us and felt utterly helpless and angry that I'd fucked up everything because I'd tightly grouped the whole company. It was the only time that day that I was really scared. Thank God we didn't get any casualties. When a Pucará decided to do the same we shot him down. The track itself and the exploding bomb dump were, I guessed, the obvious indicators for the aircraft. Rather than remain as a target for aircraft I decided to take our chances in the minefield, so we moved off into a nice reverse-slope position.

By this stage it was just coming up to last light and we heard from Chris Keeble that we weren't to exploit further because he had other moves afoot. John Crosland had by then gone round to the south-west and so we had the village encircled. This was one of John's canny moves, because they tried to land reinforcements for Goose Green down there but old John had pre-empted them and got himself between them and the village. Probably no one else had sussed out that possibility.

We began to unwind slightly. We were very low on food; more important, we had very little ammunition left; most people had run out of water and we had no warm clothing. We spent the night in this position and were very cold. Under cover of darkness we were able to bring down the bodies of Jim Barry, Lance-Corporal Smith and Corporal Sullivan. Jim, of course, shouldn't have been with us at all. He'd been picked for the Americas Cup trials in Newport and I'd give him the choice of a cruise there or a cruise in the South Atlantic. Being the sort of bloke he was, he returned from the Americas Cup team to join us without a second thought. A snowcat came forward with some ammunition and took their bodies and our other casualties back at last. It was not until then that we had any direct link with the rest of the battalion.

We entered the village the next day and my company went up

with Chris Keeble to organize the surrender. It was something of an eye-opener to see over 900 Argentinians still fully armed come out to meet our three small platoons. It seemed a little incongruous to say the least. In Goose Green the welcome was fairly rapturous – we went into the village house where everyone had been cooped up and we were given cups of tea. It was quite nice to be treated like the conquering hero for a bit. My Company HQ ended up in the farm manager's house and we were looked after very well.

We realized we had fought a major battle against fairly remarkable odds. I think that we had stuck our necks out and it had not been a controlled or typical situation at all. In saner moments we probably realized that it wasn't the sort of thing to commit a single battalion to at all but we had been committed and done it. So everyone was pleased with themselves. I think that what we had achieved as a battalion was very much a reflection of H. A more phlegmatic person probably would not have committed us to such uncertainty. But he was a real warrior and was determined to get stuck in. Not only that, but he had imbued such a faith in the battalion, in our abilities, that I don't think the idea of failure entered anyone's mind. We just assumed we would win and this did a lot for everyone's approach when things got rough – "Just a minor hiccup – soon sort it out." Anyway, the Argentinians were a lot less frightening than he could be! I think that was his major contribution; that, and not expecting anything more than he was prepared to do himself and making that clear to everyone. That's why he got killed, but what he'd set in motion of course didn't die with him. The act just kept rolling. We all knew what was expected of us and it would have taken a deliberate act at that stage to stop what he'd started. I sometimes disagreed with his military judgment, but I had no doubts about him as a man – an extraordinary personality. I just don't think the Battle of Goose Green as we know it could have ever happened without him.

Mitsuo Fuchida

Tora! Tora! Tora!

Fuchida led the strike of 353 Japanese fighters and bombers that struck Pearl Harbor on 7 December 1941 and brought the USA into the Second World War.

On the flight deck a green lamp was waved in a circle to signal "Take off!" The engine of the foremost fighter plane began to roar. With the ship still pitching and rolling, the plane started its run, slowly at first but with steadily increasing speed. Men lining the flight deck held their breath as the first plane took off successfully just before the ship took a downward pitch. The next plane was already moving forward. There were loud cheers as each plane rose into the air.

Thus did the first wave of 183 fighters, bombers, and torpedo planes take off from the six carriers. Within fifteen minutes they had all been launched and were forming up in the still-dark sky, guided only by signal lights of the lead planes. After one great circling over the fleet formation, the planes set course due south for Oahu Island and Pearl Harbor. It was 0615.

Under my direct command were 49 level bombers. About 500 meters to my right and slightly below me were 40 torpedo planes. The same distance to my left, but about 200 meters above me, were 51 dive bombers, and flying cover for the formation there were 43 fighters. These other three groups were led by Lieutenant Commanders Murata, Takahashi, and Itaya, respectively.

We flew through and over the thick clouds which were at 2000 meters, up to where day was ready to dawn. And the clouds began gradually to brighten below us after the brilliant sun burst into the eastern sky. I opened the cockpit canopy and looked back at the large formation of planes. The wings glittered in the bright morning sunlight.

The speedometer indicated 125 knots and we were favored by a tail wind. At 0700 I figured that we should reach Oahu in less than an hour. But flying over the clouds we could not see the surface of the water, and, consequently, had no check on our drift. I switched on the radio-direction finder to tune in the Honolulu radio station and soon picked up some light music. By turning the antenna I found the exact direction from which the broadcast was coming and corrected our course, which had been five degrees off.

Continuing to listen to the program, I was wondering how to get below the clouds after reaching Oahu. If the island was covered by thick clouds like those below us, the level bombing would be difficult; and we had not yet had reports from the reconnaissance planes.

In tuning the radio a little finer I heard, along with the music, what seemed to be a weather report. Holding my breath, I adjusted the dial and listened intently. Then I heard it come through a second time, slowly and distinctly: "Averaging partly cloudy, with clouds mostly over the mountains. Cloud base at 3500 feet. Visibility good. Wind north, 10 knots."

What a windfall for us! No matter how careful the planning, a more favorable situation could not have been imagined. Weather conditions over Pearl Harbor had been worrying me greatly, but now with this information I could turn my attention to other problems. Since Honolulu was only partly cloudy, there must be breaks in the clouds over the island. But since the clouds over the mountains were at 1000 meters altitude, it would not be wise to attack from the northeast, flying over the eastern mountains, as previously planned. The wind was north and visibility good. It would be better to pass to the west of the island and make our approach from the south.

We had been in the air for about an hour and a half. It was time that we were seeing land, but there was only a solid layer of clouds below. All of a sudden the clouds broke, and a long white line of coast appeared. We were over Kahuku Point, the northern tip of the island, and now it was time for our deployment.

There were alternate plans for the attack: if we had surprise, the torpedo planes were to strike first, followed by the level bombers and then the dive bombers, which were to attack the air bases including Hickam and Ford Island near the anchorage. If these bases were first hit by the dive bombers, it was feared that the resultant smoke might hinder torpedo and level-bombing attacks on the ships.

On the other hand, if enemy resistance was expected, the dive bombers would attack first to cause confusion and attract enemy fire. Level bombers, coming next, were to bomb and destroy enemy anti-aircraft guns, followed by the torpedo planes which would attack the ships.

The selection of attack method was for my decision, to be indicated by signal pistol: one "black dragon" for a surprise attack, two "black dragons" if it appeared that surprise was lost. Upon either order the fighters were immediately to dash in as cover.

There was still no news from the reconnaissance planes, but I had made up my mind that we could make a surprise attack, and there-upon ordered the deployment by raising my signal pistol outside the canopy and firing one "black dragon." The time was 0740.

With this order dive bombers rose to 4000 meters, torpedo bombers went down almost to sea level, and level bombers came down just under the clouds. The only group that failed to deploy was the fighters. Flying above the rest of the formation, they seemed to have missed the signal because of the clouds. Realizing this I fired another shot toward the fighter group. This time they noticed the signal immediately and sped toward Oahu.

This second shot, however, was taken by the commander of the dive bomber group as the second of two "black dragons," signifying a non-surprise attack which would mean that his group should attack first, and this error served to confuse some of the pilots who had understood the original signal.

Meanwhile a reconnaissance report came in from *Chikuma*'s plane giving the locations of ten battleships, one heavy cruiser, and ten light cruisers in the harbor. It also reported a 14-meter wind from bearing 080, and clouds over the U.S. Fleet at 1700 meters with a scale 7 density. The *Tone* plane also reported that "the enemy fleet is not in Lahaina Anchorage." Now I knew for sure that there were no carriers in the harbor. The sky cleared as we moved in on the target and Pearl Harbor was plainly visible from the northwest valley of the island. I studied our objective through binoculars. They were there all right, all eight of them. "Notify all planes to launch attacks," I ordered my radio man who immediately began tapping the key. The order went in plain code: "*To, to, to, to. . . .*" The time was 0749.

★ ★ ★

When Lieutenant Commander Takahashi and his dive-bombing group mistook my signal, and thought we were making a non-surprise attack, his 53 planes lost no time in dashing forward. His command was divided into two groups: one led by himself which headed for Ford Island and Hickam Field, the other, led by Lieutenant Sakamoto, headed for Wheeler Field.

The dive bombers over Hickam Field saw heavy bombers lined up on the apron. Takahashi rolled his plane sharply and went into a dive, followed immediately by the rest of his planes, and the first bombs fell at Hickam. The next places hit were Ford Island and Wheeler Field. In a very short time huge billows of black smoke were rising from these bases. The lead torpedo planes were to have started their run to the Navy Yard from over Hickam, coming from south of the bay entrance. But the sudden burst of bombs at Hickam surprised Lieutenant Commander Murata who had understood that his torpedo planes were to have attacked first. Hence he took a short cut lest the smoke from those bases cover up his targets. Thus the first torpedo was actually launched some five minutes ahead of the scheduled 0800. The time of each attack was as follows:

0755 Dive bombers at Hickam and Wheeler
0757 Torpedo planes at battleships
0800 Fighters strafing air bases
0805 Level bombers at battleships

After issuance of the attack order, my level bomber group kept east of Oahu going past the southern tip of the island. On our left was the Barbers Point airfield, but, as we had been informed, there were no planes. Our information indicated that a powerful anti-aircraft battery was stationed there, but we saw no evidence of it.

I continued to watch the sky over the harbor and activities on the ground. None but Japanese planes were in the air, and there were no indications of air combat. Ships in the harbor still appeared to be asleep, and the Honolulu radio broadcast continued normally. I felt that surprise was now assured, and that my men would succeed in their missions.

Knowing that Admirals Nagumo, Yamamoto, and the General Staff were anxious about the attack, I decided that they should be informed. I ordered the following message sent to the fleet: "We have succeeded in making a surprise attack. Request you relay this report to Tokyo." The radio man reported shortly that the message had been received by *Akagi*.

The code for a successful surprise attack was "*Tora, tora, tora* . . ." Before *Akagi*'s relay of this message reached Japan, it was received by *Nagato* in Hiroshima Bay and the General Staff in Tokyo, directly from my plane! This was surely a long-distance record for such a low-powered transmission from an airplane, and might be attributed to the use of the word. "*Tora*" as our code. There is a Japanese saying, "A tiger (*tora*) goes out 1000 *ri* (2000 miles) and returns without fail."

I saw clouds of black smoke rising from Hickam and soon thereafter from Ford Island. This bothered me and I wondered what had happened. It was not long before I saw waterspouts rising alongside the battleships, followed by more and more waterspouts. It was time to launch our level bombing attacks so I ordered my pilot to bank sharply, which was the attack signal for the planes following us. All ten of my squadrons then formed into a single column with intervals of 200 meters. It was indeed a gorgeous formation.

The lead plane in each squadron was manned by a specially trained pilot and bombardier. The pilot and bombardier of my squadron had won numerous fleet contests and were considered the best in the Japanese Navy. I approved when Lieutenant Matsuzaki asked if the lead plane should trade positions with us, and he lifted our plane a little as a signal. The new leader came forward quickly, and I could see the smiling round face of the bombardier when he saluted. In returning the salute I entrusted the command to them for the bombing mission.

As my group made its bomb run, enemy anti-aircraft guns suddenly came to life. Dark gray bursts blossomed here and there until the sky was clouded with shattering near misses which made our plane tremble. Shipboard guns seemed to open fire before the shore batteries. I was startled by the rapidity of the counter-attack which came less than five minutes after the first bomb had fallen. Were it the Japanese Fleet, the reaction would not have been so quick, because although the Japanese character is suitable for offensives, it does not readily adjust to the defensive.

Suddenly the plane bounced as if struck by a huge club. "The fuselage is holed to port," reported the radio man behind me, "and a steering-control wire is damaged." I asked hurriedly if the plane was under control, and the pilot assured me that it was.

No sooner were we feeling relieved than another burst shook the plane. My squadron was headed for *Nevada*'s mooring at the northern end of battleship row on the east side of Ford Island. We were just passing over the bay entrance and it was almost time to release our bombs. It was not easy to pass through the concentrated anti-aircraft fire. Flying at only 3000 meters, it seemed that this might well be a date with eternity.

I further saw that it was not wise to have deployed in this long single-column formation. The whole level bomber group could be destroyed like ducks in a shooting gallery. It would also have been better if we had approached the targets from the direction of Diamond Head. But here we were at our targets and there was a job to be done.

It was now a matter of utmost importance to stay on course, and the lead plane kept to its line of flight like a homing pigeon. Ignoring the barrage of shells bursting around us, I concentrated on the bomb loaded under the lead plane, pulled the safety bolt from the bomb release lever and grasped the handle. It seemed as if time was standing still.

Again we were shaken terrifically and our planes were buffeted about. When I looked out the third plane of my group was abeam of us and I saw its bomb fall! That pilot had a reputation for being careless. In training his bomb releases were poorly timed, and he had often been cautioned.

I thought, "That damn fellow has done it again!" and shook my fist in his direction. But I soon realized that there was something wrong with his plane and he was losing gasoline. I wrote on a small blackboard, "What happened?" and held it toward his plane. He explained. "Underside of fuselage hit."

Now I saw his bomb cinch lines fluttering wildly, and sorry for having scolded him, I ordered that he return to the carrier. He answered, "Fuel tank destroyed, will follow you," asking permission to stay with the group. Knowing the feelings of the pilot and crew, I gave permission, although I knew it was useless to try taking that crippled and bombless plane through the enemy fire. It was nearly time for bomb release when we ran into clouds which obscured the target, and I made out the round face of the lead bombardier who was waving his hands back and forth to indicate that we had passed the release point. Banking slightly we turned right toward Honolulu, and I studied the anti-aircraft fire, knowing that we would have to run through it again. It was now concentrated on the second squadron.

While circling for another try, I looked toward the area in which the bomb from the third plane had fallen. Just outside the bay entrance I saw a large water ring close by what looked like a destroyer. The ship seemed to be standing in a floating dock, attached to both sides of the entrance like a gate boat. I was suddenly reminded of the midget submarines which were to have entered the bay for a special attack.

At the time of our sortie I was aware of these midget submarines, but knew nothing of their characteristics, operational objectives, force organization, or the reason for their participation in the attack. In *Akagi*, Commander Shibuya, a staff officer in charge of submarine operations, had explained that they were to penetrate the harbor the night before our attack; but, no matter how good an opportunity might arise, they were not to strike until after the planes had done so.

Even now the submarines were probably concealed in the bay, awaiting the air attack. Had the entrance been left open, there would have been some opportunity for them to get out of the harbor. But in light of what I had just seen there seemed little chance of that, and, feeling now the bitterness of war, I vowed to do my best in the assigned mission.

While my group was circling over Honolulu for another bombing attempt, other groups made their runs, some making three tries before succeeding. Suddenly a colossal explosion occurred in battleship row. A huge column of dark red smoke rose to 1000 feet and a stiff shock wave reached our plane. I called the pilot's attention to the spectacle, and he observed, "Yes, Commander, the powder magazine must have exploded. Terrible indeed!" The attack was in full swing, and smoke from fires and explosions filled most of the sky over Pearl Harbor.

My group now entered on a bombing course again. Studying battleship row through binoculars, I saw that the big explosion had been on *Arizona*. She was still flaming fiercely and her smoke was covering *Nevada*, the target of my group. Since the heavy smoke would hinder our bomber accuracy, I looked for some other ship to attack. *Tennessee*, third in the left row, was already on fire; but next in row was *Maryland*, which had not yet been attacked. I gave an order changing our target to this ship, and once again we headed into the anti-aircraft fire. Then came the "ready" signal and I took a firm grip on the bomb release handle, holding my breath and staring at the bomb of the lead plane.

Pilots, observers, and radio men all shouted, "Release!" on seeing the bomb drop from the lead plane, and all the others let go their bombs. I immediately lay flat on the floor to watch the fall of bombs through a peephole. Four bombs in perfect pattern plummeted like devils of doom. The target was so far away that I wondered for a moment if they would reach it. The bombs grew smaller and smaller until I was holding my breath for fear of losing them. I forgot everything in the thrill of watching the fall toward the target. They become small as poppy seeds and finally disappeared just as tiny white flashes of smoke appeared on and near the ship.

Livy

Hannibal at Cannae

The Battle of Cannae, in 216 BC, is as famous for Hannibal's tactics as for its role in Roman history – it was an early, and near-perfect, example of the "pincer movement".

WHILST time was thus being wasted in disputes instead of deliberation, Hannibal withdrew the bulk of his army, who had been standing most of the day in order of battle, into camp. He sent his Numidians, however, across the river to attack the parties who were getting water for the smaller camp. They had hardly gained the opposite bank when with their shouting and uproar they sent the crowd flying in wild disorder, and galloping on as far as the outpost in front of the rampart, they nearly reached the gates of the camp. It was looked upon as such as insult for a Roman camp to be actually terrorised by irregular auxiliaries that one thing, and one thing alone, held back the Romans from instantly crossing the river and forming their battle line – the supreme command that day rested with Paulus.

The following day Varro, whose turn it now was, without any consultation with his colleague, exhibited the signal for battle and led his forces drawn up for action across the river. Paulus followed, for though he disapproved of the measure, he was bound to support it. After crossing, they strengthened their line with the force in the smaller camp and completed their formation. On the right, which was nearest to the river, the Roman cavalry were posted, then came the infantry: on the extreme left were the cavalry of the allies, their infantry were between them and the Roman legions. The javelin men with the rest of the light-armed auxiliaries formed the front line. The consuls took their stations on the wings, Terentius Varro on the left, Æmilius Paulus on the right.

As soon as it grew light Hannibal sent forward the Balearics and the other light infantry. He then crossed the river in person and as each division was brought across he assigned it its place in the line. The Gaulish and Spanish horse he posted near the bank on the left wing in front of the Roman cavalry; the right wing was assigned to the Numidian troopers. The centre consisted of a strong force of infantry, the Gauls and Spaniards in the middle, the Africans at either end of them. You might fancy that the Africans were for the most part a body of Romans from the way they were armed, they were so completely equipped with the arms, some of which they had taken at the Trebia, but the most part at Trasumennus. The Gauls and Spaniards had shields almost of the same shape; their swords were totally different, those of the Gauls being very long and without a point, the Spaniard, accustomed to thrust more than to cut, had a short handy sword, pointed like a dagger. These nations, more than any other, inspired terror by the vastness of their stature and their frightful appearance: the Gauls were naked above the waist, the Spaniards had taken up their position wearing white tunics embroidered with purple, of dazzling brilliancy. The total number of infantry in the field was 40,000, and there were 10,000 cavalry. Hasdrubal was in command of the left wing, Maharbal of the right; Hannibal himself with his brother Mago commanded the centre. It was a great convenience to both armies that the sun shone obliquely on them, whether it was that they had purposely so placed themselves, or whether it happened by accident, since the Romans faced the north, the Carthaginians the south. The wind, called by the inhabitants the Vulturnus, was against the Romans, and blew great clouds of dust into their faces, making it impossible for them to see in front of them.

When the battle shout was raised the auxiliaries ran forward, and the battle began with the light infantry. Then the Gauls and Spaniards on the left engaged the Roman cavalry on the right; the battle was not at all like a cavalry fight, for there was no room for manœuvering, the river on the one side and the infantry on the other hemming them in, compelled them to fight face to face. Each side tried to force their way straight forward, till at last the horses were standing in a closely pressed mass, and the riders seized their opponents and tried to drag them from their horses. It had become mainly a struggle of infantry, fierce but short, and the Roman cavalry was repulsed and fled. Just as this battle of the cavalry was finished, the infantry became engaged, and as long as the Gauls and Spaniards kept their ranks unbroken, both sides were equally matched in strength and courage. At length after long and repeated efforts the Romans closed up their ranks, echeloned their front, and by the sheer weight of their deep column bore down the division of the enemy which was stationed in front of Hannibal's line, and was too thin and weak to resist the pressure. Without a moment's pause they followed up their broken and hastily retreating foe till they took to headlong flight. Cutting their way through the mass of fugitives, who offered no resistance, they penetrated as far as the Africans who were stationed on both wings, somewhat further back than the Gauls and Spaniards who had formed the advanced centre. As the latter fell back the whole front became level, and as they continued to give ground it became concave and crescent-shaped, the Africans at either end forming the horns. As the Romans rushed on incautiously between them, they were enfiladed by the two wings, which extended and closed round them in the rear. On this, the Romans, who had fought one battle to no purpose, left the Gauls and Spaniards, whose rear they had been slaughtering, and commenced a fresh struggle with the Africans. The contest was a very one-sided one, for not only were they hemmed in on all sides, but wearied with the previous fighting they were meeting fresh and vigorous opponents.

By this time the Roman left wing, where the allied cavalry were fronting the Numidians, had become engaged, but the fighting was slack at first owing to a Carthaginian stratagem. About 500 Numidians, carrying, besides their usual arms and missiles, swords concealed under their coats of mail, rode out from their own line with their shields slung behind their backs as though they were deserters, and suddenly leaped from their horses and flung their shields and javelins at the feet of their enemy. They were received into their ranks, conducted to the rear, and ordered to remain quiet. While the battle was spreading to the various parts of the field they remained quiet, but when the eyes and minds of all were wholly taken up with the fighting they seized the large Roman shields which were lying everywhere amongst the heaps of slain and commenced a furious attack upon the rear of the Roman line. Slashing away at backs and hips, they made a great slaughter and a still greater panic and confusion. Amidst the rout and panic in one part of the field and the obstinate but hopeless struggle in the other, Hasdrubal, who was in command of that arm, withdrew some Numidians from the centre of the right wing, where the fighting was feebly kept up, and sent them in pursuit of the fugitives, and at the same time sent the Spanish and Gaulish horse to the aid of the Africans, who were by this time more wearied by slaughter than by fighting.

Paulus was on the other side of the field. In spite of his having been seriously wounded at the commencement of the action by a bullet from a sling, he frequently encountered Hannibal with a compact body of troops, and in several places restored the battle. The Roman cavalry formed a bodyguard round him, but at last, as he became too weak to manage his horse, they all dismounted. It is stated that when some one reported to Hannibal that the consul had ordered his men to fight on foot, he remarked, "I would rather he handed them over to me bound hand and foot." Now that the victory of the enemy was no longer doubtful this struggle of the dismounted cavalry was such as might be expected when men preferred to die where they stood rather than flee, and the victors, furious at them for delaying the victory, butchered without mercy those whom they could not dislodge. They did, however, repulse a few survivors exhausted with their exertions and their wounds. All were at last scattered, and those who could regained their horses for flight. Cn. Lentulus, a military tribune, saw, as he rode by, the consul covered with blood sitting on a boulder. "Lucius Æmilius," he said, "the one man whom the gods must hold guiltless of this day's disaster, take this horse while you have still some strength left, and I can lift you into the saddle and keep by your side to protect you. Do not make this day of battle still more fatal by a consul's death, there are enough tears and mourning without that." The consul replied: "Long may you live to do brave deeds, Cornelius, but do not waste in useless pity the few moments left in which to escape from the hands of the enemy. Go, announce publicly to the senate that they must fortify Rome and make its defence strong before the victorious enemy approaches, and tell Q. Fabius privately that I have ever remembered his precepts in life and in death. Suffer me to breathe my last among my slaughtered soldiers, let me not have to defend myself again when I am no longer consul, or appear as the accuser of my colleague and protect my own innocence by throwing the guilt on another." During this conversation a crowd of fugitives came suddenly upon them, followed by the enemy, who, not knowing who the consul was, overwhelmed him with a shower of missiles. Lentulus escaped on horseback in the rush.

Then there was flight in all directions; 7000 men escaped to the smaller camp, 10,000 to the larger, and about 2000 to the village of Cannae. These latter were at once surrounded by Carthalo and his cavalry, as the village was quite unfortified. The other consul, who either by accident or design had not joined any of these bodies of fugitives, escaped with about fifty cavalry to Venusia; 45,500 infantry, 2700 cavalry – almost an equal proportion of Romans and allies – are said to have been killed. Amongst the number were both the quaestors attached to the consuls, L. Atilius and L. Furius Bibulcus, twenty-nine military tribunes, several ex-consuls, ex-praetors, and ex-ædiles (amongst them are included Cn. Servilius Geminus and M. Minucius, who was Master of the Horse the previous year and, some years before that, consul), and in addition to these, eighty men who had either been senators or filled offices qualifying them for election to the senate and who had volunteered for service with the legions. The prisoners taken in the battle are stated to have amounted to 3000 infantry and 1500 cavalry.

Edward Creasy

The Battle that Saved the West

"The events that rescued our ancestors of Britain and our neighbours of Gaul from the . . . yoke of the Koran" was Gibbon's evaluation of Charles Martel's defeat of the Saracens at Tours in AD 732.

Although three centuries had passed away since the Germanic conquerors of Rome had crossed the Rhine, never to repass that frontier stream, no settled system of institutions or government, no amalgamation of the various races into our people, no uniformity of language or habits had been established in the country at the time when Charles Martel was called to repel the menacing tide of Saracenic invasion from the south. Gaul was not yet France. In that, as in other provinces of the Roman empire of the west, the dominion of the Caesars had been shattered as early as the fifth century, and barbaric kingdoms and principalities had promptly arisen on the ruins of the Roman power. But few of these had any permanency, and none of them consolidated the rest, or any considerable number of the rest, into one coherent and organized civil and political society. The great bulk of the population still consisted of the conquered provincials, that is to say, of Romanized Celts, of a Gallic race which had long been under the dominion of the Caesars, and had acquired, together with no slight infusion of Roman blood, the language, the literature, the laws, and the civilization of Latium. Among these, and dominant over them, roved or dwelt the German victors; some retaining nearly all the rude independence of their primitive national character, others softened and disciplined by the aspect and contact of the manners and institutions of civilized life; for it is to be borne in mind that the Roman empire in the west was not crushed by any sudden avalanche of barbaric invasion. The German conquerors came across the Rhine, not in enormous hosts, but in bands of a few thousand warriors at a time. The conquest of a province was the result of an infinite series of partial local invasions, carried on by little armies of this description. The victorious warriors either retired with their booty, or fixed themselves in the invaded district, taking care to keep sufficiently concentrated for military purposes, and ever ready for some fresh foray, either against a rival Teutonic band, or some hitherto unassailed city of the provincials. Gradually, however, the conquerors acquired a desire for permanent landed possessions. They lost somewhat of the restless thirst for novelty and adventure which had first made them throng beneath the banner of the boldest captains of their tribe, and leave their native forests for a roving military life on the left bank of the Rhine. They were converted to the Christian faith, and gave up with their old creed much of the coarse ferocity which must have been fostered in the spirits of the ancient warriors of the north by a mythology which promised, as the reward of the brave on earth, an eternal cycle of fighting and drunkenness in heaven.

But, although their conversion and other civilizing influences operated powerfully upon the Germans in Gaul, and although the Franks (who were originally a confederation of the Teutonic tribes that dwelt between the Rhine, the Maine, and the Weser) established a decisive superiority over the other conquerors of the province, as well as over the conquered provincials, the country long remained a chaos of uncombined and shifting elements. The early princes of the Merovingian dynasty were generally occupied in wars against other princes of their house, occasioned by the frequent subdivisions of the Frank monarchy and the ablest and best of them had found all their energies tasked to the utmost to defend the barrier of the Rhine against the pagan Germans who strove to pass that river and gather their share of the spoils of the empire.

The conquests which the Saracens effected over the southern and eastern provinces of Rome were far more rapid than those achieved by the Germans in the north, and the new organizations of society which the Moslems introduced were summarily and uniformly enforced. Exactly a century passed between the death of Mohammed and the date of the battle of Tours. During that century the followers of the Prophet had torn away half the Roman empire; and, besides their conquests over Persia, the Saracens had overrun Syria, Egypt, Africa, and Spain, in an uncheckered and apparently irresistible career of victory. Nor, at the commencement of the eighth century of our era, was the Mohammedan world divided against itself, as it subsequently became. All these vast regions obeyed the caliph; throughout them all, from the Pyrenees to the Oxus, the name of Mohammed was invoked in prayer, and the Koran revered as the book of the law.

It was under one of their ablest and most renowned commanders, with a veteran army, and with every apparent advantage of time, place, and circumstance, that the Arabs made their great effort at the conquest of Europe north of the Pyrenees. The victorious Moslem soldiery in Spain,

> A countless multitude;
> Syrian, Moor, Saracen, Greek renegade,
> Persian, and Copt, and Tartar, in one bond
> Of erring faith conjoined – strong in the youth
> And heat of zeal – a dreadful brotherhood.

were eager for the plunder of more Christian cities and shrines, and full of fanatic confidence in the invincibility of their arms.

> Nor were the chiefs
> Of victory less assured, by long success
> Elate, and proud of that o'erwhelming strength
> Which, surely they believed, as it had rolled
> Thus far uncheck'd, would roll victorious on,
> Till, like the Orient, the subjected West
> Should bow in reverence at Mohammed's name;
> And pilgrims from remotest Arctic shores
> Tread with religious feet the burning sands
> Of Araby and Mecca's stony soil.
>
> Southey's *Roderick*

It is not only by the modern Christian poet, but by the old Arabian chroniclers also, that these feelings of ambition and arrogance are attributed to the Moslems who had overthrown the Visigoth power in Spain. And their eager expectations of new wars were excited to the utmost on the reappointment by the caliph of Abderrahman Ibn Abdillah Alghafeki to the government of that country, in AD729, which restored them a general who had signalized his skill and prowess during the conquests of Africa and Spain, whose ready valor and generosity had made him the idol of the troops, who had already been engaged in several expeditions into Gaul, so as to be well acquainted with the national character and tactics of the Franks, and who was known to thirst, like a good Moslem, for revenge for the slaughter of some detachments of the True Believers, which had been cut off on the north of the Pyrenees.

In addition to his cardinal military virtues, Abderrahman is described by the Arab writers as a model of integrity and justice. The first two years of his second administration in Spain were occupied in severe reforms of the abuses which under his predecessors had crept into the system of government, and in extensive preparations for his intended conquest in Gaul. Besides the troops which he collected from his province, he obtained from Africa a large body of chosen Berber cavalry, officered by Arabs of proved skill and valor; and in the summer of 732, he crossed the Pyrenees at the head of an army which some Arab writers rate at eighty thousand strong, while some of the Christian chroniclers swell its numbers to many hundreds of thousands more. Probably the Arab account diminishes, but of the two keeps nearer to the truth. It was from this formidable host, after Eudes, the Count of Aquitaine, had vainly striven to check it, after many strong cities had fallen before it, and half the land had been overrun, that Gaul and Christendom were at last rescued by the strong arm of Prince Charles, who acquired a surname, like that of the war-god of his forefathers' creed, from the might with which he broke and shattered his enemies in the battle.

The Merovingian kings had sunk into absolute insignificance, and had become mere puppets of royalty before the eighth century. Charles Martel, like his father, Pepin Heristal, was Duke of the Austrasian Franks, the bravest and most thoroughly Germanic part of the nation, and exercised, in the name of the titular king, what little paramount authority the turbulent minor rulers of districts and towns could be persuaded or compelled to acknowledge. Engaged with his national competitors in perpetual conflicts for power, and in more serious struggles for safety against the fierce tribes of the unconverted Frisians, Bavarians, Saxons, and Thuringians, who at that epoch assailed with peculiar ferocity the Christianized Germans on the left bank of the Rhine, Charles Martel added experienced skill to his natural courage, and he had also formed a militia of veterans among the Franks. Hallam has thrown out a doubt whether, in our admiration of his victory at Tours, we do not judge a little too much by the event, and whether there was not rashness in his risking the fate of France on the result of a general battle with the invaders. But when we remember that Charles had no standing army, and the independent spirit of the Frank warriors who followed his standard, it seems most probable that it was not in his power to adopt the cautious policy of watching the invaders, and wearing out their strength by delay. So dreadful and so widespread were the ravages of the Saracenic light cavalry throughout Gaul, that it must have been impossible to restrain for any length of time the indignant ardor of the Franks. And, even, if Charles could have persuaded his men to look tamely on while the Arabs stormed more towns and desolated more districts, he could not have kept an army together when the usual period of a military expedition had expired. If, indeed, the Arab account of the disorganization of the Moslem forces be correct, the battle was as well timed on the part of Charles, as it was, beyond all question, well fought.

The monkish chroniclers, from whom we are obliged to glean a narrative of this memorable campaign, bear full evidence to the terror which the Saracen invasion inspired, and to the agony of that great struggle. The Saracens, say they, and their king, who was called Abdirames, came out of Spain, with all their wives, and their children, and their substance, in such great multitudes that no man could reckon or estimate them. They brought with them all their armor, and whatever they had, as if they were thenceforth always to dwell in France.

> Then Abderrahman, seeing the land filled with the multitude of his army, pierces through the mountains, tramples over rough and level ground, plunders far into the country of the Franks, and smites all with the sword, insomuch that when Eudo came to battle with him at the River Garonne, and fled before him, God alone knows the number of the slain. Then Abderrahman pursued after Count Eudo, and, while he strives to spoil and burn the holy shrine at Tours, he encounters the chief of the Austrasian Franks, Charles, a man of war from his youth up, to whom Eudo had sent warning. There for nearly seven days they strive intensely, and at last they set themselves in battle array, and the nations of the North

> standing firm as a wall, and impenetrable as a zone of ice, utterly slay the Arabs with the edge of the sword.

The European writers all concur in speaking of the fall of Abderrahman as one of the principal causes of the defeat of the Arabs, who, according to one writer, after finding that their leader was slain, dispersed in the night, to the agreeable surprise of the Christians, who expected the next morning to see them issue from their tents and renew the combat. One monkish chronicler puts the loss of the Arabs at 375,000 men, while he says that only 1,007 Christians fell; a disparity of loss which he feels bound to account for by a special interposition of Providence. I have translated above some of the most spirited passages of these writers; but it is impossible to collect from them anything like a full or authentic description of the great battle itself, or of the operations which preceded and followed it.

Though, however, we may have cause to regret the meagreness and doubtful character of these narratives, we have the great advantage of being able to compare the accounts given of Abderrahman's expedition by the national writers of each side. This is a benefit which the inquirer into antiquity so seldom can obtain, that the fact of possessing it, in the case of the battle of Tours, makes us think the historical testimony respecting that great event more certain and satisfactory than is the case in many other instances, where we possess abundant details respecting military exploits, but where those details come to us from the annalist of one nation only, and where we have consequently no safeguard against the exaggemtions, the distortions, and the fictions which national vanity has so often put forth in the garb and under the title of history. The Arabian writers who recorded the conquests and wars of their countrymen in Spain have narrated also the expedition into Gaul of their great emir, and his defeat and death near Tours, in battle with the host of the Franks under King Caldus, the name into which they metamorphose Charles Martel.

They tell us how there was war between the count of the Frankish frontier and the Moslems, and how the count gathered together all his people, and fought for a time with doubtful success. "But," say the Arabian chroniclers, "Abderrahman drove them back; and the men of Abderrahman were puffed up in spirit by their repeated successes, and they were full of trust in the valor and the practice in war of their emir. So the

Moslems smote their enemies, and passed the River Garonne, and laid waste the country, and took captives without number. And that army went through all places like a desolating storm. Prosperity made these warriors insatiable. At the passage of the river, Abderrahman overthrew the count, and the count retired into his stronghold, but the Moslems fought against it, and entered it by force and slew the count; for everything gave way to their cimeters, which were the robbers of lives. All the nations of the Franks trembled at that terrible army, and they betook them to their king Caldus, and told him of the havoc made by the Moslem horsemen, and how they rode at their will through all the land of Narbonne, Toulouse, and Bordeaux, and they told the king of the death of their count. Then the king bade them be of good cheer, and offered to aid them. And in the 114th year he mounted his horse, and he took with him a host that could not be numbered, and went against the Moslems. And he came upon them at the great city of Tours. And Abderrahman and other prudent cavaliers saw the disorder of the Moslem troops, who were loaded with spoil; but they did not venture to displease the soldiers by ordering them to abandon everything except their arms and war-horses. And Abderrahman trusted in the valor of his soldiers, and in the good fortune which had ever attended him. But (the Arab writer remarks) such defect of discipline always is fatal to armies. So Abderrahman and his host attacked Tours to gain still more spoil, and they fought against it so fiercely that they stormed the city almost before the eyes of the army that came to save it; and the fury and the cruelty of the Moslems towards the inhabitants of the city were like the fury and cruelty of raging tigers. It was manifest," adds the Arab, "that God's chastisement was sure to follow such excesses; and Fortune thereupon turned her back upon the Moslems."

Near the River Owar, the two great hosts of the two languages and the two creeds were set in array against each other. The hearts of Abderrahman, his captains, and his men, were filled with wrath and pride, and they were the first to begin the fight. The Moslem horsemen dashed fierce and frequent forward against the battalions of the Franks, who resisted manfully, and many fell dead on either side, until the going down of the sun. Night parted the two armies; but in the gray of the morning the Moslems returned to the battle. Their cavaliers had soon hewn their way into the centre of the Christian host. But many of the Moslems were fearful for the safety of the spoil which they had stored in their tents, and a false cry arose in their ranks that some of the enemy were plundering the camp; whereupon several squadrons of the Moslem horsemen rode off to protect their tents. But it seemed as if they fled; and all the host was troubled. And, while Abderrahman strove to check their tumult, and to lead them back to battle, the warriors of the Franks came around him, and he was pierced through with many spears, so that he died. Then all the host fled before the enemy and many died in the flight. This deadly defeat of the Moslems, and the loss of the great leader and good cavalier, Abderrahman, took place in the hundred and fifteenth year.

It would be difficult to expect from an adversary a more explicit confession of having been thoroughly vanquished than the Arabs here accord to the Europeans. The points on which their narrative differs from those of the Christians – as to how many days the conflict lasted, whether the assailed city was actually rescued or not, and the like – are of little moment compared with the admitted great fact that there was a decisive trial of strength between Frank and Saracen, in which the former conquered. The enduring importance of the battle of Tours in the eyes of the Moslems is attested not only by the expressions of "the deadly battle" and "the disgraceful overthrow" which their writers constantly employ when referring to it, but also by the fact that no more serious attempts at conquest beyond the Pyrenees were made by the Saracens. Charles Martel, and his son and grandson, were left at leisure to consolidate and extend their power. The new Christian Roman empire of the West, which the genius of Charlemagne founded, and throughout which his iron will imposed peace on the old anarchy of creeds and races, did not indeed retain its integrity after its great ruler's death. Fresh troubles came over Europe; but Christendom, though disunited, was safe. The progress of civilization, and the development of the nationalities and governments of modern Europe, from that time forth went forward in not uninterrupted, but ultimately certain career.

John Parker

Not Men But Devils

Each year the French Foreign Legion parades the wooden hand of Captain Jean Danjou – the only relic of the officer found after he had led the Legion's celebrated action at Camerone in Mexico in 1863.

Two years before the Battle of Camerone, the new Mexican president, Benito Pablo Juarez, suspended payment of interest on the country's debts. The French King, Napoleon III, having recently purchased some of the Mexican debts from a Swiss bank, demanded military action to secure his investment. He proposed the installation of a puppet monarch, Maximilian of Hapsburg, younger brother of Franz Josef of Austria. The French would send an expeditionary force of 3,000 to support him. They were joined by moderate and less than enthusiastic contingents from Britain and Spain. The French infantry arrived early in 1862, began their march inland and were promptly driven back to the beach where they had landed. The British and the Spanish withdrew their troops after heavy losses, but Napoleon III refused to accept the humiliation of retreat.

By the year's end, the French had ferried 40,000 men across the Atlantic, falling ever deeper into another guerilla war. The Foreign Legion had not been called upon to support this intervention. It had been considered that the international forces of the French, British and Spanish regular armies would have neither need of the Legion, nor of their methods of fighting. The French army were laying siege to Mexico City, and developing large concentrations of forces at Puebla.

The Legion, languishing at Sidi-bel-Abbès, in Algeria, began to get restless. They complained they were being used simply as police troops and construction workers. Junior officers, fearing dissension and stagnation, petitioned Napoleon III, begging that they be allowed to participate in the "Mexicana Affaire". On 19 January 1863, approval was received from Paris. A regiment of march, comprising two battalions of infantry, a base company and the band left for Mexico. The Legion's 3rd Battalion would remain in Algeria to hold the garrison and train new Legion recruits for Mexico.

On 9 February, the 2,000-strong Legion contingent sailed from Oran. It consisted of 48 officers, 1,432 Legionnaires along with sundry support forces which included eight canteen managers. They arrived on 28 March at Vera Cruz, 250 kilometres from where the French army had stalled in the face of strong local resistance at Puebla. The legionnaires discovered they were not being thrust into battle. Their task was to secure and safeguard French supply lines which travelled through 120 kilometres of tropical swamplands. The French Commander-in-Chief, General Forey, made no secret of his decision to place the Legion on guard duty. "I preferred to leave foreigners, rather than French, to guard that most unhealthy area," he wrote.

As the legionnaires soon discovered, their enemy was not the Mexicans but disease. Malaria, yellow fever, typhus and many infections not yet existing in the textbooks of their medical orderlies were soon attacking the newcomers. Within weeks, sickness had taken a huge toll on the Legion's strength. So much so that when the Legion's commanding officer, Colonel Jeanningros, was summoned to muster two companies to protect a slow-moving convoy, he had difficulty in staffing it.

The convoy was of particular importance. It consisted of sixty horse-drawn wagons filled with heavy guns, ammunition, supplies and three million francs in gold pieces bound for General Foley, bogged down at Puebla. Two days later, on 29 April, Jeanningros received news from a spy that the convoy was to be ambushed, not by guerrillas but by the Mexican army, anxious to avail itself of the new weapons, and the gold. Several battalions of Mexican regular infantry were already moving into position. Jeanningros detailed the 3rd Company of the 1st Battalion of the Legion to go out on patrol and hopefully make contact with the convoy and/or track the movements of the ambush troops. The Company was a sorry sight, decimated by sickness. Only sixty-two of the original 120 were still standing, and not a single officer among them.

A member of the commander's own HQ staff, Captain Jean Danjou, volunteered to lead the Company, and he was joined by two lieutenants promoted from the ranks, Vilain, not yet thirty, and a dour veteran sergeant named Maudet who had also fought at the Crimea and Magenta. Among the NCOs was Corporal Berg, who had given up a career as an officer in the French regular army to join the Legion, and Corporal Maine, who also left the regular army at the rank of sergeant major, having decided that only in the Legion would he find what he believed was his true vocation, as a fighting soldier. With Captain Danjou at their head, and with his newly promoted officers and the Company drummer, Legionnaire Lai, immediately behind, they set off marching in the cooler temperatures of the night towards Palo Verde, stopping only for a coffee break at a post held by the battalion's Grenadier Company.

Danjou pressed on and at dawn they were making their way through the foothills and deep ravines of the Mexican mountains, dotted only with scorched and withering trees. They were already suffering from the morning heat when he called a halt as they reached the humid plains close to Palo Verde around 7 a.m. The men sat down on the parched ground, tired and hot. Out of the dusty packs came the mess tins and soon the smell of coffee rose in the morning air. Danjou, a square, tall man with a small goatee beard walked among them, fiddling occasionally with the leather strap that attached his articulated wooden left hand to his forearm. The men had hardly time to drink their coffee when from the crest behind them, a sentry reported a cloud of dust from approaching horsemen, heading from the direction they had marched.

Danjou called the two lieutenants and ordered the Company to draw their arms and move out. The barren spot they had chosen to rest was no place to meet the oncoming Mexicans and he decided to head back to the tumbledown collection of farm buildings near Camerone which they had passed earlier, about half a mile or so away. They didn't make it. A swarm of Mexican cavalry, guns firing, reached them when they were still some distance from the farm. They took up position in thick, low scrub.

"Form a square," Danjou ordered. "Fire only on orders."

The Mexicans divided into two squadrons, to attack the legionnaires from opposite sides. They approached in a controlled walk, then at fifty metres, the "Charge" order was given, and with sabres flashing the Mexicans headed in at the gallop.

Simultaneously Captain Danjou screamed, "Fire!"

The legionnaires opened fire with their first round and then waited for the second command: "Fire!" Another sixty rounds exploded into the horseflesh and the riders. The Mexicans took some heavy casualties, pulled up and turned away. Danjou barked another order: "Fire at will!"

The legionnaires, with one foot forward, and heads down fired volley after volley into the mass of men and screaming horses. In the pandemonium, the Mexicans drew back, evidently surprised at the resistance. The legionnaires took the chance to make a dash for cover in a roadside hacienda, leaving their dead where they had fallen and the mules carrying their supplies disappearing into the distance.

Within fifteen minutes of the start of a running fight, Danjou, the two officers and forty-six legionnaires reached farm buildings, surrounded by a stone quadrangle. The large rickety wooden gates were rapidly slammed shut and barred with timber. Danjou deployed his men at strategic points in the buildings, some at windows in the farmhouse, others inside and on top of the stable block, more still lining the walls of the yard. Another Legion veteran, a Polish sergeant named Morzycki, climbed to the highest point of the roof, and came back with the gloomy report that there were "hundreds of Mexicans all around us".

Meanwhile, a cavalry rider had reached the encampment of the Mexican infantry, an hour's march away, under the command of Colonel Milan. His reaction erred on the side of overkill. He ordered his three battalions to move out and head for the scene. Under a blazing sun, the Mexican cavalry had reduced the pace of the battle, sending in snipers or rushing the weakest parts of Danjou's defences. They fought for two hours with little change in the situation until, at 9.30, the Mexicans put up a white flag and sent Lieutenant Ramon Laine to the gate of the hacienda, offering to accept an honourable surrender so as to end the slaughter.

"There are 2,000 of us, and more on the way," said the lieutenant. "We guarantee you safe conduct as prisoners of war."

Danjou sent him packing.

By 11.00, he had lost another twelve men. Every quarter of an hour, crawling on his hands and knees, he made a tour of his defences, talking to his men. He asked each one to take an oath to fight unto death. One by one they took it.

At 11.30, Danjou was shot, a bullet in the throat. Lieutenant Vilain took his medals and his sword. He was now the leader of the company's remaining thirty-two men capable of continuing the fight. At 12.00 the legionnaires heard bugles and drums. For a moment they thought that their own regiment had come up to their rescue. Their hopes were quickly dashed as another 1,000 Mexicans, having secured the convoy, appeared on the horizon.

Lieutenant Vilain took a bullet in the forehead and died instantly at 2.00. Mauder took command. The heat was now overpowering and they had no food or water and were running low on ammunition. They had been fighting for nine hours and had eaten nothing since the day before. The surviving legionnaires took Maudet's command and carried on, loading and firing, their faces now black with powder and stumbling over the dead and badly wounded. The deaths mounted as the afternoon wore on and towards 4.00, the Mexicans, whose own casualties now reached 280, set fire to the sheds and straw. The Mexican colonel, Milan, made a speech to his men about national pride, and sent a new attack with his own infantry bearing modern American carbines, firing on the hacienda from all sides.

Morzycki was shot from the stable roof, along with three others. By 5.00, only Lieutenant Maudet, Corporal Maine and three legionnaires, Wenzel, Catteau and Constantin were still standing. They fought on for another fifteen minutes, huddled, choking and retching in the smoking ruins of a shed. When each man had one round of ammunition left, Maudet gave the signal.

Corporal Maine's account of those last minutes remains in the Legion archives:

> We had held the enemy at a distance but we could not hold out any longer as our bullets were almost exhausted. It was six o'clock and we had fought since the early morning. The lieutenant shouted, "Ready, Fire!" and we discharged our remaining five bullets and, he in front, we jumped forward with fixed bayonets. We were met by a formidable volley. Catteau threw himself in front of his officer to make a rampart with his body and was struck with nineteen bullets. In spite of this devotion, Lieutenant Maudet himself was hit with two bullets. Wenzel also fell wounded in the shoulder but got up immediately. There were now three of us on our feet – Wenzel, Constantin and I.

> We were about to jump over the lieutenant's body and charge when the Mexicans encircled us and held their bayonets to our chests. We thought we had breathed our last when a senior officer who was in the front rank of the soldiers ordered them to stop and with a sharp movement of his sabre, raised their bayonets which threatened us. "Now will you surrender," he called to us. I replied, "We will surrender if you will leave us our arms and treat our lieutenant who is wounded." He agreed and offered me his arm and gave the other to help Wenzel. They brought out a stretcher for the lieutenant.

The Mexican infantry colonel who saved the lives of the three remaining legionnaires was named Cambas. He took Maine and the other two like VIPs to his commander, Colonel Milan, who uttered the famous description: "Truly, these are not men but devils".

The Mexicans moved off. There were twenty seriously wounded legionnaires still lying in the hacienda, although several of them were so badly wounded they did not survive long. Meanwhile, scant rumours of the Camerone battle had reached the Legion's Colonel Jeanningros by nightfall and the following morning he set off with a relief column in search of the missing company. Twenty miles away, he came across the battle scene, which had been cleared up by the Mexicans. Only the bodies of the dead lay naked in a ditch.

It was another two days before Jeanningros could return to bury the dead according to Legion convention, or at least all that remained of the bodies after the vultures and the coyotes had had their fill. One thing that no one noticed and which had remained untouched by the animals was the wooden left hand of Captain Danjou. It was discovered by chance by a local rancher who, realizing its potential value, offered it two years later to the Legion for fifty piastres. After lengthy negotiation by letter, the money was eventually paid and the hand returned to the Legion headquarters at Sidi-bel-Abbès, to take pride of place in the Legion's *Salle d'Honneur*.

Gordon Williamson

The Short Life and Many Kills of Michael Wittmann

SS-Hauptsturmführer Michael Wittmann was the leading tank ace of the Second World War.

The greatest tank ace in history, Michael Wittmann, was born on 22 April 1914 in Vogelthal, Oberpfalz. His father, Johann Wittmann, was a local farmer and after completing his education, Michael worked for his father for a short time. In February 1934, he joined the Freiwilligearbeitsdienst or FAD. This was the forerunner of the compulsory Reichsarbeitsdienst. After six months' labour service he returned to his father's farm until October when he enlisted in the Army, joining Infanterie Regiment 19. After completing his two years of military service he was discharged in September 1936 with the rank of Gefreiter.

Wittmann volunteered for the SS on 1 April 1937 and was accepted into the élite Leibstandarte Adolf Hitler. As a trained soldier the military aspects of the training posed no problems for Wittmann. The stress on comradeship between all ranks during his SS training no doubt contributed greatly to the moulding of the young Wittmann's character.

On the outbreak of war in 1939 Wittmann was an SS-Unterscharführer and saw action in the Polish and western campaigns in an armoured unit. When Hitler's armies invaded the Soviet Union in June 1941, Wittmann was serving in the Sturmgeschütz Abteilung of the Leibstandarte. Under XIV Panzer Korps on the southern sector of the front, the Leibstandarte was in the thick of the action and Wittmann's Sturmgeschütz Abteilung saw heavy action where he quickly gained a reputation for being cool-headed and determined. Earning the Iron Cross Second Class on 12 July 1941, a few weeks later he was wounded for the first time, winning the Wound Badge in Black on 20 August 1941. The Iron Cross First Class followed shortly afterwards, on 8 September, and on 21 November 1941 he qualified for the Panzer Assault Badge.

During one engagement when he came under attack from eight enemy tanks, Wittmann coolly dispatched six of them. This sort of determined courage was not to go overlooked and Wittmann was selected to attend an officer candidates' course at the SS-Junkerschule Bad Tölz in Bavaria in July 1942. Successfully completing his training he was commissioned in December 1942 as an SS-Untersturmführer.

At the beginning of 1943 Wittmann was assigned to 13 (schwere) Kompanie of SS Panzer Regiment 1. He had exchanged his Sturmgeschütz for one of the world's most deadly weapons, the Tiger tank. He applied the same skills and determination in his command of a Tiger as he had with his Sturmgeschütz. On the first day of the great tank battle at Kursk alone, Wittmann knocked out eight enemy tanks and seven artillery pieces. By the end of that ill-fated offensive, Wittmann had added 30 enemy tanks and 28 guns to his ever-mounting score.

One one day in late autumn 1943, Wittmann destroyed ten tanks in a single engagement, bringing his score to 66 tanks destroyed. The Press release which was issued at the time of the award of the Knight's Cross of the Iron Cross to Wittmann on 14 January 1944 refers to Wittmann and his troop of Tigers preventing the breakthrough of an entire Soviet Armoured Brigade on 8/9 January 1944. Many of Wittmann's comrades felt that his award had been long delayed, but if he had waited over long for the Knight's Cross, the Oakleaves were much quicker in coming. On 31 January 1944 Wittmann received a telegram informing him that the Oakleaves had been awarded on the previous day. It read: "In thankful appreciation of your heroic actions in the battle for the future of our people, I award you, as the 380th soldier of the German Wehrmacht, the Oakleaves to the Knight's Cross of the Iron Cross. Adolf Hitler." Wittmann was also granted a much-deserved promotion to SS-Obersturmführer. In April 1944, he became Commander of 1 Kompanie schwere SS-Panzer Abteilung 501.

It was during the Allied invasion of Normandy in June 1944 that Wittmann was to earn his place in military history. His name has become so much associated with his formidable weapon that one can hardly mention the Tiger tank without thinking of this great ace.

The Tiger tank, designed by Professor Ferdinand Porsche, was a formidable weapon. Its low speed and poor manoeuvrability made it a rather inadequate offensive weapon, but in defence, as a form of mobile pillbox, it was nearly invincible. None of the standard Allied tanks in Normandy was a match for this monster. Its frontal armour was all but impenetrable. In the hands of an ace such as Wittmann, the Tiger was truly an awesome weapon. Allied tank troops would become almost paranoid in their fear of this terrible beast.

Stationed at Beauvais at the time of the invasion, schwere SS-Panzer Abteilung 501 did not reach the front until 12 June. On 13 June Wittmann took a force of four Tigers and a Panzer Mk IV on a reconnaissance of the battle area. At Villers-Bocage he spotted a number of armoured vehicles advancing towards the German lines. Wittmann entered the town from behind the British vehicles and quickly knocked out three of four Cromwell tanks he encountered. Passing through the town he encountered Sherman Firefly tanks armed with the fearsome 17-pounder gun. Seeing that he was outnumbered, Wittmann withdrew through the town only to find that the remaining Cromwell had been stalking *him*. Two shots at almost point-blank range from the Cromwell merely bounced off the Tiger's thick armour plating and Wittmann coolly eliminated the Cromwell.

After rejoining the rest of his force and replenishing his ammunition, Wittmann moved against a British armoured column at the far end of the town. Approaching under cover of a wood, he knocked out the lead and tail vehicles and then calmly proceeded to destroy the entire column of about 25 armoured vehicles. He then withdrew through Villers-Bocage with the rest of his force, but British tanks were lying in ambush and succeeded in getting in a flank shot against his Tiger, disabling it. Wittmann and his crew were able to escape on foot. The British poured petrol over Wittmann's Tiger and set it on fire to prevent its recovery.

Wittmann's actions on that day prevented the British armoured thrust which threatened to encircle Panzer Lehr Division. Its commander, General Fritz Bayerlein, immediately recommended Wittmann for the award of the Swords to his Knight's Cross. The original recommendation document for the Swords was signed by the Commander of the Leibstandarte Adolf Hitler, SS-Obergruppenführer und Panzer General der

Waffen-SS, Sepp Dietrich, and included the following report on the action:

Headquarters 1 SS Panzer Korps Korps Field HQ 13 June 1944

Leibstandarte

On 12 June 1944, SS-Obersturmführer Wittmann was ordered to secure the left flank of the Korps near Villers-Bocage, because of growing reports that British armoured forces had broken through and were pushing to the south and south-east. Wittmann set off at the appointed time with 6 Tigers. In the night of 12/13 June 1944 Wittmann's Kompanie had to change its position several times due to heavy artillery bombardments, but in the early hours of 13 June Wittmann had 5 Tigers ready for action by Hill 213 north-east of Villers-Bocage.

At 0800 hrs a report reached Wittmann that a strong column of enemy armour was proceeding along the Caen to Villers-Bocage road. Wittmann with his Tiger concealed under cover 200m south of the road knew that the British armoured battalion would be accompanied by armoured personnel carriers.

The situation required immediate action and Wittmann struck right away, firing on the move, into the British column. His swift action trapped the British column. From a distance of 80m he destroyed 4 Sherman tanks then travelled along the length of the column at a distance of only 10 to 30m and succeeded in destroying a further 15 enemy armoured vehicles. A further 6 tanks were hit and their crews forced to abandon them. The leading armoured personnel carriers were quickly destroyed. The accompanying tanks from Wittmann's Kompanie took 230 prisoners.

Wittmann then raced ahead of his Kompanie into the village of Villers-Bocage. In the town centre his tank was struck by an enemy anti-tank shell and disabled. Nevertheless he destroyed all enemy vehicles within range before abandoning his tank. He and his crew struck out on foot approximately 15km north towards the Panzer Lehr Division where he reported the action to the Divisional Staff Officer . . . Through his determined act, against an enemy deep behind his own lines, acting alone and on his own initiative with great personal gallantry, with his tank he destroyed the greater part of the British 22nd Armoured Brigade and saved the entire front of the 1 SS-Panzer Korps from the imminent danger which threatened. The Korps at the time had no reserves available.

With today's battle. Wittmann has destroyed a total of 138 enemy tanks and 132 anti-tank guns.

signed Dietrich
SS-Obergruppenführer
und Panzer General der Waffen-SS

The Swords were awarded on 22 June 1944 and a few days later Wittmann was promoted to SS-Hauptsturmführer. Offered a transfer to a non-combatant post at a training school, where the benefit of his vast experience could be passed on to future Panzer soldiers, Wittmann refused, preferring to remain with his comrades.

On 8 August the Abteilung, attached to the remnants of the battered 12th SS-Panzer Division Hitlerjugend, was ordered to capture the village of Cintheaux in order to protect the Division's flank. Engaging M4 Shermans of the Canadian 4th Armoured Division, a battle ensued which lasted several hours until Panthers from 12th SS Panzer Division arrived and Cintheaux was finally captured. That evening Wittmann was reported missing and eyewitness accounts report that he had engaged five Shermans at one time and had taken hits from all sides. Even the mighty Tiger could not withstand such punishment. Wittmann's Tiger exploded and he and his entire crew were killed. Wittmann's remains were undiscovered until 1987 when a road-widening operation uncovered his unmarked grave. He is now buried at the Soldatenfriedhof at La Cambe.

Michael Wittmann was the most successful tank commander in history and had the honour of receiving the Knight's Cross, Oakleaves and Swords all within five months. His total recorded score was 138 tanks and 132 guns in a period of less than two years.

Never the flash, daredevil type, Michael Wittmann was a quiet, thoughtful man much admired and esteemed by his comrades and very highly regarded by his superiors. His unquestionable skill and personal gallantry have assured him a well-deserved place in the annals of military history.

Mention should also be made of Wittmann's gunlayer,

Balthasar Woll. In no small way did Woll contribute to Wittmann's success, so much so that he was awarded the Knight's Cross on 16 January 1944. At the time of Wittmann's death Woll had command of his own Tiger. He survived the war.

Siegfried Sassoon

Trench Raid at Mametz Wood

The poet Siegfried Sassoon served with the Royal Welch Fusiliers on the Western Front, before becoming an outspoken opponent of the First World War.

The Germans had evidently been digging when we attacked, and had left their packs and other equipment ranged along the reverse edge of the trench. I stared about me; the smoke-drifted twilight was alive with intense movement, and there was a wild strangeness in the scene which somehow excited me. Our men seemed a bit out of hand and I couldn't see any of the responsible N.C.O.s; some of the troops were firing excitedly at the Wood; others were rummaging in the German packs. Fernby said that we were being sniped from the trees on both sides. Mametz Wood was a menacing wall of gloom, and now an outburst of rapid thudding explosions began from that direction. There was a sap from the Quadrangle to the wood, and along this the Germans were bombing. In all this confusion I formed the obvious notion that we ought to be deepening the trench. Daylight would be on us at once, and we were along a slope exposed to enfilade fire from the wood. I told Fernby to make the men dig for all they were worth, and went to the right with Kendle. The Germans had left a lot of shovels, but we were making no use of them. Two tough-looking privates were disputing the ownership of a pair of field-glasses, so I pulled out my pistol and urged them, with ferocious objurgations, to chuck all that fooling and dig. I seem to be getting pretty handy with my pistol, I thought, for the conditions in Quadrangle Trench were giving me a sort of angry impetus. In some places it was only a foot deep, and already men were lying wounded and killed by sniping. There were high-booted German bodies, too, and in the bleak beginning of daylight they seemed as much the victims of a catastrophe as the men who had attacked them. As I stepped over one of the Germans an impulse made me lift him up from the miserable ditch. Propped against the bank, his blond face was undisfigured, except by the mud which I wiped from his eyes and mouth with my coat sleeve. He'd evidently been killed while digging, for his tunic was knotted loosely about his shoulders. He didn't look to be more than eighteen. Hoisting him a little higher, I thought what a gentle face he had, and remembered that this was the first time I'd ever touched one of our enemies with my hands. Perhaps I had some dim sense of the futility which had put an end to this good-looking youth. Anyhow I hadn't expected the Battle of the Somme to be quite like this . . . Kendle, who had been trying to do something for a badly wounded man, now rejoined me, and we continued, mostly on all fours, along the dwindling trench. We passed no one until we came to a bombing post – three serious-minded men who said that no one had been further than that yet. Being of an exploring frame of mind, I took a bag of bombs and crawled another sixty or seventy yards with Kendle close behind me.

We stared across at the Wood. From the other side of the valley came an occasional rifle-shot, and a helmet bobbed up for a moment. I felt adventurous and it seemed as if Kendle and I were having great fun together. Kendle thought so too. The helmet bobbed up again.

"I'll just have a shot at him," he said, wriggling away from the crumbling bank which gave us cover.

At this moment Fernby appeared with two men and a Lewis gun. Kendle was half kneeling against some broken ground; I remember seeing him push his tin hat back from his forehead and then raise himself a few inches to take aim. After firing once he looked at us with a lively smile; a second later he fell sideways. A blotchy mark showed where the bullet had hit him just above the eyes.

The circumstances being what they were, I had no justification for feeling either shocked or astonished by the sudden extinction of Lance-Corporal Kendle. But after blank awareness that he was killed, all feelings tightened and contracted to a single intention – to "settle that sniper" on the other side of the valley. If I had stopped to think, I shouldn't have gone at all. As it was, I discarded my tin hat and equipment, slung a bag of bombs across my shoulder, abruptly informed Fernby that I was going to find out who *was* there, and set off at a down hill double. While I was running I pulled the safety-pin out of a Mills bomb; my right hand being loaded, I did the same for my left. I mention this because I was obliged to extract the second safety-pin with my teeth, and the grating sensation reminded me that I was half way across and not so reckless as I had been when I started. I was even a little out of breath as I trotted up the opposite slope. Just before I arrived at the top I slowed up and threw my two bombs. Then I rushed at the bank, vaguely expecting some sort of scuffle with my imagined enemy. I had lost my temper with the man who had shot Kendle; quite unexpectedly, I found myself looking down into a well-conducted trench with a great many Germans in it. Fortunately for me, they were already retreating. It had not occurred to them that they were being attacked by a single fool; and, Fernby, with presence of mind which probably saved me, had covered my advance by traversing the top of the trench with a Lewis gun. I slung a few more bombs, but they fell short of the clumsy field-grey figures, some of whom half turned to fire their rifles over the left shoulder as they ran across the open toward the wood, while a crowd of jostling helmets vanished along the trench. Idiotically elated, I stood there with my finger in my right ear and emitted a series of "view holloas" (a gesture which ought to win the approval of people who still regard war as a form of outdoor sport) . . .

Little Fernby's anxious face awaited me, and I flopped down beside him with an outburst of hysterical laughter. When he'd heard my story he asked whether we oughtn't to send a party across to occupy the trench, but I said that the Germans would be bound to come back quite soon. Moreover, my rapid return had attracted the attention of a machine gun which was now firing on gaily along the valley from a position in front of the wood. In my excitement I had forgotten about Kendle. The sight of his body gave me a bit of a shock. His face had gone a bluish colour; I told one of the bombers to cover it with something. Then I put on my web-equipment and its attachments, took a pull at my water-bottle, for my mouth had suddenly become unbearably dry, and set off on my return journey, leaving Fernby to look after the bombing post. It was now six o'clock in the morning.

Alternatively crouching and crawling, I worked my way back. I passed the young German whose body I had rescued from disfigurement a couple of hours before. He was down in the mud again, and someone had trodden on his face. It disheartened me to see him, though his body had now lost all touch with life and was part of the wastage of the war. He and Kendle had cancelled one another out in the process called "attrition of man-power".

Further along I found one of our men dying slowly with a hole in his forehead. His eyes were open and he breathed with a horrible snoring sound. Close by him knelt two of his former mates; one of them was hacking at the ground with an entrenching tool while the other scooped the earth out of the trench with his hands. They weren't worrying about souvenirs now.

Byron Farwell

Rorke's Drift

On 22 January 1879, the Zulus of King Cetewayo massacred a British column at Isandhlwana. Only 350 men out of nearly 1,800 escaped alive. Confident of another easy victory, the Zulus then descended on the British mission station at nearby Rorke's Drift. Eleven Victoria Crosses were won in the ensuing stand against a Zulu impi.

Rorke's Drift had been churned into a muddy quagmire by the passing army and the continued movement of oxen and supply wagons. A mission station-farm was located about a quarter of a mile from the drift on the Natal side of the river and this had been turned into a field hospital and supply centre. On the morning of 22 January there were thirty-six men in the hospital, together with a surgeon, a chaplain, and one orderly; eighty-four men of B Company of the 2nd Battalion of the 24th Regiment and a company of Natal Kaffirs were there to guard the crossing; there was also an engineer officer who helped wagons to cross the river, and a few casuals.

Neither of the two regular officers entitled to hold a command (the surgeon-major did not count) was regarded as oustanding. At least neither of them had ever done anything remarkable in their careers up to this point. The senior of the two was black-bearded Lieutenant John Rouse Merriot Chard, the Royal Engineer officer. Commissioned at the age of twenty-one he had served for more than eleven years without ever seeing action or receiving a promotion.

Lieutenant Gonville Bromhead was in charge of B company of the 24th at Rorke's Drift. His brother, Major Charles Bromhead, was in the same regiment, as was natural, for members of the Bromhead family had served in the 24th Regiment for more than 120 years. Charles was regarded as a brilliant officer; he had been in the Ashanti War with Wolseley and was now on staff duty in London. But Gonville, thirty-three years old with nearly twelve years of service, had been a lieutenant for eight years; he was not so bright and was almost totally deaf. He ought not to have been in the army at all. That he was left behind and assigned to the dull job of watching the river crossing was probably due to the natural reluctance of Pulleine to allow him to command a company in battle.

It was about the middle of the afternoon before Chard and Bromhead learned from two volunteer officers of the Natal Kaffirs of the disaster at Isandhlwana and of their own danger. There were no defences at all at Rorke's Drift, but Chard decided that it would be impossible to bring away all the sick and injured men in hospital so they must do what they could to make the mission defensible. Using wagons, biscuit boxes, bags of mealies and existing walls, they managed to enclose the house, barn and kraal. Fortunately, the buildings were of stone, as was the wall around the kraal, but the house, now being used as a hospital, had a thatched roof, making it vulnerable. All the sick and injured who were well enough to shoot were given rifles and ammunition and Chard counted on having about 300 men to defend his little improvised fort.

A few refugees from Isandhlwana reached Rorke's Drift, but most continued their flight. The mounted natives who had been stationed at the drift and the native contingent with Chard all fled, together with their colonial officers and non-commissioned officers. Chard was left with only 140 men, including the patients from the hospital, to man his 300 yard perimeter. Late in the afternoon a man came racing down the hill in back of the station shouting "Here they come, black as hell and thick as grass!" And a Zulu impi of 4,000 warriors now descended on Rorke's Drift.

The soldiers were still carrying biscuit boxes and mealie bags to the walls when the Zulus, with their black and white cowhide shields and with assegais flashing in the sun, came running into view. Boxes and bags were dropped, rifles and cartridge pouches were seized and the soldiers ran to man the barricades. Rifles crashed as the defenders fired into the black masses of Zulu warriors that swept down on them. The Zulus had a deadly open space to cross and took terrible casualties – but they came on in waves. The soldiers could not shoot fast enough and as the Zulus swept around the walls of the hospital there were hand-to-hand fights along the makeshift barricades, bayonets against assegais, the Zulus mounting the bodies of their own dead and wounded to grab at the rifle barrels and jab at the soldiers.

The men of the 24th had already been enraged before the Zulus arrived by the sight of their native allies and colonial volunteers deserting them. One soldier had even put a bullet into the retreating back of a European non-commissioned officer of the Native Contingent. They were in a fighting mood, and now with the Zulus upon them they fought with a frenzy.

Some of the Zulus who were armed with rifles crouched behind boulders on the rocky slopes behind the mission and fired at the backs of the defenders on the far wall. Fortunately their shooting was erratic, and they did little damage. The steady marksmanship of the soldiers was better; one private downed eight Zulus with eight cartridges during the first charge. The soldiers had found plenty of ammunition among the stores in the barn and the chaplain circulated among them distributing handfuls of fresh cartridges.

There was wild, vicious room-to-room fighting when the Zulus broke into the hospital; the sick and wounded, together with a few men from B Company, held them off with desperate courage until they set fire to the thatched roof. Meanwhile, Chard was trying to withdraw his men into a narrower perimeter encompassing only the barn, kraal and the yard in front of the barn. Into this area the men who had escaped from the hospital, the freshly wounded and Chard's remaining effectives retreated and continued the fight. It was dark now, but the Zulus still came on and by the light of the burning hospital the fight went on.

In rush after rush the Zulus pressed back the soldiers. The kraal, which had been defended by bayonets and clubbed rifles when there was no time to reload, had at last to be abandoned. Rifles had now been fired so often and so fast that the barrels burned the fingers and the fouled guns bruised and battered the shoulders and frequently jammed. The wounded cried for water and the canteens were empty, but Chard led a sally over the wall to retrieve the two-wheeled water cart that stood in the yard by the hospital. It was about four o'clock in the morning before the Zulu attacks subsided, but even then flung assegais continued to whistle over the walls.

It seems nearly incredible that even brave, disciplined British soldiers could have sustained such determined attacks by men equally brave and in such numbers. But they did. By morning, Chard and Bromhead had about eighty men still standing. Fifteen had been killed, two were dying and most were wounded. When dawn broke over the hills, the soldiers looked over the walls and braced their tired, wounded bodies for another charge. Their faces were blackened and their eyes were red; their bodies ached and their nerves were stretched taut from the strain. But the Zulus were gone. Around them were hundreds of black corpses; a few wounded Zulus could be seen retreating painfully over a hill; the ground was littered with the debris of battle: Zulu shields and assegais; British helmets, belts and other accoutrements; broken wagons, biscuit boxes and mealie bags, and the cartridge cases of the 20,000 rounds of ammunition the defenders had fired. Chard sent out some cautious patrols, but there were no signs of the enemy in the immediate vicinity. The soldiers cleared away some dead Zulus from the cook house and began to make tea.

About seven-thirty the Zulus suddenly appeared again. Chard called his men and they manned the walls, but the Zulus simply sat down on a hill out of rifle range. They, too, were exhausted, and they had not eaten for more than two days. They had no desire to renew the fight. Besides, the leader of the impi had disobeyed the order of Cetewayo by crossing the Buffalo River into Natal and he was doubtless considering how he would explain to his chief the costly night of savage fighting outside the boundaries of Zululand. While Chard and his men grimly watched, the Zulus rose and wearily moved off over the hills.

Later in the morning, some mounted infantry rode up and soon after Chelmsford appeared with what was left of his main force. He had hoped that some portion of his troops had been able to retreat from Isandhlwana to Rorke's Drift, but he found only the survivors of those who had been left there. With the remnants of his column and the handful of men from Rorke's Drift, he sadly retreated into Natal.

Eleven Victoria Crosses were awarded the defenders of Rorke's Drift: the most ever given for a single engagement. There might have been even more, but posthumous awards were not then made. Both Chard and Bromhead received the medal. There were many Welshmen in the 24th (which later became the South Wales Borderers), and among the eighty-four men of

B Company there were five men named Jones and five named Williams; two of the Joneses and one Williams won the Victoria Cross. Private William's real name, however, was John Williams Fielding; he had run away from home to enlist and had changed his name so that his father, a policeman, would not find him.

Private Frederick Hitch, twenty-four, also won a Victoria Cross and survived his years of service to enjoy the wearing of it as a commissionaire. The bad luck of his regiment seemed to pursue Hitch and his medal, even to the grave. One day while Hitch was in his commissionaire's uniform and wearing his medals a thief snatched the Victoria Cross from his chest. It was never seen again. King Edward VII eventually gave him another to replace it, but when Hitch died in 1913 this one, too, had disappeared. Fifteen years later it turned up in an auction room; his family bought it and it is now in the museum of his old regiment. Mounted on his tomb in Chiswick cemetery was a bronze replica of the Victoria Cross. In 1968 thieves stole that.

Lieutenant Chard finally received his first promotion – to brevet major, becoming the first officer in the Royal Engineers ever to skip the rank of captain. He was also invited to Balmoral where Queen Victoria gave him a gold signet ring. He served for another eighteen years in Cyprus, India and Singapore, but he received only one more promotion. In 1897 cancer of the tongue caused him to retire and he died three months later.

Bromhead was also promoted to captain and brevet major, though he never rose any higher. He, too, was invited to Balmoral by the Queen, but being on a fishing trip when the invitation arrived he missed the occasion. He died in 1891 at the age of forty-six in Allahabad, still in the 24th Regiment.

Bromhead and Chard were fortunate in a sense when their moment for glory arrived: fighting with their backs to the wall, they had only to show the kind of stubborn bravery and simple leadership for which the British officer was conditioned and which he was best equipped to display. No great decision or military genius was required of them. But this is not to detract from their feat of courage and the British army was rightly proud of them.

James McCudden

Death of an Ace

A flight commander in 56 Squadron RAF, McCudden was an eyewitness to the classic dogfight of the First World War, when Rhys-Davids of 56 Squadron brought down the German ace Werner Voss on 23 September 1917.

On the evening of the 23rd I led my patrol from the aerodrome and crossed the lines at Bixschoote at 8,000 feet as there was a very thick wall of clouds up at 9,000 feet. As soon as we crossed over Hunland I noted abnormal enemy activity, and indeed there seemed to be a great many machines of both sides about. This was because every machine that was up was between 9,000 feet and the ground instead of as usual from 20,000 feet downwards.

We flew south from Houthoulst Forest, and although there were many Huns about they were all well over. Archie was at his best this evening, for he had us all silhouetted against a leaden sky, and we were flying mostly at 7,000 feet. When over Gheluvelt, I saw a two-seater coming north near Houthem. I dived, followed by my patrol, and opened fire from above and behind the D.F.W., whose occupants had not seen me, having been engrossed in artillery registration. I fired a good burst from both guns, a stream of water came from the D.F.W.'s centre-section, and then the machine went down in a vertical dive and crashed to nothing, north-east of Houthem.

We went north, climbing at about 6,000 feet. A heavy layer of grey clouds hung at 9,000 feet, and although the visibility was poor for observation, the atmosphere was fairly clear in a horizontal direction. Away to the east one could see clusters of little black specks, all moving swiftly, first in one direction and then another. Farther north we could see formations of our own machines, Camels, Pups, S.E.'s, Spads and Bristols, and lower down in the haze our artillery R.E.8's.

We were just on the point of engaging six Albatros Scouts away to our right, when we saw ahead of us, just above Poelcappelle, an S.E. half spinning down closely pursued by a silvery blue German triplane at very close range. The S.E. certainly looked very unhappy, so we changed our minds about attacking the six V-strutters, and went to the rescue of the unfortunate S.E.

The Hun triplane was practically underneath our formation now, and so down we dived at a colossal speed. I went to the right, Rhys-Davids to the left, and we got behind the triplane together. The German pilot saw us and turned in a most disconcertingly quick manner, not a climbing nor Immelmann turn, but a sort of flat half spin. By now the German triplane was in the middle of our formation, and its handling was wonderful to behold. The pilot seemed to be firing at all of us simultaneously, and although I got behind him a second time, I could hardly stay there for a second. His movements were so quick and uncertain that none of us could hold him in sight at all for any decisive time.

I now got a good opportunity as he was coming towards me nose on, and slightly underneath, and had apparently not seen me. I dropped my nose, got him well in my sight, and pressed both triggers. As soon as I fired up came his nose at me, and I heard clack-clack-clack-clack, as his bullets passed close to me and through my wings. I distinctly noticed the red-yellow flashes from his parallel Spandau guns. As he flashed by me I caught a glimpse of a black head in the triplane with no hat on at all.

By this time a red-nosed Albatros Scout had arrived, and was apparently doing its best to guard the triplane's tail, and it was well handled too. The formation of six Albatros Scouts which we were going to attack at first stayed above us, and were prevented from diving on us by the arrival of a formation of Spads, whose leader apparently appreciated our position, and kept the six Albatroses otherwise engaged.

The triplane was still circling round in the midst of six S.E.'s, who were all firing at it as opportunity offered, and at one time I noted the triplane in the apex of a cone of tracer bullets from at least five machines simultaneously, and each machine had two guns. By now the fighting was very low, and the red-nosed Albatros had gone down and out, but the triplane still remained. I had temporarily lost sight of the triplane whilst changing a drum of my Lewis gun, and when I next saw him he was very low, still being engaged by an S.E. marked I, the pilot being Rhys-Davids. I noticed that the triplane's movements were very erratic, and then I saw him go into a fairly steep dive and so I continued to watch, and then saw the triplane hit the ground and disappear into a thousand fragments, for it seemed to me that it literally went to powder.

Strange to say, I was the only pilot who witnessed the triplane crash, for even Rhys-Davids, who finally shot it down, did not see its end.

It was now quite late, so we flew home to the aerodrome, and as long as I live I shall never forget my admiration for that German pilot, who single-handed fought seven of us for ten minutes, and also put some bullets through all of our machines. His flying was wonderful, his courage magnificent, and in my opinion he is the bravest German airman whom it has been my privilege to see fight.

We arrived back at the mess, and at dinner the main topic was the wonderful fight, We all conjectured that the enemy pilot must be one of the enemy's best, and we debated as to whether it was Richthofen or Wolff or Voss. The tri-plane fell in our lines, and the next morning we had a wire from the Wing saying that the dead pilot was found wearing the Boelcke collar and his name was Werner Voss. He had the "Ordre Pour le Mérite."

Rhys-Davids came in for a shower of congratulations, and no one deserved them better, but as the boy himself said to me, "Oh, if I could only have brought him down alive," and his remark was in agreement with my own thoughts.

Ross Munro

The Canadians at Dieppe

On 19 August 1942 5,000 Canadians were committed to a seaborne raid on German-occupied Dieppe in France.

Even before we put to sea some had an ominous feeling about what was ahead of them on the other side of the Channel. Nobody said anything but many were wondering how the security had been in the time since 7 July. Did the Germans know the Canadians were going to France and were they waiting? This was the question being asked in many minds.

They were puzzled, too, why the raid had been decided upon so suddenly. They would have liked more time to adjust themselves.

I shared most of their mental discomfort. For the first hour or so I ran over the plan and studied my maps and photographs and was surprised I had forgotten so much of the detail. I found misgivings growing in my mind. This seemed somewhat haphazard, compared with the serene way in which the cancelled raid was mounted.

The final Dieppe plan was altered only slightly from the one prepared for July. British Commandos were assigned to tasks on the flanks previously allotted to paratroopers.

. . . It was one of the finest evenings of the summer. The sea was smooth, the sky was clear and there was the slightest of breezes. The ships cleared and the Royals went to dinner before making their final preparations. In the wardroom, the officers sat around the tables and dined in Navy style, as the last sunshine poured through the open portholes. We had a good meal and everyone ate hungrily, for on the way to the boats all we had had was haversack fare – a few bully-beef sandwiches.

The Royals officers were in good spirits at dinner. Looking around the table you would never have thought that they were facing the biggest test of their lives. They joked and bantered across the tables and renewed old friendships with the naval officers whom they had known in "practice Dieppe" training days.

. . . We were about ten miles from the French coast and until now there hadn't been a hitch in the plan. The minefield was behind us. The boats filled with infantrymen were lowered as the *Emma* stopped and anchored. Nobody spoke. Silence was the strict order but as our boat, which was the largest of the landing craft and was jammed with about eighty soldiers, pushed off from the *Emma*, a veteran sailor leaned over and in a stage whisper said, "Cheerio, lads, all the best; give the bastards a walloping." Then we were drifting off into the darkness and our coxswain peered through the night to link up with the rest of our assault flotilla.

. . . Eyes were accustomed to the darkness now and we could discern practically all our little craft; the sea was glossy with starlight.

The boats plunged along, curling up white foam at their bows and leaving a phosphorescent wake that stood out like diamonds on black velvet.

We were about seven or eight miles from Dieppe when the first alarm shook us. To our left there was a streak of tracer bullets – light blue and white dots in the night – and the angry clatter of automatic guns. This wasn't according to plan and everyone in that boat of ours tightened up like a drum. We kept our heads down behind the steel bulwark of our little craft, but it was so crowded there that even to crouch was crowding someone beside you. I sat on a cartful of 3-inch mortar bombs. More tracer bullets swept across ahead of us and some pinged off our steel sides. A big sailor by my side rigged his Lewis gun through a slit at the stern of our boat and answered with a few short bursts. A blob in the night that was an enemy ship – an armed trawler or more likely an E-boat – was less than two hundred yards away. It was firing at half a dozen craft including ours, which was in the lead at that time. From other directions came more German tracer. There might have been four ships intercepting us.

There wasn't much we could do. There isn't any armament on these assault craft to engage in a naval action against E-boats or trawlers. Our support craft didn't seem to be about at that particular time. It looked as if we were going to be cut up piecemeal by this interception; our flotilla already had been broken up from the close pattern of two columns we had held before the attack.

I blew up my lifebelt a little more. A few more blasts of tracer whistled past and then there was a great flash and a bang of gunfire behind us. In the flash we could see one of our destroyers speeding up wide-open to our assistance. It fired a dozen rounds at the enemy ships and they turned and disappeared towards the French coast. They probably went right into Dieppe harbour and spread the word that British landing craft were heading in. . . . Our coxswain tried to take us in to one section of the beach and it proved the wrong spot. Before he grounded he swung the craft out again and we fumbled through the smoke to the small strip of sand which was the Puits beach. The smoke was spotty and the last thirty yards was in the clear. Geysers from artillery shells or mortar bombs shot up in our path. Miraculously we weren't hit by any of them. The din of the German ack-ack guns and machine-guns on the cliff was so deafening you could not hear the man next to you shout.

The men in our boat crouched low, their faces tense and grim. They were awed by this unexpected blast of German fire, and it was their initiation to frightful battle noises. They gripped their weapons more tightly and waited for the ramp of our craft to go down.

We bumped on the beach and down went the ramp and out poured the first infantrymen. They plunged into about two feet of water and machine-gun bullets laced into them. Bodies piled up on the ramp. Some staggered to the beach and fell. Bullets were splattering into the boat itself, wounding and killing our men.

I was near the stern and to one side. Looking out the open bow over the bodies on the ramp, I saw the slope leading a short way up to a stone wall littered with Royals casualties. There must have been sixty or seventy of them, lying sprawled on the green grass and the brown earth. They had been cut down before they had a chance to fire a shot.

A dozen Canadians were running along the edge of the cliff towards the stone wall. They carried their weapons and some were firing as they ran. But some had no helmets, some were already wounded, their uniforms torn and bloody. One by one they were cut down and rolled down the slope to the sea.

I don't know how long we were nosed down on that beach. It

may have been five minutes. It may have been twenty. On no other front have I witnessed such a carnage. It was brutal and terrible and shocked you almost to insensibility to see the piles of dead and feel the hopelessness of the attack at this point.

There was one young lad crouching six feet away from me. He had made several vain attempts to rush down the ramp to the beach but each time a hail of fire had driven him back. He had been wounded in the arm but was determined to try again. He lunged forward and a streak of red-white tracer slashed through his stomach.

I'll never forget his anguished cry as he collapsed on the blood-soaked deck: "Christ, we gotta beat them; we gotta beat them!" He was dead in a few minutes.

. . . For the rest of that morning one lost all sense of time and developments in the frantic events of the battle. Although the Puits landing had obviously failed and the headland to the east of Dieppe would still be held by the Germans, I felt that the main attack by three infantry battalions and the tanks had possibly fared better on the beach in front of the town.

Landing craft were moving along the coast in relays and the destroyers were going in perilously close to hit the headlands with shell-fire. I clambered from one landing craft to another to try to learn what was going on. Several times we were bombed too closely by long, black German planes that sailed right through our flak and our fighter cover.

Smoke was laid by destroyers and our planes along the sea and on the beach. Finally the landing craft in which I was at the time, with some naval ratings, touched down on the sloping pebble main beach which ran about sixty-yards at that point to a high sea wall and the Esplanade, with the town beyond.

Smoke was everywhere and under its cover several of our ratings ran on to the beach and picked up two casualties by the barbed wire on the beach, lugging them back to the boat. I floundered through the loose shale to the sea-wall. There was heavy machine-gun fire down the beach towards the Casino. A group of men crouched twenty yards away under the shelter of the sea-wall.

The tobacco factory was blazing fiercely. For a moment there was no firing. It was one of those brief lulls you get in any battle. I thought our infantry were thick in the town but the Esplanade looked far too bare and empty.

There was no beach organization as there should have been.

Some dead lay by the wall and on the shale. The attack here had not gone as planned either. A string of mortar bombs whanged on the Esplanade. The naval ratings waved and I lunged back to the boat as the beach battle opened up again. In choking smoke we pulled back to the boat pool.

. . . Then the German air force struck with its most furious attack of the day. All morning long, British and Canadian fighters kept a constant patrol over the ships and the beaches, whole squadrons twisting and curling in the blue, cloud-flecked sky. Hundreds of other planes swept far over northern France, intercepting enemy fighters and bombers long before they reached Dieppe. Reconnaissance planes kept a constant lookout on the roads from Amiens and Abbeville and Rouen where reinforcements could be expected. There were air combats going on practically all morning long. It was the greatest air show since the Battle of Britain in the fall of 1940, and the R.A.F. and R.C.A.F. had overwhelming superiority. The High Command had hoped the German air force would be lured into the sky and most of the enemy strength in western Europe came up.

. . . Bullets screeched in every direction. The whole sky and sea had gone mad with the confusion of that sudden air attack, and a dozen times I clung to the bottom of the boat expecting that this moment was the last as we were cannoned or another stick of bombs churned the sea.

Several landing craft near us blew up, hit by bombs and cannon shells. There was nothing left. They just disintegrated. These craft had been trying to make the main beach again, as we had been, to take off troops on the withdrawal.

Xenophon

The March of the Ten Thousand to the Sea

Xenophon was a soldier as well as a chronicler. In 401 BC he joined the Greek mercenaries fighting under the Persian prince Cyrus the Younger. After Cyrus's death at Cunaxa, Xenophon led the Ten Thousand Greeks on their 1,500-mile retreat through hostile territory, one of the greatest of all martial endeavours.

The resolution to which they came was that they must force a passage through the hills into the territory of the Kurds; since, according to what their informants told them, when they had once passed these, they would find themselves in Armenia – the rich and large territory governed by Orontas; and from Armenia, it would be easy to proceed in any direction whatever. Thereupon they offered sacrifice, so as to be ready to start on the march as soon as the right moment appeared to have arrived. Their chief fear was that the high pass over the mountains might be occupied in advance: and a general order was issued, that after supper every one should get his kit together for starting, and repose, in readiness to follow as soon as the word of command was given.

It was now about the last watch, and enough of the night remained to allow them to cross the valley under cover of darkness; when, at the word of command, they rose and set off on their march, reaching the mountains at daybreak. At this stage of the march Cheirisophus, at the head of his own division, with the whole of the light troops, led the van, while Xenophon followed behind with the heavy infantry of the rearguard, but without any light troops, since there seemed to be no danger of pursuit or attack from the rear, while they were making their way up hill. Cheirisophus reached the summit without any of the enemy perceiving him. Then he led on slowly, and the rest of the army followed, wave upon wave, cresting the summit and descending into the villages which nestled in the hollows and recesses of the hills.

Thereupon the Carduchians abandoned their dwelling-places, and with their wives and children fled to the mountains; so there was plenty of provisions to be got for the mere trouble of taking, and the homesteads too were well supplied with a copious store of bronze vessels and utensils which the Hellenes kept their hands off, abstaining at the same time from all pursuit of the folk themselves, gently handling them, in hopes that the Carduchians might be willing to give them friendly passage through their country, since they too were enemies of the king: only they helped themselves to such provisions as fell in their way, which indeed was a sheer necessity. But the Carduchians neither gave ear, when they called to them, nor showed any other friendly sign; and now, as the last of the Hellenes descended into the villages from the pass, they were already in the dark, since, owing to the narrowness of the road, the whole day had been spent in the ascent and descent. At that instant a party of the Carduchians, who had collected, made an attack on the hindmost men, killing some and wounding others with stones and arrows – though it was quite a small body who attacked. The fact was, the approach of the Hellenic army had taken them by surprise; if, however, they had mustered in larger force at this time, the chances are that a large portion of the army would have been annihilated. As it was, they got into quarters, and bivouacked in the villages that night, while the Carduchians kept many watch-fires blazing in a circle on the mountains, and kept each other in sight all round.

But with the dawn the generals and officers of the Hellenes met and resolved to proceed, taking only the necessary number of stout baggage animals, and leaving the weaklings behind. They resolved further to let go free all the lately-captured slaves in the host; for the pace of the march was necessarily rendered slow by the quantity of animals and prisoners, and the number of non-combatants in attendance on these was excessive, while, with such a crowd of human beings to satisfy, twice the amount of provisions had to be procured and carried. These resolutions passed, they caused a proclamation by herald to be made for their enforcement.

When they had breakfasted and the march recommenced, the generals planted themselves a little to one side in a narrow place, and when they found any of the aforesaid slaves or other property still retained, they confiscated them. The soldiers yielded obedience, except where some smuggler, prompted by desire of a good-looking boy or woman, managed to make off with his prize. During this day they contrived to get along after a fashion, now fighting and now resting. But on the next day they were visited by a great storm, in spite of which they were obliged to continue the march, owing to insufficiency of provisions. Cheirisophus was as usual leading in front, while Xenophon headed the rearguard, when the enemy began a violent and sustained attack. At one narrow place after another they came up quite close, pouring in volleys of arrows and sling-stones, so that the Hellenes had no choice but to make sallies in pursuit and then again recoil, making but very little progress. Over and over again Xenophon would send an order to the front to slacken pace, when the enemy were pressing their attack severely. As a rule, when the word was so passed up, Cheirisophus slackened; but sometimes instead of slackening, Cheirisophus quickened, sending down a counter-order to the rear to follow on quickly. It was clear that there was something or other happening, but there was no time to go to the front and discover the cause of the hurry. Under these circumstances the march, at any rate in the rear, became very like a rout, and here a brave man lost his life, Cleonymus the Laconian, shot with an arrow in the ribs right through shield and corselet, as also Basias, an Arcadian, shot clean through the head.

As soon as they reached a halting-place, Xenophon, without more ado, came up to Cheirisophus, and took him to task for not having waited, "whereby," said he, "we were forced to fight and flee at the same moment; and now it has cost us the lives of two fine fellows; they are dead, and we were not able to pick up their bodies or bury them." Cheirisophus answered: "Look up there," pointing as he spoke to the mountain, "do you see how inaccessible it all is? Only this one road, which you see, going straight up, and on it all that crowd of men who have seized and are guarding the single exit. That is why I hastened on, and why I could not wait for you, hoping to be beforehand with them yonder in seizing the pass: the guides we have got say there is no other way." And Xenophon replied: "But I have got two prisoners also; the enemy annoyed us so much that we laid an ambuscade for them, which also gave us time to recover our breaths; we killed some of them, and did our best to catch one or two alive – for this very reason – that we might have guides who knew the country, to depend upon."

The two were brought up at once and questioned separately: "Did they know of any other road than the one visible?" The first said *no*: and in spite of all sorts of terrors applied to extract a better answer – *no*, he persisted. When nothing could be got out of him, he was killed before the eyes of his fellow. This latter then explained: "Yonder man said he did not know, because he has got a daughter married to a husband in those parts. I can take you," he added, "by a good road, practicable even for beasts." And when asked whether there was any point on it difficult to pass, he replied that there was a col which it would be impossible to pass unless it were occupied in advance.

Then it was resolved to summon the officers of the light infantry and some of those of the heavy infantry, and to acquaint them with the state of affairs, and ask them whether any of them were minded to distinguish themselves, and would step forward as volunteers on an expedition. Two or three heavy infantry soldiers stepped forward at once – two Arcadians, Aristonymus of Methydrium, and Agasias of Stymphalus – and in emulation of these, a third, also an Arcadian, Callimachus from Parrhasia, who said he was ready to go, and would get volunteers from the whole army to join him. "I know," he added, "there will be no lack of youngsters to follow where I lead." After that they asked, "Were there any captains of light infantry willing to accompany the expedition?" Aristeas, a Chian, who on several occasions proved his usefulness to the army on such service, volunteered.

It was already late afternoon, when they ordered the storming party to take a snatch of food and set off; then they bound the guide and handed him over to them. The agreement was, that if they succeeded in taking the summit they were to guard the position that night, and at daybreak to give a signal by bugle. At this signal the party on the summit were to attack the enemy in occupation of the visible pass, while the generals with the main body would bring up their succours; making their way up with what speed they might. With this understanding, off they set, two thousand strong; and there was a heavy downpour of rain, but Xenophon, with his rearguard, began advancing to the visible pass, so that the enemy might fix his attention on this road, and the party creeping round might, as much as possible, elude observation. Now when the rearguard, so advancing, had reached a ravine which they must cross in order to strike up the steep, at that instant the barbarians began rolling down great boulders, each a wagon load, some larger, some smaller; against the rocks they crashed and splintered flying like slingstones in every direction – so that it was absolutely out of the question even to approach the entrance of the pass. Some of the officers finding themselves baulked at this point, kept trying other ways, nor did they desist till darkness set in; and then, when they thought they would not be seen retiring, they returned to supper. Some of them who had been on duty in the rearguard had had no breakfast (it so happened). However, the enemy never ceased rolling down their stones all through the night, as was easy to infer from the booming sound.

The party with the guide made a circuit and surprised the enemy's guards seated round their fire, and after killing some, and driving out the rest, took their places, thinking that they were in possession of the height. As a matter of fact they were not, for above them, lay a breastlike hill skirted by the narrow road on which they had found the guards seated. Still, from the spot in question there was an approach to the enemy, who were seated on the pass before mentioned.

Here then they passed the night, but at the first glimpse of dawn they marched stealthily and in battle order against the enemy. There was a mist, so that they could get quite close without being observed. But as soon as they caught sight of one another, the trumpet sounded, and with a loud cheer they rushed upon the fellows, who did not wait their coming, but left the road and made off; with the loss of only a few lives however, so nimble were they. Cheirisophus and his men, catching the sound of the bugle, charged up by the well-marked road, while others of the generals pushed their way up by pathless routes, where each division chanced to be; the men mounting as they were best able, and hoisting one another up by means of their spears; and these were the first to unite with the party who had already taken the position by storm. Xenophon, with the rearguard, followed the path which the party with the guide had taken, since it was easiest for the beasts of burthen; one half of his men he had posted in rear of the baggage animals; the other half he had with himself. In their course they encountered a crest above the road, occupied by the enemy, whom they must either dislodge or be themselves cut off from the rest of the Hellenes. The men by themselves could have taken the same route as the rest, but the baggage animals could not mount by any other way than this.

Here then, with shouts of encouragement to each other, they dashed at the hill with their storming columns, not from all sides, but leaving an avenue of escape for the enemy, if he chose to avail himself of it. For a while, as the men scrambled up where each best could, the natives kept up a fire of arrows and darts, yet did not receive them at close quarters, but presently left the position in flight. No sooner, however, were the Hellenes safely past this crest, than they came in sight of another in front of them, also occupied, and deemed it advisable to storm it also. But now it struck Xenophon that if they left the ridge just taken unprotected in their rear, the enemy might re-occupy it and attack the baggage animals as they filed past, presenting a long extended line owing to the narrowness of the road by which they made their way. To obviate this, he left some officers in charge of the ridge – Cephisodorus, son of Cephisophon, an Athenian; Amphicrates, the son of Amphidemus, an Athenian; and Archagoras, an Argive exile – while he in person with the rest of the men attacked the second ridge; this they took in the same fashion, only to find that they had still a third knoll left, far the steepest of the three. This was none other than the mamelon mentioned as above the outpost, which had been captured over their fire by the volunteer storming party in the night. But when the Hellenes were close, the natives, to the astonishment of all, without a struggle deserted the knoll. It was conjectured that they had left their position from fear of being encircled and besieged, but the fact was that they, from their higher ground, had been able to see what was going on in the rear, and had all made off in this fashion to attack the rearguard.

So then Xenophon, with the youngest men, scaled up to the top, leaving orders to the rest to march on slowly, so as to allow the hindmost companies to unite with them; they were to advance by the road, and when they reached the level to ground arms. Meanwhile the Argive Archagoras arrived, in full flight, with the announcement that they had been dislodged from the first ridge, and that Cephisodorus and Amphicrates were slain, with a number of others besides, all in fact who had not jumped down the crags and so reached the rearguard. After this achievement the barbarians came to a crest facing the mamelon, and Xenophon held a colloquy with them by means of an interpreter, to negotiate a truce, and demanded back the dead bodies. These they agreed to restore if he would not burn their houses, and to these terms Xenophon agreed. Meanwhile, as the rest of the army filed past, and the colloquy was proceeding, all the people of the place had time to gather gradually, and the enemy formed; and as soon as the Hellenes began to descend from the mamelon to join the others where the troops were halted, on rushed the foe, in full force, with hue and cry. They reached the summit of the mamelon from which Xenophon was descending, and began rolling down crags. One man's leg was crushed to pieces. Xenophon was left by his shield-bearer, who carried off his shield, but Eurylochus of Lusia, an Arcadian hoplite, ran up to him, and threw his shield in front to protect both of them; so the two together beat a retreat, and so too the rest, and joined the serried ranks of the main body.

After this the whole Hellenic force united, and took up their quarters there in numerous beautiful dwellings, with an ample store of provisions, for there was wine so plentiful that they had it in cemented cisterns. Xenophon and Cheirisophus arranged to recover the dead, and in return restored the guide; afterwards they did everything for the dead, according to the means at their disposal, with the customary honours paid to good men.

Next day they set off without a guide; and the enemy, by keeping up a continuous battle and occupying in advance every narrow place, obstructed passage after passage. Accordingly, whenever the van was obstructed, Xenophon, from behind, made a dash up the hills and broke the barricade, and freed the vanguard by endeavouring to get above the obstructing enemy. Whenever the rear was the point attacked, Cheirisophus, in the same way, made a détour, and by endeavouring to mount higher than the barricaders, freed the passage for the rear rank; and in this way, turn and turn about, they rescued each other, and paid unflinching attention to their mutual needs. At times it happened that the relief party, having mounted, encountered considerable annoyance in their descent from the barbarians, who were so agile that they allowed them to come up quite close, before they turned back, and still escaped, partly no doubt because the only weapons they had to carry were bows and slings.

They were, moreover, excellent archers, using bows nearly three cubits long and arrows more than two cubits. When discharging the arrow, they drew the string by getting a purchase with the left foot planted forward on the lower end of the bow. The arrows pierced through shield and cuirass, and the Hellenes, when they got hold of them, used them as javelins, fitting them to their thongs. In these districts the Cretans were highly serviceable. They were under the command of Stratocles, a Cretan.

During this day they bivouacked in the villages which lie above the plain of the river Centrites, which is about two hundred feet broad. It is the frontier river between Armenia and the country of the Carduchians. Here the Hellenes recruited themselves, and the sight of the plain filled them with joy, for the river was but six or seven furlongs distant from the mountains of the Carduchians. For the moment then they bivouacked right happily; they had their provisions, they had also many memories of the labours that were now passed; seeing that the last seven days spent in traversing the country of the Carduchians had been one long continuous battle, which had cost them more suffering than the whole of their troubles at the hands of the king and Tissaphernes put together. As though they were truly quit of them forever, they laid their heads to rest in sweet content.

But with the morrow's dawn they espied horsemen at a certain point across the river, armed *cap-à-pie*, as if they meant to dispute the passage. Infantry, too, drawn up in line upon the banks above the cavalry, threatened to prevent them debouching into Armenia. These troops were Armenian and Mardian and Chaldaean mercenaries belonging to Orontas and Artuchas. The last of the three, the Chaldaeans, were said to be a free and brave set of people. They were armed with long wicker shields and lances. The banks before named on which they were drawn up were a hundred yards or more distant from the river, and the single road which was visible was one leading upwards and looking like a regular artificially constructed highway. At this point the Hellenes endeavoured to cross, but on their making the attempt the water proved to be more than breast-deep, and the river bed was rough with great slippery stones, and as to holding their arms in the water, it was out of the question – the stream swept them away – or if they tried to carry them over the head, the body was left exposed to the arrows and other missiles; accordingly they turned back and encamped there by the bank of the river.

At the point where they had themselves been last night, up on the mountains, they could see the Carduchians collected in large numbers and under arms. A shadow of deep despair again descended on their souls, whichever way they turned their eyes – in front lay the river so difficult to ford; over on the other side, a new enemy threatening to bar the passage; on the hills behind, the Carduchians ready to fall upon their rear should they once again attempt to cross. Thus for this day and night they halted, sunk in perplexity. But Xenophon had a dream. In his sleep he thought that he was bound in fetters, but these, of their own accord, fell from off him, so that he was loosed, and could stretch his legs as freely as he wished. So at the first glimpse of daylight he came to Cheirisophus and told him that he had hopes that all things would go well, and related to him his dream.

The other was well pleased, and with the first faint gleam of dawn the generals all were present and did sacrifice; and the victims were favourable at the first essay. Retiring from the sacrifice, the generals and officers issued an order to the troops to take their breakfasts; and while Xenophon was taking his, two young men came running up to him, for every one knew that, breakfasting or supping, he was always accessible, or that even if asleep any one was welcome to awaken him who had anything to say bearing on the business of war. What the two young men had at this time to say was that they had been collecting brushwood for fire, and had presently espied on the opposite side, in among some rocks which came down to the river's brink, an old man and some women and little girls depositing, as it would appear, bags of clothes in a cavernous rock. When they saw them, it struck them that it was safe to cross; in any case the enemy's cavalry could not approach at this point. So they stripped naked, expecting to have to swim for it, and with their long knives in their hands began crossing, but going forward crossed without being wet up to the fork. Once across they captured the clothes, and came back again.

Accordingly Xenophon at once poured out a libation himself, and bade the two young fellows fill the cup and pray to the gods, who showed to him this vision and to them a passage, to bring all other blessings for them to accomplishment. When he had poured out the libations, he at once led the two young men to Cheirisophus, and they repeated to him their story. Cheirisophus, on hearing it, offered libations also, and when they had performed them, they sent a general order to the troops to pack up ready for starting, while they themselves called a meeting of the generals and took counsel how they might best effect a passage, so as to overpower the enemy in front without suffering any loss from the men behind. And they resolved that Cheirisophus should lead the van and cross with half the army, the other half still remaining behind under Xenophon, while the baggage animals and the mob of sutlers were to cross between the two divisions.

When all was duly ordered the move began, the young men pioneering them, and keeping the river on their left. It was about four furlongs' march to the crossing, and as they moved along the bank, the squadrons of cavalry kept pace with them on the opposite side.

But when they had reached a point in a line with the ford, and the cliff-like banks of the river, they ordered arms, and first Cheirisophus himself placed a wreath upon his brows, and throwing off his cloak, resumed his arms, passing the order to all the rest to do the same, and bade the captains form their companies in open order in deep columns, some to left and some to right of himself. Meanwhile the soothsayers were slaying a victim over the river, and the enemy were letting fly their arrows and slingstones; but as yet they were out of range. As soon as the victims were favourable, all the soldiers began singing the battle hymn, and with the notes of the paean mingled the shouting of the men accompanied by the shriller chant of the women, for there were many women in the camp.

So Cheirisophus with his detachment stept in. But Xenophon, taking the most active-bodied of the rearguard, began running back at full speed to the passage facing the egress into the hills of Armenia, making a feint of crossing at that point to intercept their cavalry on the river bank. The enemy, seeing Cheirisophus's detachment easily crossing the stream, and Xenophon's men racing back, were seized with the fear of being intercepted, and fled at full speed in the direction of the road which emerged from the stream. But when they were come opposite to it they raced up hill towards their mountains. Then Lycius, who commanded the cavalry, and Aeschines, who was in command of the division of light infantry attached to Cheirisophus, no sooner saw them fleeing so lustily than they were after them, and the soldiers shouted not to fall behind, but to follow them right up to the mountains. Cheirisophus, on getting across, forbore to pursue the cavalry, but advanced by the bluffs which reached to the river to attack the enemy overhead. And these, seeing their own cavalry fleeing, seeing also the heavy infantry advancing upon them, abandoned the heights above the river.

Xenophon, as soon as he saw that things were going well on the other side, fell back with all speed to join the troops engaged in crossing, for by this time the Carduchians were well in sight, descending into the plain to attack their rear.

Cheirisophus was in possession of the higher ground, and Lycius, with his little squadron, in an attempt to follow up the pursuit, had captured some stragglers of their baggage-bearers, and with them some handsome apparel and drinking-cups. The baggage animals of the Hellenes and the mob of non-combatants were just about to cross, when Xenophon turned his troops right about to face the Carduchians. *Vis-à-vis* he formed his line, passing the order to the captains each to form his company into sections, and to deploy them into line to the left, the captains of companies and lieutenants in command of sections to advance to meet the Carduchians, while the rear leaders would keep their position facing the river. But when the Carduchians saw the rearguard so stript of the mass, and looking now like a mere handful of men, they advanced all the more quickly, singing certain songs the while. Then, as matters were safe with him, Cheirisophus sent back the peltasts and slingers and archers to join Xenophon, with orders to carry out his instructions. They were in the act of recrossing, when Xenophon, who saw their intention, sent a messenger across, bidding them wait there at the river's brink without crossing; but as soon as he and his detachment began to cross they were to step in facing him in two flanking divisions right and left of them, as if in the act of crossing; the javelin men with their javelins on the thong, and the bowmen with their arrows on the string; but they were not to advance far into the stream. The order passed to his own men was: "Wait till you are within sling-shot, and the shield rattles, then sound the paean and charge the enemy. As soon as he turns, and the bugle from the river sounds for 'the attack,' you will face to the right about, the rear rank leading, and the whole detachment falling back and crossing the river as quickly as possible, every one preserving his original rank, so as to avoid trammelling one another: the bravest man is he who gets to the other side first."

The Carduchians, seeing that the remnant left was the merest handful (for many even of those whose duty it was to remain had gone off in their anxiety to protect their beasts of burden, or their personal kit, or their mistresses), bore down upon them valorously, and opened fire with sling-stones and arrows. But the Hellenes, raising the battle hymn, dashed at them at a run, and they did not await them; armed well enough for mountain warfare, and with a view to sudden attack followed by speedy flight, they were not by any means sufficiently equipped for an engagement at close quarters. At this instant the signal of the bugle was heard. Its notes added wings to the flight of the barbarians, but the Hellenes turned right about in the opposite direction, and betook themselves to the river with what speed they might. Some of the enemy, here a man and there another, perceived, and running back to the river, let fly their arrows and wounded a few; but the majority, even when the Hellenes were well across, were still to be seen pursuing their flight. The detachment which came to meet Xenophon's men, carried away by their valour, advanced further than they had need to, and had to cross back again in the rear of Xenophon's men, and of these too a few were wounded.

The passage effected, they fell into line about midday, and marched through Armenian territory, one long plain with smooth rolling hillocks, not less than five parasangs in distance; for owing to the wars of this people with the Carduchians there were no villages near the river. The village eventually reached was large, and possessed a palace belonging to the satrap, and most of the houses were crowned with turrets; provisions were plentiful.

From this village they marched two stages – ten parasangs – until they had surmounted the sources of the river Tigris; and from this point they marched three stages – fifteen parasangs – to the river Teleboas. This was a fine stream, though not large, and there were many villages about it. The district was named Western Armenia. The lieutenant-governor of it was Tiribazus, the king's friend, and whenever the latter paid a visit, he alone had the privilege of mounting the king upon his horse. This officer rode up to the Hellenes with a body of cavalry, and sending forward an interpreter, stated that he desired a colloquy with the leaders. The generals resolved to hear what he had to say; and advancing on their side to within speaking distance, they demanded what he wanted. He replied that he wished to make a treaty with them, in accordance with which he on his side would abstain from injuring the Hellenes, if they would not burn his houses, but merely take such provisions as they needed. This proposal satisfied the generals, and a treaty was made on the terms suggested.

From this place they marched three stages – fifteen parasangs – through plain country, Tiribazus the while keeping close behind with his own forces more than a mile off. Presently they reached a palace with villages clustered round about it, which were full of supplies in great variety. But while they were encamping in the night there was a heavy fall of snow, and in the morning it was resolved to billet out the different regiments, with their generals, throughout the villages. There was no enemy in sight, and the proceeding seemed prudent, owing to the quantity of snow. In these quarters they had for provisions all the good things there are – sacrificial beasts, corn, old wines with an exquisite bouquet, dried grapes, and vegetables of all sorts. But some of the stragglers from the camp reported having seen an army, and the blaze of many watchfires in the night. Accordingly the generals concluded that it was not prudent to separate their quarters in this way, and a resolution was passed to bring the troops together again. After that they reunited, the more so that the weather promised to be fine with a clear sky; but while they lay there in open quarters, during the night down came so thick a fall of snow that it completely covered up the stacks of arms and the men themselves lying down. It cramped and crippled the baggage animals; and there was great unreadiness to get up, so gently fell the snow as they lay there warm and comfortable, and formed a blanket, except where it slipped off the sleeper's shoulders; and it was not until Xenophon roused himself to get up, and, without his cloak on, began to split wood, that quickly first one and then another got up, and taking the log away from him, fell to splitting. Thereat the rest followed suit, got up, and began kindling fires and oiling their bodies, for there was a scented unguent to be found there in abundance, which they used instead of oil. It was made from pig's fat, sesame, bitter almonds, and turpentine. There was a sweet oil also to be found, made of the same ingredients.

After this it was resolved that they must again separate their quarters and get under cover in the villages. At this news the soldiers, with much joy and shouting, rushed upon the covered houses and the provisions; but all who in their blind folly had set fire to the houses when they left them before, now paid the penalty in the poor quarters they got. From this place one night they sent off a party under Democrates, a Temenite, up into the mountains, where the stragglers reported having seen watch-fires. The leader selected was a man whose judgment might be depended upon to verify the truth of the matter. With a happy gift to distinguish between fact and fiction, he had often been successfully appealed to. He went and reported that he had seen no watchfires, but he had got a man, whom he brought back with him, carrying a Persian bow and quiver, and a sagaris or battle-axe like those worn by the Amazons. When asked "from what country he came," the prisoner answered that he was "a Persian, and was going from the army of Tiribazus to get provisions." They next asked him "how large the army was, and for what object it had been collected." His answer was that "it consisted of Tiribazus at the head of his own forces, and aided by some Chalybian and Taochian mercenaries. Tiribazus had got it together," he added, "meaning to attack the Hellenes on the high mountain pass, in a defile which was the sole passage."

When the generals heard this news, they resolved to collect the troops, and they set off at once, taking the prisoner to act as guide, and leaving a garrison behind with Sophaenetus the Stymphalian in command of those who remained in the camp. As soon as they had begun to cross the hills, the light infantry, advancing in front and catching sight of the camp, did not wait for the heavy infantry, but with a loud shout rushed upon the enemy's entrenchment. The natives, hearing the din and clatter, did not care to stop, but took rapidly to their heels. But, for all their expedition, some of them were killed, and as many as twenty horses were captured, with the tent of Tiribazus, and its contents, silver-footed couches and goblets, besides certain persons styling themselves the butlers and bakers. As soon as the generals of the heavy infantry division had learnt the news, they resolved to return to the camp with all speed, for fear of an attack being made on the remnant left behind. The recall was sounded and the retreat commenced; the camp was reached the same day.

The next day it was resolved that they should set off with all possible speed, before the enemy had time to collect and occupy the defile. Having got their kit and baggage together, they at once began their march through deep snow with several guides, and, crossing the high pass the same day on which Tiribazus was to have attacked them, got safely into cantonments. From this point they marched three desert stages – fifteen parasangs – to the river Euphrates, and crossed it in water up to the waist. The sources of the river were reported to be at no great distance. From this place they marched through deep snow over a flat country three stages – fifteen parasangs. The last of these marches was trying, with the north wind blowing in their teeth, drying up everything and benumbing the men. Here one of the seers suggested to them to do sacrifice to Boreas, and sacrifice was done. The effect was obvious to all in the diminished fierceness of the blast. But there was six feet of snow, so that many of the baggage animals and slaves were lost, and about thirty of the men themselves.

They spent the whole night in kindling fires, for there was fortunately no dearth of wood at the halting-place; only those who came late into camp had no wood. Accordingly those who had arrived a good while and had kindled fires were not for allowing these late-comers near their fires, unless they would in return give a share of their corn or of any other victuals they might have. Here then a general exchange of goods was set up. Where the fire was kindled the snow melted, and great trenches formed themselves down to the bare earth, and here it was possible to measure the depth of the snow.

Leaving these quarters, they marched the whole of the next day over snow, and many of the men were afflicted with "boulimia" (or hunger-faintness). Xenophon, who was guarding the rear, came upon some men who had dropt down, and he did not know what ailed them; but some one who was experienced in such matters suggested to him that they had evidently got boulimia; and if they got something to eat, they would revive. Then he went the round of the baggage train, and laying an embargo on any eatables he could see, doled out with his own hands, or sent off other able-bodied agents to distribute to the sufferers, who as soon as they had taken a mouthful got on their legs again and continued the march.

On and on they marched, and about dusk Cheirisophus reached a village, and surprised some women and girls who had come from the village to fetch water at the fountain outside the stockade. These asked them who they were. The interpreters answered for them in Persian: "They were on their way from the king to the satrap;" in reply to which the women gave them to understand that the satrap was not at home, but was away a parasang farther on. As it was late they entered with the water-carriers within the stockade to visit the headman of the village. Accordingly Cheirisophus and as many of the troops as were able got into cantonments there, while the rest of the soldiers – those namely who were unable to complete the march – had to spend the night out, without food and without fire; under the circumstances some of the men perished.

On the heels of the army hung perpetually bands of the enemy, snatching away disabled baggage animals and fighting with each other over the carcases. And in its track not seldom were left to their fate disabled soldiers, struck down with snow-blindness or with toes mortified by frostbite. As to the eyes, it was some alleviation against the snow to march with something black before them; for their feet, the only remedy was to keep in motion without stopping for an instant, and to loose the sandal at night. If they went to sleep with the sandals on, the thong worked into the feet, and the sandals were frozen fast to them. This was partly due to the fact that, since their old sandals had failed, they wore untanned brogues made of newly-flayed ox-hides. It was owing to some such dire necessity that a party of men fell out and were left behind, and seeing a black-looking patch of ground where the snow had evidently disappeared, they conjectured it must have been melted; and this was actually so, owing to a spring of some sort which was to be seen steaming up in a dell close by. To this they had turned aside and sat down, and were loth to go a step further. But Xenophon, with his rearguard, perceived them, and begged and implored them by all manner of means not to be left behind, telling them that the enemy were after them in large packs pursuing; and he ended by growing angry. They merely bade him put a knife to their throats; not one step farther would they stir. Then it seemed best to frighten the pursuing enemy if possible, and prevent their falling upon the invalids. It was already dusk, and the pursuers were advancing with much noise and hubbub, wrangling and disputing over their spoils. Then all of a sudden the rearguard, in the plenitude of health and strength, sprang up out of their lair and ran upon the enemy, whilst those weary wights bawled out as loud as their sick throats could sound, and clashed their spears against their shields; and the enemy in terror hurled themselves through the snow into the dell, and not one of them ever uttered a sound again.

Xenophon and his party, telling the sick folk that next day

people would come for them, set off, and before they had gone half a mile they fell in with some soldiers who had laid down to rest on the snow with their cloaks wrapped round them, but never a guard was established, and they made them get up. Their explanation was that those in front would not move on. Passing by this group he sent forward the strongest of his light infantry in advance, with orders to find out what the stoppage was. They reported that the whole army lay reposing in the same fashion. That being so, Xenophon's men had nothing for it but to bivouac in the open air also, without fire and supperless, merely posting what pickets they could under the circumstances. But as soon as it drew towards day, Xenophon despatched the youngest of his men to the sick folk behind, with orders to make them get up and force them to proceed. Meanwhile Cheirisophus had sent some of his men quartered in the village to enquire how they fared in the rear; they were overjoyed to see them, and handed over the sick folk to them to carry into camp, while they themselves continued their march forwards, and ere twenty furlongs were past reached the village in which Cheirisophus was quartered. As soon as the two divisions were met, the resolution was come to that it would be safe to billet the regiments throughout the villages; Cheirisophus remained where he was, while the rest drew lots for the villages in sight, and then, with their several detachments, marched off to their respective destinations.

It was here that Polycrates, an Athenian and captain of a company, asked for leave of absence – he wished to be off on a quest of his own; and putting himself at the head of the active men of the division, he ran to the village which had been allotted to Xenophon. He surprised within it the villagers with their headman, and seventeen young horses which were being reared as a tribute for the king, and, last of all, the headman's own daughter, a young bride only eight days wed. Her husband had gone off to chase hares, and so he escaped being taken with the other villagers. The houses were underground structures with an aperture like the mouth of a well by which to enter, but they were broad and spacious below. The entrance for the beasts of burden was dug out, but the human occupants descended by a ladder. In these dwellings were to be found goats and sheep and cattle, and cocks and hens, with their various progeny. The flocks and herds were all reared under cover upon green food. There were stores within of wheat and barley and vegetables, and wine made from barley in great big bowls; the grains of barley malt lay floating in the beverage up to the lip of the vessel, and reeds lay in them, some longer, some shorter, without joints; when you were thirsty you must take one of these into your mouth, and suck. The beverage without admixture of water was very strong, and of a delicious flavour to certain palates, but the taste must be acquired.

Xenophon made the headman of the village his guest at supper, and bade him keep a good heart; so far from robbing him of his children, they would fill his house full of good things in return for what they took before they went away; only he must set them an example, and discover some blessing or other for the army, until they found themselves with another tribe. To this he readily assented, and with the utmost cordiality showed them the cellar where the wine was buried. For this night then, having taken up their several quarters as described, they slumbered in the midst of plenty, one and all, with the headman under watch and ward, and his children with him safe in sight.

But on the following day Xenophon took the headman and set off to Cheirisophus, making a round of the villages, and at each place turning in to visit the different parties. Everywhere alike he found them faring sumptuously and merry-making. There was not a single village where they did not insist on setting a breakfast before them, and on the same table were spread half a dozen dishes at least, lamb, kid, pork, veal, fowls, with various sorts of bread, some of wheat and some of barley. When, as an act of courtesy, any one wished to drink his neighbour's health, he would drag him to the big bowl, and when there, he must duck his head and take a long pull, drinking like an ox. The headman, they insisted everywhere, must accept as a present whatever he liked to have. But he would accept nothing, except where he espied any of his relations, when he made a point of taking them off, him or her, with himself.

When they reached Cheirisophus they found a similar scene. There too the men were feasting in their quarters, garlanded with whisps of hay and dry grass, and Armenian boys were playing the part of waiters in barbaric costumes, only they had to point out by gesture to the boys what they were to do, like deaf and dumb. After the first formalities, when Cheirisophus and Xenophon had greeted one another like bosom friends, they interrogated the headman in common by means of the Persian-speaking interpreter, "What was the country?" they asked; he replied, "Armenia." And again, "For whom are the horses being bred?" "They are tribute for the king," he replied. "And the neighbouring country?" "Is the land of the Chalybes," he said; and he described the road which led to it. So for the present Xenophon went off, taking the headman back with him to his household and friends. He also made him a present of an oldish horse which he had got; he had heard that the headman was a priest of the sun, and so he could fatten up the beast and sacrifice him; otherwise he was afraid it might die outright, for it had been injured by the long marching. For himself he took his pick of the colts, and gave a colt apiece to each of his fellow-generals and officers. The horses here were smaller than the Persian horses, but much more spirited. It was here too that their friend the head-man explained to them, how they should wrap small bags or sacks round the feet of the horses and other cattle when marching through the snow, for without such precautions the creatures sank up to their bellies.

When a week had passed, on the eighth day Xenophon delivered over the guide (that is to say, the village headman) to Cheirisophus. He left the headman's household safe behind in the village, with the exception of his son, a lad in the bloom of youth. This boy was entrusted to Episthenes of Amphipolis to guard; if the headman proved himself a good guide, he was to take away his son also at his departure. They finally made his house the repository of all the good things they could contrive to get together; then they broke up their camp and commenced to march, the headman guiding them through the snow unfettered. When they had reached the third stage Cheirisophus flew into a rage with him, because he had not brought them to any villages. The headman pleaded that there were none in this part. Cheirisophus struck him, but forgot to bind him, and the end of it was that the headman ran away in the night and was gone, leaving his son behind him. This was the sole ground of difference between Cheirisophus and Xenophon during the march, this combination of ill-treatment and neglect in the case of the guide. As to the boy, Episthenes conceived a passion for him, and took him home with him, and found in him the most faithful of friends.

After this they marched seven stages at the rate of five parasangs a day, to the banks of the river Phasis, which is a hundred feet broad: and thence they marched another couple of stages, ten parasangs; but at the pass leading down into the plain there appeared in front of them a mixed body of Chalybes and Taochians and Phasians. When Cheirisophus caught sight of the enemy on the pass at a distance of about three or four miles, he ceased marching, not caring to approach the enemy with his troops in column, and he passed down the order to the others: to deploy their companies to the front, that the troops might form into line. As soon as the rearguard had come up, he assembled the generals and officers, and addressed them: "The enemy, as you see, are in occupation of the mountain pass, it is time we should consider how we are to make the best fight to win it. My opinion is, that we should give orders to the troops to take their morning meal, whilst we deliberate whether we should cross the mountains to-day or to-morrow." "My opinion," said Cleanor, "is, that as soon as we have breakfasted, we should arm for the fight and attack the enemy, without loss of time, for if we fritter away to-day, the enemy who are now content to look at us, will grow bolder, and with their growing courage, depend upon it, others more numerous will join them."

After him Xenophon spoke: "This," he said, "is how I see the matter; if fight we must, let us make preparation to sell our lives dearly, but if we desire to cross with the greatest ease, the point to consider is, how we may get the fewest wounds and throw away the smallest number of good men. Well then, that part of the mountain which is visible stretches nearly seven miles. Where are the men posted to intercept us? Except at the road itself, they are nowhere to be seen. It is much better then to try if possible to steal a point of this desert mountain unobserved, and before they know where we are, secure the prize, than to fly at a strong position and an enemy thoroughly prepared. Since it is much easier to march up a mountain without fighting than to tramp along a level when assailants are on either hand; and provided he has not to fight, a man will see what lies at his feet much more plainly even at night than in broad daylight in the midst of battle; and a rough road to feet that roam in peace may be pleasanter than a smooth surface with the bullets whistling about your ears. Nor is it so impossible, I take it, to steal a march, since it is open to us to go by night, when we cannot be seen, and to fall back so far that they will never notice us. In my opinion, however, if we make a feint of attacking here, we shall find the mountain chain all the more deserted elsewhere, since the enemy will be waiting for us here in thicker swarm.

"But what right have I to be drawing conclusions about

stealing in your presence, Cheirisophus? For you Lacedaemonians, as I have often been told, you who belong to the 'peers,' practise stealing from your boyhood up; and it is no disgrace but honourable rather to steal, except such things as the law forbids; and in order, I presume, to stimulate your sense of secretiveness, and to make you master thieves, it is lawful for you further to get a whipping, if you are caught. Now then you have a fine opportunity of displaying your training. But take care we are not caught stealing over the mountain, or we shall catch it ourselves." "For all that," retorted Cheirisophus, "I have heard that you Athenians are clever hands at stealing the public moneys; and that too though there is fearful risk for the person so employed; but, I am told, it is your best men who are most addicted to it; if it is your best men who are thought worthy to rule. So it is a fine opportunity for yourself also, Xenophon, to exhibit your education." "And I," replied Xenophon, "am ready to take the rear division, as soon as we have supped, and seize the mountain chain. I have already got guides, for the light troops laid an ambuscade, and seized some of the cut-purse vagabonds who hung on our rear. I am further informed by them that the mountain is not inaccessible, but is grazed by goats and cattle, so that if we can once get hold of any portion of it, there will be no difficulty as regards our animals – they can cross. As to the enemy, I expect they will not even wait for us any longer, when they once see us on a level with themselves on the heights, for they do not even at present care to come down and meet us on fair ground." Cheirisophus answered: "But why should you go and leave your command in the rear? Send others rather, unless a band of volunteers will present themselves." Thereupon Aristonymus the Methydrian came forward with some heavy infantry, and Aristeas the Chian with some light troops, and Nicomachus the Oetean with another body of light troops, and they made an agreement to kindle several watch-fires as soon as they held the heights. The arrangements made, they breakfasted; and after breakfast Cheirisophus advanced the whole army ten furlongs closer towards the enemy, so as to strengthen the impression that he intended to attack them at that point.

But as soon as they had supped and night had fallen, the party under orders set off and occupied the mountain, while the main body rested where they were. Now as soon as the enemy perceived that the mountain was taken, they banished all thought of sleep, and kept many watch-fires blazing through the night. But at break of day Cheirisophus offered sacrifice, and began advancing along the road, while the detachment which held the mountain advanced *pari passu* by the high ground. The larger mass of the enemy, on his side, remained still on the mountain-pass, but a section of them turned to confront the detachment on the heights. Before the main bodies had time to draw together, the detachment on the height came to close quarters, and the Hellenes were victorious and gave chase. Meanwhile the light division of the Hellenes, issuing from the plain, were rapidly advancing against the serried lines of the enemy, whilst Cheirisophus followed up with his heavy infantry at quick march. But the enemy on the road no sooner saw their higher division being worsted than they fled, and some few of them were slain, and a vast number of wicker shields were taken, which the Hellenes hacked to pieces with their short swords and rendered useless. So when they had reached the summit of the pass, they sacrificed and set up a trophy, and descending into the plain, reached villages abounding in good things of every kind.

After this they marched into the country of the Taochians five stages – thirty parasangs – and provisions failed; for the Taochians lived in strong places, into which they had carried up all their stores. Now when the army arrived before one of these strong places – a mere fortress, without city or houses, into which a motley crowd of men and women and numerous flocks and herds were gathered – Cheirisophus attacked at once. When the first regiment fell back tired, a second advanced, and again a third, for it was impossible to surround the place in full force, as it was encircled by a river. Presently Xenophon came up with the rearguard, consisting of both light and heavy infantry, whereupon Cheirisophus hailed him with the words: "In the nick of time you have come; we must take this place, for the troops have no provisions, unless we take it." Thereupon they consulted together, and to Xenophon's inquiry, "What it was which hindered their simply walking in?" Cheirisophus replied, "There is just this one narrow approach which you see; but when we attempt to pass by it they roll down volleys of stones from yonder overhanging crag," pointing up, "and this is the state in which you find yourself, if you chance to be caught;" and he pointed to some poor fellows with their legs or ribs crushed to bits. "But when they have expended their ammunition," said Xenophon, "there is nothing else, is there, to hinder our passing? Certainly, except yonder handful of fellows, there is no one in front of us that we can see; and of them, only two or three apparently are armed, and the distance to be traversed under fire is, as your eyes will tell you, about one hundred and fifty feet as near as can be, and of this space the first hundred is thickly covered with great pines at intervals; under cover of these, what harm can come to our men from a pelt of stones, flying or rolling? So then, there is only fifty feet left to cross, during a lull of stones." "Ay," said Cheirisophus, "but with our first attempt to approach the bush a galling fire of stones commences." "The very thing we want," said the other, "for they will use up their ammunition all the quicker; but let us select a point from which we shall have only a brief space to run across, if we can, and from which it will be easier to get back, if we wish."

Thereupon Cheirisophus and Xenophon set out with Callimachus the Parrhasian, the captain in command of the officers of the rearguard that day; the rest of the captains remained out of danger. That done, the next step was for a party of about seventy men to get away under the trees, not in a body, but one by one, every one using his best precaution; and Agasias the Stymphalian, and Aristonymus the Methydrian, who were also officers of the rearguard, were posted as supports outside the trees; for it was not possible for more than a single company to stand safely within the trees. Here Callimachus hit upon a pretty contrivance – he ran forward from the tree under which he was posted two or three paces, and as soon as the stones came whizzing, he retired easily, but at each excursion more than ten wagon-loads of rocks were expended. Agasias, seeing how Callimachus was amusing himself, and the whole army looking on as spectators, was seized with the fear that he might miss his chance of being first to run the gauntlet of the enemy's fire and get into the place. So, without a word of summons to his next neighbour, Aristonymus, or to Eurylochus of Lusia, both comrades of his, or to any one else, off he set on his own account, and passed the whole detachment. But Callimachus, seeing him tearing past, caught hold of his shield by the rim, and in the meantime Aristonymus the Methydrian ran past both, and after him. Eurylochus of Lusia; for they were one and all aspirants to valour, and in that high pursuit, each was the eager rival of the rest. So in this strife of honour, the four of them took the fortress, and when they had once rushed in, not a stone more was hurled from overhead.

And here a terrible spectacle displayed itself: the women first cast their infants down the cliff, and then they cast themselves after their fallen little ones, and the men likewise. In such a scene, Aeneas the Stymphalian, an officer, caught sight of a man with a fine dress about to throw himself over, and seized hold of him to stop him; but the other caught him to his arms, and both were gone in an instant headlong down the crags, and were killed. Out of this place the merest handful of human beings were taken prisoner, but cattle and asses in abundance and flocks of sheep.

From this place they marched through the Chalybes seven stages, fifty parasangs. These were the bravest men whom they encountered on the whole march, coming cheerily to close quarters with them. They wore linen corselets reaching to the groin, and instead of the ordinary "wings" or basques, a thickly-plaited fringe of cords. They were also provided with greaves and helmets, and at the girdle a short sabre, about as long as the Laconian dagger, with which they cut the throats of those they mastered, and after severing the head from the trunk they would march along carrying it, singing and dancing, when they drew within their enemy's field of view. They carried also a spear fifteen cubits long, lanced at one end. These folk stayed in regular townships, and whenever the Hellenes passed by they invariably hung close on their heels fighting. They had dwelling-places in their fortresses, and into them they had carried up their supplies, so that the Hellenes could get nothing from this district, but supported themselves on the flocks and herds they had taken from the Taochians. After this the Hellenes reached the river Harpasus, which was four hundred feet broad. Hence they marched through the Scythenians four stages – twenty parasangs – through a long level country to more villages, among which they halted three days, and got in supplies.

Passing on from thence in four stages of twenty parasangs, they reached a large and prosperous well-populated city, which went by the name of Gymnias, from which the governor of the country sent them a guide to lead them through a district hostile to his own. This guide told them that within five days he would lead them to a place from which they could see the sea, "and," he added, "if I fail of my word, you are free to take my life." Accordingly he put himself at their head; but he no sooner set foot in the country hostile to himself than he fell to encouraging them to burn and harry the land; indeed his exhortations were so earnest, it was plain that it was for this he had come, and not out of the good-will he bore the Hellenes.

On the fifth day they reached the mountain, the name of which was Theches. No sooner had the men in front ascended it and caught sight of the sea than a great cry arose, and Xenophon, with the rearguard, catching the sound of it, conjectured that another set of enemies must surely be attacking in front; for they were followed by the inhabitants of the country, which was all aflame; indeed the rearguard had killed some and captured others alive by laying an ambuscade; they had taken also about twenty wicker shields, covered with the raw hides of shaggy oxen.

But as the shout became louder and nearer, and those who from time to time came up, began racing at the top of their speed towards the shouters, and the shouting continually recommenced with yet greater volume as the numbers increased, Xenophon settled in his mind that something extraordinary must have happened, so he mounted his horse, and taking with him Lycius and the cavalry, he galloped to the rescue. Presently they could hear the soldiers shouting and passing on the joyful word, *The sea! the sea!*

Thereupon they began running, rearguard and all, and the baggage animals and horses came galloping up. But when they had reached the summit, then indeed they fell to embracing one another – generals and officers and all – and the tears trickled down their cheeks. And on a sudden, some one, whoever it was, having passed down the order, the soldiers began bringing stones and erecting a great cairn, whereon they dedicated a host of untanned skins, and staves, and captured wicker shields, and with his own hand the guide hacked the shields to pieces, inviting the rest to follow his example. After this the Hellenes dismissed the guide with a present raised from the common store, to wit, a horse, a silver bowl, a Persian dress, and ten darics; but what he most begged to have were their rings, and of these he got several from the soldiers. So, after pointing out to them a village where they would find quarters, and the road by which they would proceed towards the land of the Macrones, as evening fell, he turned his back upon them in the night and was gone.

From this point the Hellenes marched through the country of

the Macrones in three stages – ten parasangs – and on the first day they reached the river, which formed the boundary between the land of the Macrones and the land of the Scythenians. Above them, on their right, they had a country of the sternest and ruggedest character, and on their left another river, into which the frontier river discharges itself, and which they must cross. This was thickly fringed with trees which, though not of any great bulk, were closely packed. As soon as they came up to them, the Hellenes proceeded to cut them down in their haste to get out of the place as soon as possible. But the Macrones, armed with wicker shields and lances and hair tunics, were already drawn up to receive them immediately opposite the crossing. They were cheering one another on, and kept up a steady pelt of stones into the river, though they failed to reach the other side or do any harm.

At this juncture one of the light infantry came up to Xenophon; he had been, he said, a slave at Athens, and he wished to tell him that he recognised the speech of these people. "I think," said he, "this must be my native country, and if there is no objection I will have a talk with them." "No objection at all," replied Xenophon, "pray talk to them, and ask them first, who they are." In answer to this question they said, "they were Macrones." "Well, then," said he, "ask them why they are drawn up in battle and want to fight with us." They answered, "Because you are invading our country." The generals bade him say: "If so, it is with no intention certainly of doing it or you any harm: but we have been at war with the king, and are now returning to Hellas, and all we want is to reach the sea." The others asked, "Were they willing to give them pledges to that effect?" They replied: "Yes, they were ready to give and receive pledges to that effect." Then the Macrones gave a barbaric lance to the Hellenes, and the Hellenes a Hellenic lance to them: "for these," they said, "would serve as pledges," and both sides called upon the gods to witness.

After the pledges were exchanged, the Macrones fell to vigorously hewing down trees and constructing a road to help them across, mingling freely with the Hellenes and fraternising in their midst, and they afforded them as good a market as they could, and for three days conducted them on their march, until they had brought them safely to the confines of the Colchians. At this point they were confronted by a great mountain chain, which however was accessible, and on it the Colchians were drawn up for battle. In the first instance, the Hellenes drew up opposite in line of battle, as though they were minded to assault the hill in that order; but afterwards the generals determined to hold a council of war, and consider how to make the fairest fight.

Accordingly Xenophon said: "I am not for advancing in line, but advise to form into column of sections. To begin with, the line," he urged, "would be scattered and thrown into disorder at once; for we shall find the mountain full of inequalities, it will be pathless here and easy to traverse there. The mere fact of first having formed in line, and then seeing the line thrown into disorder, must exercise a disheartening effect. Again, if we advance several deep, the enemy will none the less overlap us, and turn their superfluous numbers to account as best they like; while, if we march in shallow order, we may fully expect our line to be cut through and through by the thick rain of missiles and rush of men, and if this happens anywhere along the line, the whole line will equally suffer. No; my notion is to form columns by companies, covering ground sufficient with spaces between the companies to allow the last companies of each flank to be outside the enemy's flanks. Thus we shall with our extreme companies be outside the enemy's line, and the best men at the head of their columns will lead the attack, and every company will pick its way where the ground is easy; also it will be difficult for the enemy to force his way into the intervening spaces, when there are companies on both sides; nor will it be easy for him to cut in twain any individual company marching in column. If, too, any particular company should be pressed, the neighbouring company will come to the rescue, or if at any point any single company succeed in reaching the height, from that moment not one man of the enemy will stand his ground."

This proposal was carried, and they formed into columns by companies. Then Xenophon, returning from the right wing to the left, addressed the soldiers. "Men," he said, "these men whom you see in front of you are the sole obstacles still interposed between us and the haven of our hopes so long deferred. We will swallow them up whole, without cooking, if we can."

The several divisions fell into position, the companies were formed into columns, and the result was a total of something like eighty companies of heavy infantry, each company consisting on an average of a hundred men. The light infantry and bowmen were arranged in three divisions – two outside to support the left and the right respectively, and the third in the centre – each division consisting of about six hundred men.

Before starting, the generals passed the order to offer prayer; and with the prayer and battle hymn rising from their lips they commenced their advance. Cheirisophus and Xenophon, and the light infantry with them, advanced outside the enemy's line to right and left, and the enemy, seeing their advance, made an effort to keep parallel and confront them, but in order to do so, as he extended partly to right and partly to left, he was pulled to pieces, and there was a large space or hollow left in the centre of his line. Seeing them separate thus, the light infantry attached to the Arcadian battalion, under command of Aeschines, an Acarnanian, mistook the movement for flight, and with a loud shout rushed on, and these were the first to scale the mountain summit; but they were closely followed up by the Arcadian heavy infantry, under command of Cleanor of Orchomenus.

When they began running in that way, the enemy stood their ground no longer, but betook themselves to flight, one in one direction, one in another, and the Hellenes scaled the hill and found quarters in numerous villages which contained supplies in abundance. Here, generally speaking, there was nothing to excite their wonderment, but the numbers of beehives were indeed astonishing, and so were certain properties of the honey. The effect upon the soldiers who tasted the combs was, that they all went for the nonce quite off their heads, and suffered from vomiting and diarrhoea, with a total inability to stand steady on their legs. A small dose produced a condition not unlike violent drunkenness, a large one an attack very like a fit of madness, and some dropped down, apparently at death's door. So they lay, hundreds of them, as if there had been a great defeat, a prey to the cruellest despondency. But the next day, none had died; and almost at the same hour of the day at which they had eaten they recovered their senses, and on the third or fourth day got on their legs again like convalescents after a severe course of medical treatment.

From this place they marched on two stages – seven parasangs – and reached the sea at Trapezus, a populous Hellenic city on the Euxine Sea, a colony of the Sinopeans, in the territory of the Colchians. Here they halted for about thirty days in the villages of the Colchians, which they used as a base of operations to ravage the whole territory of Colchis. The men of Trapezus supplied the army with a market, entertained them, and gave them, as gifts of hospitality, oxen and wheat and wine. Further, they negotiated with them in behalf of their neighbours the Colchians, who dwelt in the plain for the most part, and from these folk also came gifts of hospitality in the shape of cattle. And now the Hellenes made preparation for the sacrifice which they had vowed, and a sufficient number of cattle came in for them to offer thank-offerings for safe guidance to Zeus the Saviour, and to Heracles, and to the other gods, according to their vows. They instituted also a gymnastic contest on the mountain side, just where they were quartered, and chose Dracontius, a Spartan (who had been banished from home when a lad, having unintentionally slain another boy with a blow of his dagger), to superintend the course, and be president of the games.

As soon as the sacrifices were over, they handed over the hides of the beasts to Dracontius, and bade him lead the way to his racecourse. He merely waved his hand and pointed to where they were standing, and said, "There, this ridge is just the place for running, anywhere, everywhere." "But how," it was asked, "will they manage to wrestle on the hard scrubby ground?" "Oh! worse knocks for those who are thrown," the president replied. There was a mile race for boys, the majority being captive lads; and for the long race more than sixty Cretans competed; there was wrestling, boxing, and the pankration. Altogether it was a beautiful spectacle. There was a large number of entries, and the emulation, with their companions, male and female, standing as spectators, was immense. There was horse-racing also; the riders had to gallop down a steep incline to the sea, and then turn and come up again to the altar, and on the descent more than half rolled head over heels, and then back they came toiling up the tremendous steep, scarcely out of a walking pace. Loud were the shouts, the laughter, and the cheers.

Hilary St George Saunders

The Drop

The drop by the British 1st Airborne Division at Arnhem in Holland was intended to capture a bridge over the Rhine that would allow Allied armour to roll into the industrial heartland of Nazi Germany. Unfortunately for the 1st Airborne, the 9th and 10th SS Panzer Divisions were refitting in the area. Arnhem, like Dunkirk, was a failure but one made heroic by the sheer bravery of those who fought it – the "Red Berets" of the Parachute Regiment.

AT 09.30 on 17 September, a Sunday, the 1st Battalion[1] left Grimsthorpe Castle in lorries bound for Grantham airfield. Mile after mile went by, and no dispatch riders appeared to cancel the operation. Parachutes had been drawn an hour before – a member of "T" Company resolutely refusing one marked "Dummy" – and harness fitted, and the men were in the highest spirits. "It was a perfect summer day. All the planes were lined up in the bright sunshine . . . the men lay on the ground beside them, resting on their parachutes, eating haversack rations." With the order to put on parachutes appeared the battalion joker, "Guv." Beech, the physical training sergeant, who walked down to the line of men wearing "his well-known opera hat which he kept taking off *à la* Winston Churchill", and bowing to left and right.

The 2nd Battalion boarded its aircraft in similar conditions at Saltby; and so did the 3rd Battalion.

The journey was uneventful. "Once in the air," writes one who dropped with the 1st Battalion that day, "one could see an endless line of Dakotas behind. One could catch glimpses of fighters diving about round the convoy." They came from the

1 Of the Parachute Regiment.

8th U.S. Army Air Force, which provided continuous and most efficient protection, and which, together with the 9th American Army Air Force, attacked anti-aircraft guns which opened up upon the main stream of Dakotas when it reached the Dutch coast. In the van were twelve Stirlings carrying six officers and a hundred and eighty men of the 21st Independent Parachute Company under the command of Major B. A. Wilson. They constituted the pathfinder force, whose duty it was to lay out the various aids and indications on the dropping zone. "I shall always remember," says Wilson, a man then in his forties, "that first flight on that lovely Sunday morning. I sat with the pilot as we flew in over the Dutch coast. Everything looked so peaceful. There were cows feeding quietly in the fields and peasants going about their work. Not a sign of fighting or war. Not a glimpse of the enemy. I had just said to the pilot, 'This seems a pretty quiet area. Suppose we get out here', when, before he could answer, a number of shells burst round the aircraft. . . . A few minutes later he wished me good luck as I sailed down to the glorious uncertainty of the welcome I should receive as one of the first parachute troops to enter German-occupied Holland. . . . The ensuing half-hour while we waited for the main force to drop was, to say the least of it, interesting."

The interest for Wilson and his men lay in accepting the surrender of some fifteen frightened Germans. During the drop two men were hit, one in his ammunition pouches, the other in his haversack. Neither was hurt. Punctually on time came the Dakotas and a moment later the members of the pathfinder force could see "the blue field of the sky suddenly blossom with the white flowers of the parachutes". The drop of the 1st Brigade was more successful than anything which had so far been achieved by the airborne forces of either side in the war, even during an exercise. Nearly 100 per cent. arrived at the right time and place. By 15.00 hours the units were ready and prepared to move. All had prospered marvellously, but now came the first check. The Air Landing Reconnaissance Squadron could not attempt the planned *coup de main* against the bridge, for the few gliders that failed to arrive were unfortunately those carrying the transport. It fell, therefore, to the 2nd Battalion under Frost,[2] that veteran of Bruneval, North Africa and Sicily, to seize the bridge. His simple plan was for "A"

2 Lieut.-Col. J. D. Frost, D.S.O., M.C.

Company in the van to move straight to the main bridge, while "C" Company, following in its rear, was to seize the railway bridge if it was still intact, pass over it and attack the main bridge from the south. To "B" Company, coming last of all, was allotted the capture of the pontoon bridge, if it existed, and, if it did not, the seizure of some high ground called Den Brink, which controlled the entrance to Arnhem from the west. . . .

The force at the bridge under Frost's command now amounted to between three and four hundred men. With dawn came German patrols, including a latrine squad. The latrine lorry was knocked out and the others moved "somewhat aimlessly up and down the road in front of us. Presently the drivers seemed to hesitate. They had seen our ugly eyes looking at them from the windows." Bombs and machine-gun fire killed them all save two, who were captured badly wounded. Hardly had these bewildered Germans been dealt with when the look-outs on the bridge reported that a German convoy had assembled on the farther end and seemed about to rush Frost's position. There ensued, he said afterwards, "the most lovely battle you have ever seen. Sixteen half-track vehicles and armoured cars advanced. There they were, these awful Boches, with their pot helmets sticking out. When we dealt with them they smoked and burned in front of us almost to the end of the battle. I believe they belonged to the 9th Reconnaissance Squadron." They were destroyed by Hawkins grenades, an anti-tank gun and Piats. Some reached the school, where Lieutenant D. R. Simpson, M.C., Royal Engineers, with his sappers, provided them with a warm welcome. The school was in shape a square horseshoe, the ends of the two arms being about ten yards from the road. As the German vehicles went by "Corporal Simpson and Sapper Perry, whose conduct that day was outstanding, stood up and fired straight into the half-tracks". The driver of one half-track, seeing what had happened to those who had preceded him, pulled out to the right along the asphalt path running beneath the windows of the school. "His vehicle did not get far before it was hit; its crew climbed out and sought the cover of bushes, but were killed before reaching them."

Much heartened by this small but not insignificant victory, Frost and his men continued to hold their positions under a shower of light shells and mortar bombs, which grew slowly but steadily heavier and began to cause casualties. They were treated by Captain James Logan, D.S.O., Royal Army Medical Corps, of the 16th Parachute Field Ambulance, whose labours were as skilful as they were indefatigable. This fire came mostly from the north and east, from somewhere, that is, in Arnhem. Frost's main fear at that time was that the Germans would obtain a foothold on the southern end of the bridge, and to prevent this he strengthened the number of light and medium machine-guns established in the upper stories of the houses he had occupied during the night. Their fire was returned by 20-mm. and 40-mm. guns, effective weapons which presently set several buildings on fire and knocked down others. Unfortunately the houses which gave the best field of fire over the bridge were made of wood, and these began to burn. . . .

By the end of the first twenty-four hours' fighting the position was this: Frost, with a mixed force, including the 2nd Battalion, was holding the northern end of the bridge and had successfully repelled all attacks; his casualties, however, were increasing, and he was in urgent need of reinforcements. These the 1st and 3rd Battalions were trying to supply and in so doing had been fought almost to a standstill. The strength of the 1st had fallen to about a hundred men, that of the 3rd was little better. . . .

The drops and the glider landings of the second lift were as successful as those of the day before. Once more the R.A.F. and American crews had performed their task with skill. On this occasion one of them showed that type of resolution which makes a man faithful unto death. Over the dropping zone a Dakota with sixteen parachutists on board was hit and set on fire. "Suddenly a little orange flame appeared on the port wing," notes a witness. "I watched the plane gradually lose height and counted the bodies baling out. They all came out, although the last two were too low for comfort. But the crew stayed in the plane and flew straight, the flames getting larger and larger, till eventually it flew into the ground" . . .

As the autumn day waned the 10th Battalion found itself unable to hold on any longer and began slowly to withdraw. Queripel,[3] cut off with a small party of men, took cover in a ditch. In addition to the wound in his face, he had now been wounded in

3 The late Captain L. E. Queripel, V.C.

both arms. He and his men lined the ditch to cover the withdrawal of the remainder of the battalion. By then they were short both of weapons and ammunition, having but a few Mills bombs, rifles and their personal pistols. German infantry were very near and more than once their stick bombs landed in the ditch, only to be flung back in their faces by the vigilant Queripel. The position became more and more untenable, but he waited until the last moment before he ordered those of his men still alive to leave while he covered their withdrawal with the aid of such grenades as remained. "That was the last occasion on which he was seen." His gallantry earned him a posthumous Victoria Cross. . . .

In this heavy fighting the brigadier and those with him took their full share. "My brigade headquarters," wrote Hackett[4] months later when he had returned from Holland, "with its clerks, signallers, Intelligence section and batmen, was holding the centre of our line as a unit. They were a splendid lot. The signallers were mostly Cheshire yeomen, the clerks were also 'foundation members' for the most part and in the close-quarter fighting in the woods on 19 and particularly 20 September did brilliantly under Staff-Sergeant Pearson, the chief clerk, one of the bravest men at really hand-to-hand fighting and one of the soundest in the brigade. . . . I found myself on 20 September as 'a broken-down cavalryman' (Urquhart's[5] phrase) leading little bayonet rushes in the very dirty stuff the brigade had to contend with before we made contact with the division, and I was impressed with the stout hearts and accurate grenade throwing of the brigade Intelligence section, particularly after the Intelligence officer (Captain Blundell) was shot and killed at about twenty yards range on the same morning."

Before recounting the last stand at Oosterbeek, what happened to Frost and the 2nd Battalion must first be recorded. That Monday night, the 18th, which began in flame and smoke from the burning houses, gradually grew quieter, until soon after midnight "there was absolute silence, or so it seemed to me," said Frost, "for some hours". The commander of the defence

4 Brig. J. W. Hackett.
5 Major-General R. E. Urquhart, Commanding British 1st Airborne Division.

was able at last to snatch some sleep. Up till then he had had but half an hour, and had sustained himself with cups of tea and an occasional nip of whisky. Before dawn he had had to issue an order bringing sniping to an end, for ammunition was running low and would have to be kept for warding off the attacks which the enemy was bound soon to launch with increasing severity. The bridge was still covered by the guns of the 1st Air Landing Light Regiment under Lieutenant-Colonel W. F. K. Thompson, Royal Artillery, but these were their only support. "It became more and more difficult to move," recounts Frost, "for the Boche were tightening their grip, though they made no effort to close with us. By then the number of wounded was very great, but the number of killed small."

The men at the bridge held on throughout that day, buoyed up by rumours, first that the 1st and 3rd Battalions were at hand, and then in the later afternoon by the news that the South Staffordshires and the 11th Battalion were fighting their way towards them. It was a day of heavy mortaring and shelling by tanks which had crept up to a position close to the river bank. Towards noon Captain A. Franks went out against them, and scored three hits with the last three Piat bombs. The German tanks clattered away out of range and did not return. At dusk, however, a Tiger tank appeared and shelled in turn each house still held by the parachutists. Among the casualties caused by this fire was Father Egan, M.C., who had served with the brigade from the outset, and Major A. D. Tatham-Warter. They were both hit, but both remained with those still fighting and refused to go below to the cellars.

During this day the conduct of Trooper Bolton of the 1st Air Landing Reconnaissance Squadron was particularly noteworthy for the calnmess with which he manned his Bren gun and refused to be parted from it. "He hated the thought of anyone using it but himself," says Captain Bernard Briggs, the staff captain at brigade headquarters, who had been at the bridge from the beginning, "and would wake from a cat-nap at any moment and leap to it ready to fire". Lieutenant P. J. Burnett, of the brigade headquarters defence platoon, showed much courage and ingenuity when he succeeded in destroying "a troublesome tank single-handed with grenades". He was to earn the Medaille Militaire Willemsworde, the Dutch Victoria Cross.

Night fell and it seemed to Frost, looking uneasily over his shoulder, that the whole town of Arnhem was on fire, including two large churches. "I never saw anything more beautiful than those burning buildings."

By now the defenders of the bridge were being driven from the houses as they caught fire. Their method of moving from one to another was, whenever possible, to "mousehole" their way from house to house in conditions which grew steadily worse. During this tedious dusty method of moving from one position to another Lieutenant Simpson succeeded in disabling a tank close to the house in which he was posted. Its crew got out and "crept along the wall till they came to a halt beneath the window where I was crouching. I dropped a grenade on them and that was that. I held it for two seconds before I let it drop."

Two things were of particular concern: the lack of water, and the breakdown of the wireless sets, which made it impossible to keep in touch with the rest of the division except by means of the civilian telephone lines. These, manned by the Dutch Resistance, continued to play a part to the end, the operators paying for their fortitude with their lives. Frost had no continuous means of communicating with the battalions who he still hoped were on their way to his relief, but could sometimes speak with divisional headquarters. Perhaps the reinforcements were not very distant. They might even be within earshot. "During a lull we yelled 'Waho Mahommed'," says Briggs, "hoping there would be some reply. But none came. Then we tore down wallpaper to make a megaphone six feet long, through which we shouted words and epithets that could only be British." But there was still no reply.

Dawn on Wednesday, the 20th, shone on Frost, still clinging with difficulty to the north-west end of the bridge, but able to prevent the Germans crossing it. But now his personal good fortune was to desert him. During the morning he was badly wounded in the leg, and Major C. H. F. Gough, M.C., Reconnaissance Squadron, assumed command, but still referred major decisions to Frost, while Tatham-Warter, "whose conduct was exemplary even amid so much gallantry", took over what remained of the 2nd Battalion. In reporting these changes to the divisional commander at Hartestein, Gough referred to himself as "the man who goes in for funny weapons", so that no German or collaborator listening-in on the town exchange which he was using would be able to identify him.

The area occupied by the parachute troops grew smaller and smaller, though they continued to control the approaches to the bridge. Conspicuous among them at this stage was Lieutenant Grayburn[6] of "A" Company. Early in the action, in leading the unsuccessful attack on the south end of the bridge, he had been hit in the shoulder, but continued to lead his men and was the last to withdraw. He then established his platoon in a very exposed house whose position was vital to the defence. In this he held out until 19 September, when it was set on fire, having repelled all attacks, including those made by tanks and self-propelled guns. Re-forming his depleted force, he was still able to maintain the defence and on 20 September led a series of fighting patrols, whose activities so galled the enemy that tanks were brought up again. Only then did Grayburn retreat and, even so, was still able to strike back. At the head of another patrol he drove off the enemy, thus allowing others to remove the fuses from the demolition charges which the Germans had succeeded in placing under the bridge. In so doing he was again wounded, but still would not leave the fight. Eventually, that evening he was killed by the fire of a tank. In his conduct "he showed an example of devotion to duty which can seldom have been equalled", and was awarded a posthumous Victoria Cross.

By the evening of that day all the buildings near the bridge had been burnt down except the U-shaped school. This now caught fire and all attempts to put it out failed. Captain J. Logan, D.S.O., the medical officer, who, with Captain D. Wright, M.C., had been tireless in tending the wounded, therefore informed Gough that he must surrender them if he wished to save them from being burnt or roasted alive in the cellars. Just after dark, under a flag of truce, the enemy picked up many of the wounded, including Frost, who had been expecting his fate and had thrown away his badges. A moment before, Wicks, his batman, had taken leave of him and gone back to the fight. He, too, was soon afterwards badly wounded.

Gough and those still unwounded continued to resist. Though ammunition was practically at an end, they nevertheless succeeded in delivering an attack at dawn on 21 September in an attempt to retake some of the houses. It failed and what remained of the 2nd Battalion scattered in small parties in an endeavour to find their way to the XXX Corps, which they had awaited so long and so vainly. At last the bridge was once more in German hands.

6 Lieut. J. H. Grayburn, V.C.

In this action the 2nd Battalion had been wiped out; but seldom can a fighting unit of any army in any age have had so glorious an end. For thrice the length of time laid down in its orders it had held a bridge against odds which were overwhelming from the beginning. Buoyed up by hope and by frequent messages that relief or support was on the way, either at the hands of the rest of the 1st Parachute Brigade and later the 4th Brigade, or from XXX Corps moving up from Nijmegen, when that hope was deferred, the hearts of its officers and men were not sick. They continued to fight, and only ceased to fire when their ammunition was gone and their wounded, now the great majority, faced with a fearful and unnecessary death. The conduct of the 2nd Battalion at the Bridge at Arnhem is more than an inspiration or an example; it was the quintessence of all those qualities which the parachute soldier must possess and display if he is to justify his training and the trust reposed in him. So great a spirit in evidence every moment of those three September days and nights can be overcome only by weight of numbers. That, and that alone, was the cause of their glorious defeat.

Anthony Deane-Drummond

Return Ticket

Deane-Drummond was dropped at Arnhem where he was captured – and then escaped.

THE game was up and I was a prisoner again. Once more I went through the indignity of being searched by the enemy, but on this occasion I at least had no weapons to surrender. I had to let go my sten gun and pistol while swimming across the Rhine.

No longer was I a free man and the anti-climax suddenly made me remember my hunger, and how every bone and muscle in my body ached for rest. Wearily, so wearily, I was marched down the road to the Company H.Q., watched by sleepy-eyed Germans from slit trenches dug into the verge.

We stopped at a farmhouse and I was shown into a room, after pushing aside a blackout blanket which had been nailed over the doorway. Inside, a hot aroma of unwashed bodies, the acrid stench of stale German tobacco smoke and seasoned sausage combined to stifle my nostrils. A hurricane lamp turned low gave the only light, and in the gloom my smarting eyes could now see gently-heaving bodies wrapped up in greatcoats lying all over the floor, with mounds of equipment taking up every vacant space. The only sounds were wheezes and snores except for the faint noise of a conversation in German coming from a next-door room.

One of my escort of three pushed by and, after mumbling what I took to be swear words, woke up one of the prostrate Germans. He turned out to be an N.C.O. and was soon kicking the other bodies to life, who grunted in a dialect I did not understand, and then stood up and stretched. A piece of paper passed hands and I was off again out of the house with three new guards to Battalion H.Q. which was about a mile away.

This time everything was much more orderly, and after a

German sentry had examined the piece of paper carried by my escort, we went down some steps into a cellar whose roof had been chocked up with large baulks of timber. A clean-shaven, middle-aged German subaltern sat at a table with a lamp on one corner. He motioned me to sit down and said in broken English:

"I must to you questions ask. You will answer."

"Oh."

"Your name, please?"

I told him.

"What day you jumped?"

"I can't say."

"How many more are you?"

"I can't say."

His eyes seemed to bulge a bit behind his glasses, and an angry flush spread up his neck.

"O.K. You no speak. We will see."

He ended with some instructions in German and I was shown outside into the back of an open *Volkswagen* car, in which I was driven along the road towards Arnhem.

We crossed the Rhine using the main bridge for which so many lives had been sacrificed. I could see many marks of the bloody fighting which had taken place as we threaded our way in and out of shell holes and burnt-out German tanks. Smoke was still coming from the ruins of the buildings on the north side of the river.

We sped on through deserted streets to the outskirts of the town and stopped outside a newly-built church which had sentries posted all round it. I was told to get out and wait inside. There I found the church full of newly-captured prisoners of war standing in little groups everywhere. In one corner I could see a few officers, none of whom I knew, and I learned from them that the division was now fighting inside a small perimeter round Oosterbeek, a suburb about three miles from the centre of Arnhem.

In another corner I saw Lance-Corporal Turner and three others who had shared the lavatory[1] with me. They, too, had been captured that morning in various places not far from where I had been taken. All touch had been lost while swimming across, and Turner had been caught while trying to find a hiding-place in a farmhouse. Daylight had come before he had reached a point anywhere near the railway bridge.

1 During the house-to-house fighting at Arnhem.

We all looked pretty scruffy in that church. I had a five-days' growth of beard, not having had a chance to shave, and many were like me. All had the slightly haggard and drawn look of soldiers who have been without sleep, and seen their best friends die, not knowing when their own turn might come. Some were rummaging in their pockets or haversacks for any crumbs left over from the once-despised 48-hour concentrated ration that we all carried. Many were lying down full-length on pews fast asleep, snoring away with mouths slightly open and heads twisted at any angle.

As the morning drew on, the air in the church became warmer, and more and more of us lay down where we were on the hard tiles and went fast asleep. I followed suit after checking that all doors were guarded and there was no way of getting out.

The Germans still gave us nothing to eat and by midday we were all getting very hungry and thirsty. Some men I spoke to asked me if I could get the Germans to do something about it.

After some argument I managed to get hold of an officer who could speak some English and in a mixture of two languages I told him that we expected to be given food within an hour, or else I would see that his name was remembered after the war when the time came to deal with the war criminals who disobeyed the Geneva Convention.

He became quite angry and spluttered:

"You can all think yourselves lucky to be alive and you will get food when it pleases us. Anyhow, what do you know about Geneva Conventions?"

"You would be surprised," I replied, "but food we must have, and it is your responsibility to provide."

"Let me tell you, Herr Major, I have just received orders to march you all to a prison near here run by the S.S. I am sure they will feed you."

With a glint in his eye, he turned on his heel, and five minutes later we were on the march with guards on all sides. For two miles we went through the suburbs and saw very few civilians, one or two of whom were brave enough to wave and smile as we went by.

Eventually we arrived at a house on the outskirts of Arnhem in another suburb called Velp. This was used as a prisoner-of-war cage and was guarded by an under-strength company of fifty-five men. It was a typical large suburban house, about twenty yards back from the main road and with exactly similar ones on either side. Two monkey-puzzle trees stood on the front lawn.

Inside the house were about five hundred all ranks of the Division, whose spirits were high except for the ignominy of being prisoners. Here I met Freddie Gough, Tony Hibbert, and many others. I learned all their news and told them mine. The Germans fed us on tins of lard and coarse brown bread, but we were not fussy and I wolfed my share down. I had not had a really square meal since leaving England, my last being breakfast on the 17th, and to-day was the 22nd. What months it all seemed and yet it was only five days.

I heard that the bridge had been captured by the Germans soon after dawn on the 21st, when nearly all the original defenders were killed or wounded, and all ammunition had been expended. Colonel Johnny Frost had himself been wounded, and for the last twenty-four hours Freddy Gough had been in command. For three days and nights this gallant force had held out against overwhelming odds, including tanks, which came up and gradually knocked down or set on fire every house that was being used for the defence. Some of these tanks had been stalked on foot and blown up with grenades. Fighting patrols had gone out every night to drive the Germans out of houses which overlooked the bridge. Deeds of heroism were done which are matchless in the history of the British Army, but received little publicity at the time because nobody returned to tell the tale. The Division had been ordered to hold the bridge for forty-eight hours until the arrival of the Second Army. It had been held for seventy-two hours by six hundred men, but unfortunately to no avail.

We now realized what a failure the whole operation had been but we still hoped that the Division could hold on where it was and provide Second Army with a bridgehead from which the advance could continue. Many were our speculations on what was happening to the rest of our units still fighting, but our hearts were heavy and we could not help thinking about ourselves and our present plight.

I remember the latrines inside the house were hopelessly inadequate for the numbers of men, and some deep trenches had been dug at the bottom of the garden at the back of the house. In this garden were growing carrots and onions, and we quickly dug these up and distributed them on the basis of half a carrot or onion per head. It took some of the hunger away.

All this time I was looking for ways out of the house or garden. I was determined to escape and not be a prisoner longer than I could help. Now would be the time and it would be infinitely easier than later on.

Some of the officers were already saying that they would leave trying to escape till they arrived at the German prison camp. It would all be "laid on" there. It is so easy to put off action till to-morrow and all this sort of talk was so reminiscent of my experiences in Italy. I told everybody I saw that their one and only chance of getting away would be before they left Holland. The farther they went back along the evacuation channels, the more difficult would escape become. I think they believed me, but most of them could not see any possible way out with any hope of success. When I started looking over the whole house and the garden there were many smiles cast in my direction. It was not possible to get away, they said, they had already been over the place with a fine-tooth comb. The trouble was that most of them were numbed by the anti-climax of being prisoners, and they did not realize that small though the chances of getting away were at the moment, they would be better now than at any future date.

I reasoned that the cage would only be temporary and would last as long as the Division did. From all accounts this would not be long, so one solution would be to hide up in the house itself till the Germans left and then to get out. Again it was just possible that the Second Army would continue their advance through the Division's bridgehead, and then the area would be liberated.

I could not see any way to escape that gave a better than fifty-fifty chance of success, so I looked everywhere for a hiding place that would hold me for two or three days. The only possible place seemed to be a wall-cupboard in one of the ground floor rooms, which had a flush-fitting concealed door. The whole door was covered with the same sort of wallpaper as that of the rest of the room, and was difficult to see except on close examination. The cupboard was about four feet across, twelve inches deep, and about seven feet high. Its interior was divided horizontally by adjustable shelves, but by removing the shelves I was able to stand inside in tolerable comfort. Fastening the door was a problem. The cupboard was fitted with the normal type of mortice lock let into the thickness of the door, with a keyhole on the outside complete with key. By unscrewing the lock, and turning it back to front, the keyhole came on the inside of the door and I was able to lock myself in. A piece of wall-paper, torn from another part of the room and pasted over the outside keyhole, helped to conceal the cupboard's presence.

The next job was to lay in a stock of water and food. All I had was my waterbottle, and I found an old two-pound jam jar that I also filled up. A one-pound tin of lard and half a small loaf of bread completed all the provisioning I could do. Some of the officers very kindly offered to give me their waterbottles, but I refused. They would need them for their own escape, which, I reminded them, they must try to make or be a prisoner for the rest of the war.

Little did I think that I would be confined to my cramped little cupboard for thirteen days and nights before getting out. I thought that the limit of my endurance would be reached after three or four days, because I did not start off in the best condition for an endurance test. The Germans came round on the evening of the 22nd to take all names, and in order to avoid a record being taken I started standing in my cupboard. Pole squatting is, I believe, a time-honoured sport in the U.S.A. I cannot recommend cupboard-standing to anybody who wants to try out something new. I stood first on one leg, then on the other; I leaned on one shoulder and then on the other. There was no room to sit down because the cupboard was too shallow. I managed to sleep all right, although occasionally my knees would give way and would drop forward against the door, making a hammer-like noise. Every bone in my body ached, and I felt quite light-headed from lack of food, water and rest.

The day after I locked myself in the cupboard the Germans turned the room into an interrogation centre. Every officer and man going through that cage was first interrogated in the room where my cupboard was. It was certainly an interesting experience, which I believe had never before been rivalled, though I scarcely appreciated its uniqueness at the time. We in the army had always been instructed that if ever we were made prisoner, the only information that we should give would be our army number, rank and name. The Germans knew this, of course, but tried every guile to get more information. The usual trick was to pretend that they were filling out a card for the Red Cross, and ask a series of innocuous questions until the prisoner was at ease, when a question of military importance would suddenly pop up. It was a surprising thing to me that very few officers or men gave only their number, rank and name. Almost everybody gave a little additional harmless information, such as the address of their parents or wives, or whether they were regular soldiers or had been in the T.A. before the war.

Only two gave away military information. One was a captain in the Glider Pilots, and another was the batman to a company commander of the leading battalion of relieving Second Army. This battalion had assaulted across the Rhine opposite the division's perimeter in order to allow the successful withdrawal of the division. These two men, who shall be nameless, gave all the information they knew or were capable of giving. Luckily neither was in possession of any real military secrets and no great harm was done, except to my pride. The officer talked so much, and seemed so promising a source of information, that he was given lunch just in front of my cupboard door. What agonies of mind and tummy! To hear all this coming out, and to smell what seemed to be a delicious meal only a few yards from my hiding place. I nearly burst out of the cupboard on several occasions to stop the wretch giving information. I think I would have done so if he had started to say anything serious. Luckily he did not know much and I kept my peace and exercised self-control over my mental anguish.

The questioning went on for several days, four or five I think, and by night the room was used as sleeping quarters for the German guard. I had no chances to get out at all, but as I had lasted so far, I resolved to try to remain a little longer. My luck must come to my rescue. It had always done so up till now.

Little by little I eked out my rations of water and bread. Four mouthfuls of water every four or five hours and just a bite or two of bread. The water was the chief shortage, and after nine or ten days I could not eat any more bread because my mouth was so dry. For the benefit of the curious, I was able to direct my urine through a paper funnel into one corner of the cupboard where there was a gap in the floorboards to allow some pipes to pass down to the cellar. It interested me to see that I continued to pass water in spite of drinking practically nothing. I did not feel the need to do anything more solid during the whole time, perhaps because there was nothing in my tummy. My system started to function again quite normally as soon as I started to eat when I got out. My only legacy was a series of bad boils, followed by styes which persisted for about a year afterwards.

It was now 5 October, 1944, and the thirteenth day of my

voluntary confinement. My water was nearly at an end, and the cramp in my muscles hurt acutely most of the time. Patience and caution were now finished and I told myself that I would have to make an attempt to escape that evening or fail in the effort.

The room outside my cupboard was still full of Germans but provided no new prisoners came in that evening there would be a good chance of the whole guard leaving the room empty for half an hour or so at sunset. On the previous evening they had all cleared out of the room and hung over the garden wall adjoining the main road outside my window, to watch the passers-by in the twilight. I suppose it is a world-wide habit to come out of the houses on a warm evening for a breather before going back inside for the night. The only thing that might spoil it would be new prisoners; but there had not been any last night, so with any luck I would get away to-night.

I slowly shifted my weight from one leg to the other, and leaned alternately on my right shoulder and then my left. By now, shifting my position had become almost automatic, and no longer required any thought or even consciousness. My mouth was dry as a bone, but I had already had both my dawn and midday mouthfuls. My evening one was not due for another two hours yet. To-night I would take three mouthfuls of water. What bliss this promised to be!

It was due to get dark about 7.30 p.m. or 8 p.m., and I hoped the room would clear by about 7 p.m. I would then have to hide up in the bushes near the house for an hour, till it was really dark, before it would be possible to move round to the back of the house and get away.

The minutes slowly crept by while I waited anxiously, my ears taut for the sound of the Germans leaving the room. Occasionally one of them would go in or out, but I could hear snores from two or three having an after-lunch nap. At about 6 o'clock I pulled on my boots and smock and gathered all my equipment. Dressing in that cupboard was a work of art, and to avoid making a noise it was three-quarters of an hour before I was ready. While I was dressing I heard two Germans stumble out of the room, but I was fairly certain that there were one or two more. Sure enough, by their grunts and the bumping of boots on the floor, I heard two more get up and go out talking about a *fräulein*.

The time had come. Cautiously I unlocked my door. There might still be the odd squarehead making up arrears of sleep. I opened the door an inch and had a quick look round. Damnation take it, there, not six feet away, was a solitary German soldier sleeping with his hands crossed over his tummy and his mouth wide open. As I had to walk across the floor and open the big French windows, which were both noisy operations, I decided to give him another half-hour.

A few troops came clattering into the building with a couple of girls, all talking at the tops of their voices. I heard them go upstairs and enter the room directly over my head, and they soon had quite a merry party going with songs and a gramophone, and an occasional girlish giggle or scream. I was in luck. They were probably not expecting any prisoners to-night, and if the noise increased as the wine flowed I should have no worries about covering up squeaks as I opened a window.

The noise upstairs woke my sleeping soldier after about twenty minutes, and he got up and walked out. This was my chance and, taking a couple of mouthfuls of water, I gently pushed the door open again. This time the room was empty. I could see the guards lining the garden fence on the main road and not ten yards away. My plan was to get the window open and then wait for a lorry or tank to go by before slipping out and into the shrubs growing almost under the sill. The Germans would be most unlikely to look back towards the house when anything interesting was passing.

I was in luck and no sooner had I opened the windows when a large truck went clattering by. This was my cue, and I was quickly out and had dropped into the shrubbery. My luck held good on the thirteenth day in that Dutch cupboard.

I quickly crawled into the bushes where it was thickest at the corner of the house, and concealed myself as best I could with dead leaves. From where I was I could see eight or ten of the guard idly leaning against the garden fence a few yards away and could hear them chatting unconcernedly about the war in general and their sweethearts at home. From the window above came occasional strains of gramophone records and the semi-delighted, semi-frightened squeals from the not-too-particular girls.

Kenneth Poolman

Zepp Sunday

Zeppelin raids on England began in 1915, and initially the airships seemed impervious to everything the Royal Flying Corps could hit them with. Not until September 1916, when RFC pilots were issued with new incendiary-explosive ammunition for their machine guns, was a Zeppelin shot down. The first "kill" went to Leefe Robinson.

At all points round the coast Zeppelins were coming in – at Skegness, over the Wash, at Wells, at Bacton, at Cromer and at Mundesley – and there were more Zepps still over the sea. Already the bombs were dropping. The long, broken night of violence had begun.

On a lonely field at Sutton's Farm, one of the R.F.C. stations guarding London, Lieutenant Robinson sprawled on his camp-bed beside his aircraft in the canvas hangar. William Leefe Robinson, at this time twenty-one years old, was born at Tollidetta, South Coory, Southern India, on July 14th, 1895, the youngest of a family of seven. His father, Horace Robinson, was in the Indian Civil Service.

At the age of fourteen he was sent to St. Bees School on the Cumberland coast, where he was a popular and outstanding house and rugger captain. He entered Sandhurst ten days after war had broken out, having crammed at Bournemouth in company with a son of Rudyard Kipling. By the end of the year he had been commissioned and gazetted to the Worcester Regiment. Transferring to the Royal Flying Corps, he was wounded in the arm over Lille while flying as an observer and turned his convalescence into a successful attempt at gaining his wings as a pilot. He had been a Home Defence pilot ever since. After a frustrating fight with a Zeppelin in May he had sworn that next time it would be "either the Zepp or I".

Leefe Robinson was a little too good, those who envied him might have thought, to be true. He was a young man of great charm, tall, athletic, handsome, with the kind of good looks associated with that figure of fun and tragedy, the Edwardian blade. It was easy to imagine the gay airman – "Robby" to his acquaintances, "Robin" to intimate friends – in straw boater and striped blazer punting down the Thames or the Isis, less easy to picture him in a smelly cockpit. He had fair, wavy hair. His mouth was rather feminine, his chin far from craggy, and his moustache no better than a good try. But his eyes gave a better clue to his character. They were blue and clear and very steady.

At eight minutes past eleven the telephone beside Robinson's bed tinkled and a drawling voice gave the order, "Take air-raid action."

No. 39 Squadron's standing order was for each flight of two aircraft at its various fields to send up one machine at two-hourly intervals. It was a system arranged to throw as wide a net as possible across the path of Zeppelins making for London.

Leefe Robinson's machine, B.E. 2c No. 2092, was quickly trundled out of the hangar and positioned for take-off. Robinson himself, an enthusiast to the last detail, supervised fuelling and final engine and airframe checks in the clammy, drifting ground fog that covered the tiny aerodrome.

The fog was quite thick and for a moment he wondered whether to cry off, but he thought that the air was probably reasonably clear a few yards off the ground and decided not to telephone. Around him in the shrouding mist he could see mechanics lighting the new, more efficient Money flares along the flight path. These wicks of asbestos and paraffin devised by the C.O., Major Higgins, burned longer and far brighter than the old buckets of petrol and shone through mist and fog. They would show him his landfall when he came down again. Still very cold, he flapped his arms vigorously across his chest and climbed stiffly up into the cockpit of the B.E.

He switched on the cockpit lighting. That was another new device for the night fliers. Not only were the dials lit up but the figures on them were daubed with luminous paint in case the lights failed. He checked the Lewis gun pointing up to the dim stars above and in front of his helmeted head, and saw that he had a full quota of drums. They all contained the new explosive-incendiary bullets mixed alternatively with tracer for aiming in the dark. He had all the gadgets tonight, he thought. If he couldn't do something with these marvels of science he ought to be sent back to the trenches.

He was ready. He pulled his goggles down over his eyes, and coughed, feeling the gritty fog in his mouth and throat. He stuck his head over the side.

"Petrol switches on."

"Petrol switches on." The mechanic echoed him hollowly through the vacuum in the mist.

"Suck in."

"Suck in."

"Contact."

"Contact."

The mechanic swung the heavy prop. The engine was cold and it took three attempts to start it.

He warmed up the engine, checked his instruments, then – "Chocks away."

"Chocks away." Mechanics jerked the blocks away, Robinson opened the throttle, and the machine was bumping over the grass. In a few minutes he was clear, and the mist had fallen away below him.

As Leefe Robinson was climbing out of the mist to begin his lonely watch under the stars a single Zeppelin which had crept in unheralded and crossed the Foulness Sands was over Coggeshall, halfway between Chelmsford and Colchester. Soon afterwards she turned west, passed over Saffron Walden, groping her way uncertainly on, and near Great Chesterford went round in a circle to try to find her bearings. Then she resumed her course west, picked up the Great Northern Railway line at Royston in Hertfordshire, and followed it as it bent south-west to Hitchin, to Luton, to the north-west fringes of London.

Luton heard her drone in the night about one o'clock. She was turning in towards the capital all the time now, the first of the raiders to approach. At London Colney, south of St. Albans, she dropped three high-explosive and three incendiary bombs to get her sights set and at the same time give her the extra buoyancy she would need to out-range the London guns. Five minutes later, heading due east, she dropped two high-explosive and two incendiaries in a wood at North Mimms. At one-twenty-eight she dropped an explosive and two incendiaries at Littleheath, where she cut a gas main and damaged two houses, then turned first north, then north-west, dropping three more bombs as she went.

At one-thirty-five she dropped two explosives and seven incendiaries on the Stud Farm at Clayhill, where a row of stables was set on fire and three valuable yearlings destroyed. She then steered westwards and dropped three incendiaries at Cockfosters, near the Enfield Isolation Hospital. At one-forty-five she crossed the Great Northern Railway line south of Hadley Wood, dropped two bombs, and turned and re-crossed the railway, going east. At ten minutes past two she dropped three bombs in a field at Southgate and turned south over Wood Green, then east.

As the lone airship passed just south of Alexandra Palace she was picked up by the Finsbury Park and Victoria Park lights. The Finsbury Park gun at once opened fire on her and other guns in the North and Central London defences followed. The guns were so accurate, especially the Temple House gun, that they forced her to turn off to the north-east over Tottenham.

At twelve minutes past two she was over Edmonton and dropped six high-explosives there. One of these fell in the grounds of Ely's explosive works, but it did not go off. Two minutes later, retreating north, she dropped two high-explosives at Ponders End, where she broke tramway and telephone wires and badly damaged the road and a water main. She bombed Enfield Highway and had just released twelve high-explosives at Forty Hill and Turkey Street when an aircraft dived upon her, coming from the south-east.

At ten minutes past one Leefe Robinson's lonely, uneventful watch had been suddenly and dramatically relieved. Somewhere south-east of Woolwich he had spotted an airship caught by two searchlights heading north-east towards the Gravesend-Tilbury area. The clouds had gathered and thickened in that part and he saw that the searchlights were having a difficult time keeping up with the raider. By this time he had managed to climb to 12,900 feet, and he steered in the direction of the Zeppelin and the few scattered shrapnel blossoms in the sky below it.

He flew on for ten minutes, slowly gaining on the Zepp. He judged it to be about 200 feet below and decided to hold on to his height advantage rather than risk a fast dive at the target, knowing he could never out-climb a Zepp once he had fallen below her. A moment later he was regretting his caution. A thicker belt of cloud rolled up, the Zeppelin plunged into it, and aeroplane and gunners were all left groping blindly. They never picked her up again, and after another quarter of an hour Leefe Robinson returned to his patrol, chagrined at losing the Zeppelin, but too generally cold and cramped to feel very angry at his error in tactics.

About ten minutes to two he noticed a fiery glow in the north-east. Thinking that where there was fire there might be a Zepp, he headed for it. He had been flying for fifteen minutes when one of the searchlights flicking to and fro out of the blackness below him halted and fixed on the long, dark shape of the Foulness Zeppelin as it came nosing out of a smudge of cloud like an Ouse chub from the shadows of a mud-bank prowling for food. Robinson had had one lesson, and he did not make the same mistake again. Engine full on, he put his nose down and steered straight for the Zepp. As he drew closer he noticed that the night tracer shells were all curving either too high above or too far below the airship's limelit hull.

A special constable standing on high ground in south-east London thought that the Zepp caught in the searchlights looked like a bar of polished steel about the thickness of an engine piston-rod. For a time it seemed to hang there motionless while the shells burst all round it. Then it made to turn as if in the direction of the coast, but a shell exploded near its nose and it swung round in the opposite direction. Its tail dipped and it made to climb, and a shell burst right over it. It sank lower, and three shells burst below, behind and in front of it. The Zepp seemed to wriggle from side to side, unable to escape. Then it seemed as if a black shadow passed between the constable's vision and the brilliant light so far up in the sky. Then he looked again. The Zepp was gone and the little shining rod of steel had disappeared. The firing ceased and the searchlights swung round the sky.

As the lights appeared to lose the Zeppelin there was a moment of intense silence in all the streets of London, then a great sigh of disappointment.

In this lull, when a second raider seemed about to escape, and her captain probably thought he was clear, Leefe Robinson's B.E. bucked and cavorted in the fierce blast of shrapnel near misses as he dived for the bow of the airship. Mackay from the R.F.C. station at North Weald, who had taken off at one o'clock, had seen the Zepp held in a searchlight and given chase. When he was within a mile of the airship he caught a brief glimpse of a B.E. heading at full speed for the bows of the Zepp.

In the early days of air defence, when the defending pilots' lack of success had first caused a great outcry from press and public, a dramatic instruction had appeared in the orders of one squadron, running thus, "If the aeroplane fails to stop the airship by the time all ammunition is expended, and the airship is still heading for London, then the pilot must decide to sacrifice himself and his machine and ram the airship at the utmost speed." The B.E. that Mackay saw fleetingly was going so fast to the bull's-eye that he thought the pilot meant to ram.

Leefe Robinson was diving to position himself beneath the giant hull. Blind attacks from above had been proved useless. The new idea was to rake the underbelly of the airship with bullets. He levelled off 800 feet underneath the nose and flew straight and level the length of the Zeppelin immediately beneath its control and engine cars, spraying it with explosive-incendiary. He used up one drum on this pass, then looked behind him, but could see no apparent result. When he had changed drums he climbed, turned, and flew right down the great looming flank of the roaring monster, emptying another drum into it. Still no flame or flicker of fire upon her. She plunged on like some huge Moby Dick trailing the toothpick lances of her attackers.

When he had reloaded again he tried a different tactic. This time he positioned himself behind the stern of the airship. He was only about 500 feet from her and concentrated the whole of the third drum on one spot on the belly near the monstrous tail fins. He saw the incandescent tracer hitting the hull. She seemed utterly invulnerable. The last round left the muzzle and the Lewis fell silent.

Just then he saw a sullen red glow inside the envelope of the Zepp in the spot he had been firing at. In a few seconds the whole after part was ablaze. Flame a hundred feet high whooshed up in front of his eyes, roaring and furnace hot.

In a street in Dulwich heads were popping out of all the windows. People were moving about in the street, calling to each other.

A man called out, "It's half-past two, dear!"

Then a woman screamed, "Look, look!"

High up and miles away in the sky a tongue of fire had appeared. Like a firework at a carnival a lick of flame ran round the sausage shape of a Zepp. Its body shone silver in the light.

In Walthamstow they saw the searchlights blink out. There was no need for them. The whole capital was illuminated by a vast, elongated, yellow flaming torch. In one narrow street a woman cried:

"Thank God. She's done for!"

A child's voice shrieked, "We've got her!"

In East Ham a boy of twelve stood among gaping men and women in pyjamas and nightdresses and saw what seemed like a furnace door suddenly opened in the night and a mass of red-hot coals exposed. Flames burst out all round the lighted body of the Zepp. There were loud cheers. Railway engines blew fancy shrieking salutes on their whistles. He thought it was just like Children's Empire Day.

All over south-east London they watched the ball of fire burning in the sky to the north. The ball swelled in size. Suddenly there was a great white burst of fire. The whole of London was illuminated as if being photographed by a giant flashlight. The dome of St. Paul's, the towers of Westminster, the silver serpent of the river, all stood out, and for the brief second of that awful flash a ghastly panorama of London was thrown on to the black screen of the night.

From Staines to Southend people gasped at the terrible vision in the sky. Then cheering broke out like a great forest fire spreading over the capital. In many a street voices began "God Save the King". Children screamed wildly, men and women danced and shouted hysterically. A spontaneous upsurge of something like New Year's Eve and Mafeking Night knocked into one and multiplied many times gripped London.

For a moment the doomed Zepp retained straight and level flight like a great bar of iron fresh from the furnace. Then it could be seen descending slowly in a shallow dive, nose down.

Below it Leefe Robinson fought the controls in a desperate effort to dodge the vast blazing mass which everywhere filled his vision. He managed with only a second to spare to side-slip out of harm's way. The crackling, roaring thing fell past him and it was like feeling the terrifying heat of the flames of hell on his face. He was not cold for a long time after that. Stunned and shaken, he watched the awful wreck sink slowly away out of the sky forever.

Exhilaration surged through him. In a burst of joy he grabbed his Very light pistol and fired off round after round of red cartridges, dropped a parachute flare, and put the B.E. into a wild loop. The people below saw the little red stars of victory and the brilliant snowflake falling and cheered their unknown champion.

Second Lieutenant Frederick Sowrey, who had taken off from Sutton's Farm half an hour after his flight commander, saw a small yellow flame in the sky and thought immediately that Robinson's engine had caught fire. Then the distant flame turned red, glowed, and grew larger, and he knew it was a Zepp burning. Mackay in his B.E. a few hundred yards away had time to notice that the whole vault of the sky from horizon to horizon was bathed red with fire, and the clouds below glared an unpleasant flesh pink. He afterwards chased a departing Zepp himself north-east of Hainault and was about to open fire when she disappeared in cloud. Hunt from the R.F.C. station at Hainault Farm got even closer to Leefe Robinson's victim than Mackay, so close in fact that he was about to fire into her when he saw her begin to burn. He broke off, and in the spreading pool of light sighted another a short distance away, well illuminated by the flames. But the glare was so fierce that it blinded him, and when he had recovered the second Zeppelin had left the zone of light. Later he chased a third airship, but lost her too.

The wild relief of seeing a nightmare Zepp destroyed in flames purged Londoners' deepest fears of the menace. It had the reverse effect upon the other airships pressing in from the rim of London.

L.16 was steering for London in the wake of the Foulness Zeppelin and only a short distance behind her. When L.16's commander saw the ship ahead on fire he turned at once and steered north at his maximum speed. Hunt in his B.E. saw him and gave chase until he lost him in the clouds.

When Captain Ernst Lehmann arrived over the London suburbs in the L.Z.98 the continued din of bombs and gunfire told him that several other Zeppelins were already in action. He saw the spectacle as an endless sea of houses lying under a silvery fog pierced by the flashes of bomb-bursts and blazing fires. The other airships were hidden, and the searchlights were groping about the sky, their conical rays passing through each other like bodiless ghosts.

He felt as if he were sitting in a theatre box, with the brightly lit stage before him and the darkened auditorium below. A whole hour went by before he ventured to come over the city.

Steering from one cloud bank to another to avoid the guns and lights, he dropped his bombs on the docks. Over the Thames, searchlights picked him up and shrapnel drove him away to the north-east.

He was in the chartroom bending over the maps to plot their course home when he heard his Staff Officer, von Gemmingen, cry out wildly. Lehmann looked behind him and saw a bright ball of flame in the sky. He knew that the blazing meteor on the further rim of the city could only be one of their ships. He, too, lost no time in heading for the coast.

Peterson in the L.32 had just fixed his course by the reservoir at Tring and was heading for London when he saw the great fire in the sky some miles ahead of him near the city. The sight knocked the heart out of him, and he very soon turned away and steered for home, unloading his bombs on Hertford as he fled.

The light-headed crowds watched the Zepp, hanging in the heavens like an enormous serpent writhing in a framework of erupting flame, continue to fall away northwards. Sometimes the crimson flame took on a bluish tinge, sometimes it became streaked with yellow. The whole of South London stared until its eyes ached. When the glowing mass finally sank out of sight behind the roof tops people fancied they could smell burning wood and fabric. In Dulwich a man with a pedantic turn of mind rushed into the house and brought out a copy of the *Daily Mail*, which he triumphantly proceeded to read in the street by the glare in the sky. He was fourteen miles away from the dying Zepp.

The whole of Chiswick, even further away, was lit up as if by a gigantic bonfire on Guy Fawkes Night. The blazing airship was seen distinctly at Gravesend, thirty miles away. The glow in the sky shed its light on places forty, even fifty miles distant.

Theodore Roosevelt

The Rough Riders at Santiago

An assistant secretary of the Navy, Roosevelt raised the "Rough Riders" to fight in the war against the Spanish in Cuba in 1898. Following the "Rough Riders" brief but illustrious campaign Roosevelt returned to the USA a national hero and shortly afterwards became the 26th President of the USA.

ON June 30th we received orders to hold ourselves in readiness to march against Santiago, and all the men were greatly overjoyed, for the inaction was trying. The one narrow road, a mere muddy track along which the army was encamped, was choked with the marching columns. As always happened when we had to change camp, everything that the men could not carry, including, of course, the officers' baggage, was left behind.

About noon the Rough Riders struck camp and drew up in column beside the road in the rear of the First Cavalry. Then we sat down and waited for hours before the order came to march, while regiment after regiment passed by, varied by bands of tatterdemalion Cuban insurgents, and by mule-trains with ammunition. Every man carried three days' provisions. We had succeeded in borrowing mules sufficient to carry along the dynamite gun and the automatic Colts.

At last, toward mid-afternoon, the First and Tenth Cavalry, ahead of us, marched, and we followed. The First was under the command of Lieutenant-Colonel Veile, the Tenth under Lieutenant-Colonel Baldwin. Every few minutes there would be a stoppage in front, and at the halt I would make the men sit or lie down beside the track, loosening their packs. The heat was intense as we passed through the still, close jungle, which formed a wall on either hand. Occasionally we came to gaps or open spaces, where some regiment was camped, and now and then one of these regiments, which apparently had been left out of its proper place, would file into the road, breaking up our line of march. As a result, we finally found ourselves following merely the tail of the regiment ahead of us, an infantry regiment being thrust into the interval. Once or twice we had to wade streams. Darkness came on, but we still continued to march. It was about eight o'clock when we turned to the left and climbed El Poso hill, on whose summit there was a ruined ranch and sugar factory, now, of course, deserted. Here I found General Wood, who was arranging for the camping of the brigade. Our own arrangements for the night were simple. I extended each troop across the road into the jungle, and then the men threw down their belongings where they stood and slept on their arms. Fortunately, there was no rain. Wood and I curled up under our rain-coats on the saddle-blankets, while his two aides, Captain A. L. Mills and Lieutenant W. N. Ship, slept near us. We were up before dawn and getting breakfast. Mills and Ship had nothing to eat, and they breakfasted with Wood and myself, as we had been able to get some handfuls of beans, and some coffee and sugar, as well as the ordinary bacon and hardtack.

We did not talk much, for though we were in ignorance as to precisely what the day would bring forth, we knew that we should see fighting. We had slept soundly enough, although, of course, both Wood and I during the night had made a round of the sentries, he of the brigade, and I of the regiment; and I suppose that, excepting among hardened veterans, there is always a certain feeling of uneasy excitement the night before the battle.

Mills and Ship were both tall, fine-looking men, of tried courage, and thoroughly trained in every detail of their profession; I remember being struck by the quiet, soldierly way they were going about their work early that morning. Before noon one was killed and the other dangerously wounded.

General Wheeler was sick, but with his usual indomitable pluck and entire indifference to his own personal comfort, he kept to the front. He was unable to retain command of the cavalry division, which accordingly devolved upon General Samuel Sumner, who commanded it until mid-afternoon, when the bulk of the fighting was over. General Sumner's own brigade fell to Colonel Henry Carroll. General Sumner led the advance with the cavalry, and the battle was fought by him and by General Kent, who commanded the infantry division, and whose foremost brigade was led by General Hawkins.

As the sun rose the men fell in, and at the same time a battery of field-guns was brought up on the hill-crest just beyond, between us and toward Santiago. It was a fine sight to see the great horses straining under the lash as they whirled the guns up the hill and into position.

Our brigade was drawn up on the hither side of a kind of half basin, a big band of Cubans being off to the left. As yet we had received no orders, except that we were told that the main fighting was to be done by Lawton's infantry division, which was to take El Caney, several miles to our right, while we were simply to make a diversion. This diversion was to be made mainly with the artillery, and the battery which had taken position immediately in front of us was to begin when Lawton began.

It was about six o'clock that the first report of the cannon from El Caney came booming to us across the miles of still jungle. It was a very lovely morning, the sky of cloudless blue, while the level, shimmering rays from the just-risen sun brought into fine relief the splendid palms which here and there towered above the lower growth. The lofty and beautiful mountains hemmed in the Santiago plain, making it an amphitheatre for the battle.

Immediately our guns opened, and at the report great clouds of white smoke hung on the ridge crest. For a minute or two there was no response. Wood and I were sitting together, and Wood remarked to me that he wished our brigade could be moved somewhere else, for we were directly in line of any return fire aimed by the Spaniards at the battery. Hardly had he spoken when there was a peculiar whistling, singing sound in the air, and immediately afterward the noise of something exploding over our heads. It was shrapnel from the Spanish batteries. We sprung to our feet and leaped on our horses. Immediately afterward a second shot came which burst directly above us; and then a third. From the second shell one of the shrapnel bullets dropped on my wrist, hardly breaking the skin, but raising a bump about as big as a hickory-nut. The same shell wounded four of my regiment, one of them being Mason Mitchell, and two or three of the regulars were also hit, one losing his leg by a great fragment of shell. Another shell exploded right in the middle of the Cubans, killing and wounding a good many, while the remainder scattered like guinea-hens. Wood's lead horse was also shot through the lungs. I at once hustled my regiment over the crest of the hill into the thick underbrush, where I had no little difficulty in getting them together again into column.

Meanwhile the firing continued for fifteen or twenty minutes, until it gradually died away. As the Spaniards used smokeless powder, their artillery had an enormous advantage over ours, and, moreover, we did not have the best type of modern guns, our fire being slow.

As soon as the firing ceased, Wood formed his brigade, with my regiment in front, and gave me orders to follow behind the First Brigade, which was just moving off the ground. In column of fours we marched down the trail toward the ford of the San Juan River. We passed two or three regiments of infantry, and were several times halted before we came to the ford. The First Brigade, which was under Colonel Carroll – Lieutenant-Colonel Hamilton commanding the Ninth Regiment, Major Wessels the Third, and Captain Kerr the Sixth – had already crossed and was marching to the right, parallel to, but a little distance from, the river. The Spaniards in the trenches and block-houses on top of the hills in front were already firing at the brigade in desultory fashion. The extreme advance of the Ninth Cavalry was under Lieutenants McNamee and Hartwick. They were joined by General Hawkins, with his staff, who was looking over the ground and deciding on the route he should take his infantry brigade.

Our orders had been of the vaguest kind, being simply to march to the right and connect with Lawton – with whom, of course, there was no chance of our connecting. No reconnaissance had been made, and the exact position and strength of the Spaniards was not known. A captive balloon was up in the air at this moment, but it was worse than useless. A previous proper reconnaissance and proper look-out from the hills would have given us exact information. As it was, Generals Kent, Sumner, and Hawkins had to be their own reconnaissance, and they fought their troops so well that we won anyhow.

I was now ordered to cross the ford, march half a mile or so to the right, and then halt and await further orders; and I promptly hurried my men across, for the fire was getting hot, and the captive balloon, to the horror of everybody, was coming down to the ford. Of course, it was a special target for the enemy's fire. I got my men across before it reached the ford. There it partly collapsed and remained, causing severe loss of life, as it indicated the exact position where the Tenth and the First Cavalry, and the infantry, were crossing.

As I led my column slowly along, under the intense heat, through the high grass of the open jungle, the First Brigade was to our left, and the firing between it and the Spaniards on the hills grew steadily hotter and hotter. After a while I came to a sunken lane, and as by this time the First Brigade had stopped and was engaged in a stand-up fight, I halted my men and sent back word for orders. As we faced toward the Spanish hills my regiment was on the right with next to it and a little in advance the First Cavalry, and behind them the Tenth. In our front the Ninth held the right, the Sixth the centre, and the Third the left; but in the jungle the lines were already overlapping in places. Kent's infantry were coming up, farther to the left.

Captain Mills was with me. The sunken lane, which had a wire fence on either side, led straight up toward, and between, the two hills in our front, the hill on the left, which contained heavy block-houses, being farther away from us than the hill on our right, which we afterward grew to call Kettle Hill, and which was surmounted merely by some large ranch buildings or haciendas, with sunken brick-lined walls and cellars. I got the men as well sheltered as I could. Many of them lay close under the bank of the lane, others slipped into the San Juan River and crouched under its hither bank, while the rest lay down behind the patches of bushy jungle in the tall grass. The heat was intense, and many of the men were already showing signs of exhaustion. The sides of the hills in front were bare; but the country up to them was, for the most part, covered with such dense jungle that in charging through it no accuracy of formation could possibly be preserved.

The fight was now on in good earnest, and the Spaniards on the hills were engaged in heavy volley firing. The Mauser bullets drove in sheets through the trees and the tall jungle grass, making a peculiar whirring or rustling sound; some of the bullets seemed to pop in the air, so that we thought they were explosive; and, indeed, many of those which were coated with brass did explode, in the sense that the brass coat was ripped off, making a thin plate of hard metal with a jagged edge, which inflicted a ghastly wound. These bullets were shot from a 45-calibre rifle carrying smokeless powder, which was much used by the guerillas and irregular Spanish troops. The Mauser bullets themselves made a small clean hole, with the result that the wound healed in a most astonishing manner. One or two of our men who were shot in the head had the skull blown open, but elsewhere the wounds from the minute steel-coated bullet, with its very high velocity, were certainly nothing like as serious as those made by the old large-calibre, low-power rifle. If a man was shot through the heart, spine, or brain he was, of course, killed instantly; but very few of the wounded died – even under the appalling conditions which prevailed, owing to the lack of attendance and supplies in the field-hospitals with the army.

While we were lying in reserve we were suffering nearly as much as afterward when we charged. I think that the bulk of the Spanish fire was practically unaimed, or at least not aimed at any particular man, and only occasionally at a particular body of men; but they swept the whole field of battle up to the edge of the river, and man after man in our ranks fell dead or wounded, although I had the troopers scattered out far apart, taking advantage of every scrap of cover.

Devereux was dangerously shot while he lay with his men on the edge of the river. A young West Point cadet, Ernest Haskell, who had taken his holiday with us as an acting second lieutenant, was shot through the stomach. He had shown great coolness and gallantry, which he displayed to an even more marked degree after being wounded, shaking my hand and saying: "All right, Colonel, I'm going to get well. Don't bother about me, and don't let any man come away with me." When I shook hands with him, I thought he would surely die; yet he recovered.

The most serious loss that I and the regiment could have suffered befell just before we charged. Bucky O'Neill was strolling up and down in front of his men, smoking his cigarette, for he was inveterately addicted to the habit. He had a theory that an officer ought never to take cover – a theory which was, of course, wrong, though in a volunteer organization the officers should certainly expose themselves very fully, simply for the effect on the men; our regimental toast on the transport running, "The officers; may the war last until each is killed, wounded, or promoted." As O'Neill moved to and fro, his men begged him to lie down, and one of the sergeants said, "Captain, a bullet is sure to hit hit you." O'Neill took his cigarette out of his mouth, and blowing out a cloud of smoke laughed and said, "Sergeant, the Spanish bullet isn't made that will kill me." A little later he discussed for a moment with one of the regular officers the direction from which the Spanish fire was coming. As he turned on his heel a bullet struck him in the mouth and came out at the back of his head; so that even before he fell his wild and gallant soul had gone out into the darkness.

My orderly was a brave young Harvard boy, Sanders, from the quaint old Massachusetts town of Salem. The work of an orderly on foot, under the blazing sun, through the hot and matted jungle, was very severe, and finally the heat overcame him. He dropped; nor did he ever recover fully, and later he died from fever. In his place I summoned a trooper whose name I did not know. Shortly afterward, while sitting beside the bank, I directed him to go back and ask whatever general he came across if I could not advance, as my men were being much cut up. He stood up to salute and then pitched forward across my knees, a bullet having gone through his throat, cutting the carotid.

When O'Neill was shot, his troop, who were devoted to him, were for the moment at a loss whom to follow. One of their number, Henry Bardshar, a huge Arizona miner, immediately attached himself to me as my orderly, and from that moment he was closer to me, not only in the fight, but throughout the rest of the campaign, than any other man, not even excepting the color-sergeant, Wright.

Captain Mills was with me; gallant Ship had already been killed. Mills was an invaluable aide, absolutely cool, absolutely unmoved or flurried in any way.

I sent messenger after messenger to try to find General Sumner or General Wood and get permission to advance, and was just about making up my mind that in the absence of orders I had better "march toward the guns," when Lieutenant-Colonel Dorst came riding up through the storm of bullets with the welcome command "to move forward and support the regulars in the assault on the hills in front." General Sumner had obtained authority to advance from Lieutenant Miley, who was representing General Shafter at the front, and was in the thick of the fire. The General at once ordered the first brigade to advance on the hills, and the second to support it. He himself was riding his horse along the lines, superintending the fight. Later I overheard a couple of my men talking together about him. What they said illustrates the value of a display of courage among the officers in hardening their soldiers; for their theme was how, as they were lying down under a fire which they could not return, and were in consequence feeling rather nervous, General Sumner suddenly appeared on horseback, sauntering by quite unmoved; and, said one of the men, "That made us feel all right. If the General could stand it, we could."

The instant I received the order I sprang on my horse and then my "crowded hour" began. The guerillas had been shooting at us from the edges of the jungle and from their perches in the leafy trees, and as they used smokeless powder, it was almost impossible to see them, though a few of my men had from time to time responded. We had also suffered from the hill on our right front, which was held chiefly by guerillas, although there were also some Spanish regulars with them, for we found their dead. I formed my men in column of troops, each troop extended in open skirmishing order, the right resting on the wire fences which bordered the sunken lane. Captain Jenkins led the first squadron, his eyes literally dancing with joyous excitement.

I started in the rear of the regiment, the position in which the colonel should theoretically stay. Captain Mills and Captain McCormick were both with me as aides; but I speedily had to send them off on special duty in getting the different bodies of men forward. I had intended to go into action on foot as at Las Guasimas, but the heat was so oppressive that I found I should be quite unable to run up and down the line and superintend matters unless I was mounted; and, moreover, when on horseback, I could see the men better and they could see me better.

A curious incident happened as I was getting the men started forward. Always when men have been lying down under cover for some time, and are required to advance, there is a little hesitation, each looking to see whether the others are going forward. As I rode down the line, calling to the troopers to go forward, and rasping brief directions to the captains and lieutenants, I came upon a man lying behind a little bush, and I ordered him to jump up. I do not think he understood that we were making a forward move, and he looked up at me for a moment with hesitation, and I again bade him rise, jeering him and saying: "Are you afraid to stand up when I am on horseback?" As I spoke, he suddenly fell forward on his face, a bullet having struck him and gone through him lengthwise. I suppose the bullet had been aimed at me; at any rate, I, who was on horseback in the open, was unhurt, and the man lying flat on the ground in the cover beside me was killed. There were several pairs of brothers with us; of the two Nortons one was killed; of the two McCurdys one was wounded.

I soon found that I could get that line, behind which I personally was, faster forward than the one immediately in front of it, with the result that the two rearmost lines of the regiment began to crowd together; so I rode through them both, the better to move on the one in front. This happened with every line in succession, until I found myself at the head of the regiment.

Both lieutenants of B Troop from Arizona had been exerting themselves greatly, and both were overcome by the heat; but Sergeants Campbell and Davidson took it forward in splendid shape. Some of the men from this troop and from the other Arizona troop (Bucky O'Neill's) joined me as a kind of fighting tail.

The Ninth Regiment was immediately in front of me, and the First on my left, and these went up Kettle Hill with my regiment. The Third, Sixth, and Tenth went partly up Kettle Hill (following the Rough Riders and the Ninth and First), and partly between that and the block-house hill, which the infantry were assailing. General Sumner in person gave the Tenth the order to charge the hills; and it went forward at a rapid gait. The three regiments went forward more or less intermingled, advancing steadily and keeping up a heavy fire. Up Kettle Hill Sergeant George Berry, of the Tenth, bore not only his own regimental colors but those of the Third, the color-sergeant of the Third having been shot down; he kept shouting, "Dress on the colors, boys, dress on the colors!" as he followed Captain Ayres, who was running in advance of his men, shouting and waving his hat. The Tenth Cavalry lost a greater proportion of its officers than any other regiment in the battle – eleven out of twenty-two.

By the time I had come to the head of the regiment we ran into the left wing of the Ninth Regulars, and some of the First Regulars, who were lying down; that is, the troopers were lying down, while the officers were walking to and fro. The officers of the white and colored regiments alike took the greatest pride in seeing that the men more than did their duty; and the mortality among them was great.

I spoke to the captain in command of the rear platoons, saying that I had been ordered to support the regulars in the attack upon the hills, and that in my judgment we could not take these hills by firing at them, and that we must rush them. He answered that his orders were to keep his men lying where they were, and that he could not charge without orders. I asked where the Colonel was, and as he was not in sight, said, "Then I am the ranking officer here and I give the order to charge" – for I did not want to keep the men longer in the open suffering under a fire which they could not effectively return. Naturally the captain hesitated to obey this order when no word had been received from his own Colonel. So I said, "Then let my men through, sir," and rode on through the lines, followed by the grinning Rough Riders, whose attention had been completely taken off the Spanish bullets, partly by my dialogue with the regulars, and partly by the language I had been using to themselves as I got the lines forward, for I had been joking with some and swearing at others, as the exigencies of the case seemed to demand. When we started to go through, however, it proved too much for the regulars, and they jumped up and came along, their officers and troops mingling with mine, all being delighted at the chance. When I got to where the head of the left wing of the Ninth was lying, through the courtesy of Lieutenant Hartwick, two of whose colored troopers threw down the fence, I was enabled to get back into the lane, at the same time waving my hat, and giving the order to charge the hill on our right front. Out of my sight, over on the right, Captains McBlain and Taylor, of the Ninth, made up their minds independently to charge at just about this time; and at almost the same moment Colonels Carroll and Hamilton, who were off, I believe, to my left, where we could see neither them nor their men, gave the order to advance. But of all this I knew nothing at the time. The whole line, tired of waiting, and eager to close with the enemy, was straining to go forward; and it seems that different parts slipped the leash at almost the same moment. The First Cavalry came up the hill just behind, and partly mixed with my regiment and the Ninth. As already said, portions of the Third, Sixth, and Tenth followed, while the rest of the members of these three regiments kept more in touch with the infantry on our left.

By this time we were all in the spirit of the thing and greatly excited by the charge, the men cheering and running forward between shots, while the delighted faces of the foremost officers, like Captain C. J. Stevens, of the Ninth, as they ran at the head of their troops, will always stay in my mind. As soon as I was in the line I galloped forward a few yards until I saw that the men were well started, and then galloped back to help Goodrich, who was in command of his troop, get his men across the road so as to attack the hill from that side. Captain Mills had already thrown three of the other troops of the regiment across this road for the same purpose. Wheeling around, I then again galloped toward the hill, passing the shouting, cheering, firing men, and went up the lane, splashing through a small stream; when I got abreast of the ranch buildings on the top of Kettle Hill, I turned and went up the slope. Being on horseback I was, of course, able to get ahead of the men on foot, excepting my orderly, Henry Bardshar, who had run ahead very fast in order to get better shots at the Spaniards, who were now running out of the ranch buildings. Sergeant Campbell and a number of the Arizona men, and Dudley Dean, among others, were very close behind. Stevens, with his platoon of the Ninth, was abreast of us; so were McNamee and Hartwick. Some forty yards from the top I ran into a wire fence and jumped off Little Texas, turning him loose. He had been scraped by a couple of bullets, one of which nicked my elbow, and I never expected to see him again. As I ran up to the hill, Bardshar stopped to shoot, and two Spaniards fell as he emptied his magazine. These were the only Spaniards I actually saw fall to aimed shots by any one of my men, with the exception of two guerillas in trees.

Almost immediately afterward the hill was covered by the troops, both Rough Riders and the colored troopers of the Ninth, and some men of the First. There was the usual confusion, and afterward there was much discussion as to exactly who had been on the hill first. The first guidons planted there were those of the three New Mexican troops, G, E, and F, of my regiment, under their Captains, Llewellen, Luna, and Muller, but on the extreme right of the hill, at the opposite end from where we struck it, Captains Taylor and McBlain and their men of the Ninth were first up. Each of the five captains was firm in the belief that his troop was first up. As for the individual men, each of whom honestly thought he was first on the summit, their name was legion. One Spaniard was captured in the buildings, another was shot as he tried to hide himself, and a few others were killed as they ran.

Among the many deeds of conspicuous gallantry here performed, two, both to the credit of the First Cavalry, may be mentioned as examples of the others, not as exceptions. Sergeant Charles Karsten, while close beside Captain Tutherly, the squadron commander, was hit by a shrapnel bullet. He continued on the line, firing until his arm grew numb; and he then refused to go to the rear, and devoted himself to taking care of the wounded, utterly unmoved by the heavy fire. Trooper Hugo Brittain, when wounded, brought the regimental standard forward, waving it to and fro, to cheer the men.

No sooner were we on the crest than the Spaniards from the line of hills in our front, where they were strongly intrenched, opened a very heavy fire upon us with their rifles. They also opened upon us with one or two pieces of artillery, using time fuses which burned very accurately, the shells exploding right over our heads.

On the top of the hill was a huge iron kettle, or something of the kind, probably used for sugar refining. Several of our men took shelter behind this. We had a splendid view of the charge on the San Juan block-house to our left, where the infantry of Kent, led by Hawkins, were climbing the hill. Obviously the proper thing to do was to help them, and I got the men together and started them volley-firing against the Spaniards in the San Juan block-house and in the trenches around it. We could only see their heads; of course this was all we ever could see when we were firing at them in their trenches. Stevens was directing not only his own colored troopers, but a number of Rough Riders; for in a mêlée good soldiers are always prompt to recognize a good officer, and are eager to follow him.

We kept up a brisk fire for some five or ten minutes; meanwhile we were much cut up ourselves. Gallant Colonel Hamilton, than whom there was never a braver man, was killed, and equally gallant Colonel Carroll wounded. When near the summit Captain Mills had been shot through the head, the bullet destroying the sight of one eye permanently and of the other temporarily. He would not go back or let any man assist him, sitting down where he was and waiting until one of the men brought him word that the hill was stormed. Colonel Veile planted the standard of the First Cavalry on the hill, and General Sumner rode up. He was fighting his division in great form, and was always himself in the thick of the fire. As the men were much excited by the firing, they seemed to pay very little heed to their own losses.

Suddenly, above the cracking of the carbines, rose a peculiar drumming sound, and some of the men cried, "The Spanish machine-guns!" Listening, I made out that it came from the flat ground to the left, and jumped to my feet, smiting my hand on my thigh, and shouting aloud with exultation, "It's the Gatlings, men, our Gatlings!" Lieutenant Parker was bringing his four Gatlings into action, and shoving them nearer and nearer the front. Now and then the drumming ceased for a moment; then it would resound again, always closer to San Juan hill, which Parker, like ourselves, was hammering to assist the infantry attack. Our men cheered lustily. We saw much of Parker after that, and there was never a more welcome sound than his Gatlings as they opened. It was the only sound which I ever heard my men cheer in battle.

The infantry got nearer and nearer the crest of the hill. At last we could see the Spaniards running from the rifle-pits as the Americans came on in their final rush. Then I stopped my men for fear they should injure their comrades, and called to them to charge the next line of trenches, on the hills in our front, from which we had been undergoing a good deal of punishment. Thinking that the men would all come, I jumped over the wire fence in front of us and started at the double; but, as a matter of fact, the troopers were so excited, what with shooting and being shot, and shouting and cheering, that they did not hear, or did not heed me; and after running about a hundred yards I found I had only five men along with me. Bullets were ripping the grass all around us, and one of the men, Clay Green, was mortally wounded; another, Winslow Clark, a Harvard man, was shot first in the leg and then through the body. He made not the slightest murmur, only asking me to put his water canteen where he could get at it, which I did; he ultimately recovered. There was no use going on with the remaining three men, and I bade them stay where they were while I went back and brought up the rest of the brigade. This was a decidedly cool request, for there was really no possible point in letting them stay there while I went back; but at the moment it seemed perfectly natural to me, and apparently so to them, for they cheerfully nodded, and sat down in the grass, firing back at the line of trenches from which the Spaniards were shooting at them. Meanwhile, I ran back, jumped over the wire fence, and went over the crest of the hill, filled with anger against the troopers, and especially those of my own regiment, for not having accompanied me. They, of course, were quite innocent of wrong-doing; and even while I taunted them bitterly for not having followed me, it was all I could do not to smile at the look of injury and surprise that came over their faces, while they cried out, "We didn't hear you, we didn't see you go, Colonel;

lead on now, we'll sure follow you." I wanted the other regiments to come too, so I ran down to where General Sumner was and asked him if I might make the charge; and he told me to go and that he would see that the men followed. By this time everybody had his attention attracted, and when I leaped over the fence again, with Major Jenkins beside me, the men of the various regiments which were already on the hill came with a rush, and we started across the wide valley which lay between us and the Spanish intrenchments. Captain Dimmick, now in command of the Ninth, was bringing it forward; Captain McBlain had a number of Rough Riders mixed in with his troop, and led them all together, Captain Taylor had been severely wounded. The long-legged men like Greenway, Goodrich, sharp-shooter Proffit, and others, outstripped the rest of us, as we had a considerable distance to go. Long before we got near them the Spaniards ran, save a few here and there, who either surrendered or were shot down. When we reached the trenches we found them filled with dead bodies in the light blue and white uniform of the Spanish regular army. There were very few wounded. Most of the fallen had little holes in their heads from which their brains were oozing; for they were covered from the neck down by the trenches.

It was at this place that Major Wessels, of the Third Cavalry, was shot in the back of the head. It was a severe wound, but after having it bound up he again came to the front in command of his regiment. Among the men who were foremost was Lieutenant Milton F. Davis, of the First Cavalry. He had been joined by three men of the Seventy-first New York, who ran up, and, saluting, said, "Lieutenant, we want to go with you, our officers won't lead us." One of the brave fellows was soon afterward shot in the face. Lieutenant Davis's first sergeant, Clarence Gould, killed a Spanish soldier with his revolver, just as the Spaniard was aiming at one of my Rough Riders. At about the same time I also shot one. I was with Henry Bardshar, running up at the double, and two Spaniards leaped from the trenches and fired at us, not ten yards away. As they turned to run I closed in and fired twice, missing the first and killing the second. My revolver was from the sunken battleship *Maine*, and had been given me by my brother-in-law, Captain W. S. Cowles, of the Navy. At the time I did not know of Gould's exploit, and supposed my feat to be unique; and although Gould had killed his Spaniard in the trenches, not very far from me, I never learned of it until weeks after. It is astonishing what a limited area of vision and experience one has in the hurly-burly of a battle.

There was very great confusion at this time, the different regiments being completely intermingled – white regulars, colored regulars, and Rough Riders. General Sumner had kept a considerable force in reserve on Kettle Hill, under Major Jackson, of the Third Cavalry. We were still under a heavy fire and I got together a mixed lot of men and pushed on from the trenches and ranch-houses which we had just taken, driving the Spaniards through a line of palm-trees, and over the crest of a chain of hills. When we reached these crests we found ourselves overlooking Santiago. Some of the men, including Jenkins, Greenway, and Goodrich, pushed on almost by themselves far ahead. Lieutenant Hugh Berkely, of the First, with a sergeant and two troopers, reached the extreme front. He was, at the time, ahead of everyone; the sergeant was killed and one trooper wounded; but the lieutenant and the remaining trooper stuck to their post for the rest of the afternoon until our line was gradually extended to include them.

While I was re-forming the troops on the chain of hills, one of General Sumner's aides, Captain Robert Howze – as dashing and gallant an officer as there was in the whole gallant cavalry division, by the way – came up with orders to me to halt where I was, not advancing farther, but to hold the hill at all hazards. Howze had his horse, and I had some difficulty in making him take proper shelter, he stayed with us for quite a time, unable to make up his mind to leave the extreme front, and meanwhile jumping at the chance to render any service, of risk or otherwise, which the moment developed.

I now had under me all the fragments of the six cavalry regiments which were at the extreme front, being the highest officer left there, and I was in immediate command of them for the remainder of the afternoon and that night. The Ninth was over to the right, and the Thirteenth Infantry afterward came up beside it. The rest of Kent's infantry was to our left. Of the Tenth, Lieutenants Anderson, Muller, and Fleming reported to me; Anderson was slightly wounded, but he paid no heed to this. All three, like every other officer, had troopers of various regiments under them; such mixing was inevitable in making repeated charges through thick jungle; it was essentially a troop commanders', indeed, almost a squad leaders', fight. The Spaniards who had been holding the trenches and the line of hills, had fallen back upon their supports and we were under a very heavy fire both from rifles and great guns. At the point where we were, the grass-covered hill-crest was gently rounded, giving poor cover, and I made my men lie down on the hither slope.

On the extreme left Captain Beck, of the Tenth, with his own troop, and small bodies of the men of other regiments, was exercising a practically independent command, driving back the Spaniards whenever they showed any symptoms of advancing. He had received his orders to hold the line at all hazards from Lieutenant Andrews, one of General Sumner's aides, just as I had received mine from Captain Howze. Finally, he was relieved by some infantry, and then rejoined the rest of the Tenth, which was engaged heavily until dark, Major Wint being among the severely wounded. Lieutenant W. N. Smith was killed. Captain Bigelow had been wounded three times.

Our artillery made one or two efforts to come into action on the firing-line of the infantry, but the black powder rendered each attempt fruitless. The Spanish guns used smokeless powder, so that it was difficult to place them. In this respect they were on a par with their own infantry and with our regular infantry and dismounted cavalry; but our only two volunteer infantry regiments, the Second Massachusetts and the Seventy-first New York, and our artillery, all had black powder. This rendered the two volunteer regiments, which were armed with the antiquated Springfield, almost useless in the battle, and did practically the same thing for the artillery wherever it was formed within rifle range. When one of the guns was discharged a thick cloud of smoke shot out and hung over the place, making an ideal target, and in a half minute every Spanish gun and rifle within range was directed at the particular spot thus indicated; the consequence was that after a more or less lengthy stand the gun was silenced or driven off. We got no appreciable help from our guns on July 1st. Our men were quick to realize the defects of our artillery, but they were entirely philosophic about it, not showing the least concern at its failure. On the contrary, whenever they heard our artillery open they would grin as they looked at one another and remark, "There go the guns again; wonder how soon they'll be shut up," and shut up they were sure to be. The light battery of Hotchkiss one-pounders, under Lieutenant J. B. Hughes, of the Tenth Cavalry, was handled with conspicuous gallantry.

On the hill-slope immediately around me I had a mixed force

composed of members of most of the cavalry regiments, and a few infantrymen. There were about fifty of my Rough Riders with Lieutenants Goodrich and Carr. Among the rest were perhaps a score of colored infantrymen, but, as it happened, at this particular point without any of their officers. No troops could have behaved better than the colored soldiers had behaved so far; but they are, of course, peculiarly dependent upon their white officers. Occasionally they produce non-commissioned officers who can take the initiative and accept responsibility precisely like the best class of whites; but this cannot be expected normally, nor is it fair to expect it. With the colored troops there should always be some of their own officers; whereas, with the white regulars, as with my own Rough Riders, experience showed that the non-commissioned officers could usually carry on the fight by themselves if they were once started, no matter whether their officers were killed or not.

At this particular time it was trying for the men, as they were lying flat on their faces, very rarely responding to the bullets, shells, and shrapnel which swept over the hill-top, and which occasionally killed or wounded one of their number. Major Albert G. Forse, of the First Cavalry, a noted Indian fighter, was killed about this time. One of my best men, Sergeant Greenly, of Arizona, who was lying beside me, suddenly said, "Beg pardon, Colonel; but I've been hit in the leg." I asked, "Badly?" He said, "Yes, Colonel; quite badly." After one of his comrades had helped him fix up his leg with a first-aid-to-the-injured bandage, he limped off to the rear.

None of the white regulars or Rough Riders showed the slightest sign of weakening; but under the strain the colored infantrymen (who had none of their officers) began to get a little uneasy and to drift to the rear, either helping wounded men, or saying that they wished to find their own regiments. This I could not allow, as it was depleting my line, so I jumped up, and walking a few yards to the rear, drew my revolver, halted the retreating soldiers, and called out to them that I appreciated the gallantry with which they had fought and would be sorry to hurt them, but that I should shoot the first man who, on any pretence whatever, went to the rear. My own men had all sat up and were watching my movements with utmost interest; so was Captain Howze. I ended my statement to the colored soldiers by saying: "Now, I shall be very sorry to hurt you, and you don't know whether or not I will keep my word, but my men can tell you that I always do"; whereupon my cow-punchers, hunters, and miners solemnly nodded their heads and commented in chorus, exactly as if in a comic opera, "He always does; he always does!"

This was the end of the trouble, for the "smoked Yankees" – as the Spaniards called the colored soldiers – flashed their white teeth at one another, as they broke into broad grins, and I had no more trouble with them, they seeming to accept me as one of their own officers. The colored cavalry-men had already so accepted me; in return, the Rough Riders, although for the most part Southwesterners, who have a strong color prejudice, grew to accept them with hearty good-will as comrades, and were entirely willing, in their own phrase, "to drink out of the same canteen." Where all the regular officers did so well, it is hard to draw any distinction; but in the cavalry division a peculiar meed of praise should be given to the officers of the Ninth and Tenth for their work, and under their leadership the colored troops did as well as any soldiers could possibly do.

In the course of the afternoon the Spaniards in our front made the only offensive movement which I saw them make during the entire campaign; for what were ordinarily called "attacks" upon our lines consisted merely of heavy firing from their trenches and from their skirmishers. In this case they did actually begin to make a forward movement, their cavalry coming up as well as the marines and reserve infantry, while their skirmishers, who were always bold, redoubled their activity. It could not be called a charge, and not only was it not pushed home, but it was stopped almost as soon as it began, our men immediately running forward to the crest of the hill with shouts of delight at seeing their enemies at last come into the open. A few seconds' firing stopped their advance and drove them into the cover of the trenches.

They kept up a very heavy fire for some time longer, and our men again lay down, only replying occasionally. Suddenly we heard on our right the peculiar drumming sound which had been so welcome in the morning, when the infantry were assailing the San Juan block-house. The Gatlings were up again! I started over to inquire, and found that Lieutenant Parker, not content with using his guns in support of the attacking forces, had thrust them forward to the extreme front of the fighting-line, where he was handling them with great effect. From this time on, throughout the fighting, Parker's Gatlings were on the right of my regiment, and his men and mine fraternized in every way. He kept his pieces at the extreme front, using them on every occasion until the last Spanish shot was fired. Indeed, the dash and efficiency with which the Gatlings were handled by Parker was one of the most striking features of the campaign; he showed that a first-rate officer could use machine-guns, on wheels, in battle and skirmish, in attacking and defending trenches, alongside of the best troops, and to their great advantage.

As night came on, the firing gradually died away. Before this happened, however, Captains Morton and Boughton, of the Third Cavalry, came over to tell me that a rumor had reached them to the effect that there had been some talk of retiring and that they wished to protest in the strongest manner. I had been watching them both, as they handled their troops with the cool confidence of the veteran regular officer, and had been congratulating myself that they were off toward the right flank, for as long as they were there, I knew I was perfectly safe in that direction. I had heard no rumor about retiring, and I cordially agreed with them that it would be far worse than a blunder to abandon our position.

To attack the Spaniards by rushing across open ground, or through wire entanglements and low, almost impassable jungle, without the help of artillery, and to force unbroken infantry, fighting behind earthworks and armed with the best repeating weapons, supported by cannon, was one thing; to repel such an attack ourselves, or to fight our foes on anything like even terms in the open, was quite another thing. No possible number of Spaniards coming at us from in front could have driven us from our position, and there was not a man on the crest who did not eagerly and devoutly hope that our opponents would make the attempt, for it would surely have been followed, not merely by a repulse, but by our immediately taking the city. There was not an officer or a man on the firing-line, so far as I saw them, who did not feel this way.

As night fell, some of my men went back to the buildings in our rear and foraged through them, for we had now been fourteen hours charging and fighting without food. They came across what was evidently the Spanish officers' mess, where their dinner was still cooking, and they brought it to the front in high glee. It was evident that the Spanish officers were living well, however the Spanish rank and file were faring. There were three big iron pots, one filled with beef-stew, one with boiled rice, and one with boiled peas; there was a big demijohn of rum (all along the trenches which the Spaniards held were empty wine and liquor bottles); there were a number of loaves of rice-bread; and there were even some small cans of preserves and a few salt fish. Of course, among so many men, the food, which was equally divided, did not give very much to each, but it freshened us all.

Soon after dark, General Wheeler, who in the afternoon had resumed command of the cavalry division, came to the front. A very few words with General Wheeler reassured us about retiring. He had been through too much heavy fighting in the Civil War to regard the present fight as very serious, and he told us not to be under any apprehension, for he had sent word that there was no need whatever of retiring, and was sure we would stay where we were until the chance came to advance. He was second in command; and to him more than to any other one man was due the prompt abandonment of the proposal to fall back – a proposal which, if adopted, would have meant shame and disaster.

Shortly afterward General Wheeler sent us orders to intrench. The men of the different regiments were now getting in place again and sifting themselves out. All of our troops who had been kept at Kettle Hill came forward and rejoined us after nightfall. During the afternoon Greenway, apparently not having enough to do in the fighting, had taken advantage of a lull to explore the buildings himself, and had found a number of Spanish intrenching tools, picks, and shovels, and these we used in digging trenches along our line. The men were very tired indeed, but they went cheerfully to work, all the officers doing their part.

Crockett, the ex-Revenue officer from Georgia, was a slight man, not physically very strong. He came to me and told me he didn't think he would be much use in digging, but that he had found a lot of Spanish coffee and would spend his time making coffee for the men, if I approved. I did approve very heartily, and Crockett officiated as cook for the next three or four hours until the trench was dug, his coffee being much appreciated by all of us.

So many acts of gallantry were performed during the day that it is quite impossible to notice them all, and it seems unjust to single out any; yet I shall mention a few, which it must always be remembered are to stand, not as exceptions, but as instances of what very many men did. It happened that I saw these myself. There were innumerable others, which either were not seen at all, or were seen only by officers who happened not to mention them; and, of course, I know chiefly those that happened in my own regiment.

Captain Llewellen was a large, heavy man, who had a grown-up son in the ranks. On the march he had frequently carried the load of some man who weakened, and he was not feeling well on the morning of the fight. Nevertheless, he kept at the head of his troop all day. In the charging and rushing, he not only became very much exhausted, but finally fell, wrenching himself terribly, and though he remained with us all night, he was so sick by morning that we had to take him behind the hill into an improvised hospital. Lieutenant Day, after handling his troop with equal gallantry and efficiency, was shot, on the summit of Kettle Hill. He was hit in the arm and was forced to go to the rear, but he would not return to the States, and rejoined us at the front long before his wound was healed. Lieutenant Leahy was also wounded, not far from him. Thirteen of the men were wounded and yet kept on fighting until the end of the day, and in some cases never went to the rear at all, even to have their wounds dressed. They were Corporals Waller and Fortescue and Trooper McKinley of Troop E; Corporal Roades of Troop D; Troopers Albertson, Winter, McGregor, and Ray Clark of Troop F; Troopers Bugbee, Jackson, and Waller of Troop A; Trumpeter McDonald of Troop L; Sergeant Hughes of Troop B; and Trooper Gievers of Troop G. One of the Wallers was a cow-puncher from New Mexico, the other the champion Yale high-jumper. The first was shot through the left arm so as to paralyze the fingers, but he continued in battle, pointing his rifle over the wounded arm as though it had been a rest. The other Waller, and Bugbee, were hit in the head, the bullets merely inflicting scalp wounds. Neither of them paid any heed to the wounds except that after nightfall each had his head done up in a bandage. Fortescue I was at times using as an extra orderly. I noticed he limped, but supposed that his foot was skinned. It proved, however, that he had been struck in the foot, though not very seriously, by a bullet, and I never knew what was the matter until the next day I saw him making wry faces as he drew off his bloody boot, which was stuck fast to the foot. Trooper Rowland again distinguished himself by his fearlessness.

For gallantry on the field of action Sergeants Dame, Ferguson, Tiffany, Greenwald, and, later on, McIlhenny, were promoted to second lieutenancies, as Sergeant Hayes had already been. Lieutenant Carr, who commanded his troop, and behaved with great gallantry throughout the day, was shot and severely wounded at nightfall. He was the son of a Confederate officer; his was the fifth generation which, from father to son, had fought in every war of the United States. Among the men whom I noticed as leading in the charges and always being nearest the enemy, were the Pawnee, Pollock, Simpson of Texas, and Dudley Dean. Jenkins was made major, Woodbury Kane, Day, and Frantz captains, and Greenway and Goodrich first lieutenants, for gallantry in action, and for the efficiency with which the first had handled his squadron, and the other five their troops – for each of them, owing to some accident to his superior, found himself in command of his troop.

Dr. Church had worked quite as hard as any man at the front in caring for the wounded; as had Chaplain Brown. Lieutenant Keyes, who acted as adjutant, did so well that he was given the position permanently. Lieutenant Coleman similarly won the position of quartermaster.

We finished digging the trench soon after midnight, and then the worn-out men laid down in rows on their rifles and dropped heavily to sleep. About one in ten of them had blankets taken from the Spaniards. Henry Bardshar, my orderly, had procured one for me. He, Goodrich, and I slept together. If the men without blankets had not been so tired that they fell asleep anyhow, they would have been very cold, for, of course, we were all drenched with sweat, and above the waist had on nothing but our flannel shirts, while the night was cool, with a heavy dew. Before anyone had time to wake from the cold, however, we were all awakened by the Spaniards, whose skirmishers suddenly opened fire on us. Of course, we could not tell whether or not this was the forerunner of a heavy attack, for our Cossack posts were responding briskly. It was about three o'clock in the morning, at which time men's courage is said to be at the lowest ebb; but the cavalry division was certainly free from any weakness in that direction. At the alarm everybody jumped to his feet and the stiff, shivering, haggard men, their eyes only half-opened, all clutched their rifles and ran forward to the trench on the crest of the hill.

The sputtering shots died away and we went to sleep again.

But in another hour dawn broke and the Spaniards opened fire in good earnest. There was a little tree only a few feet away, under which I made my head-quarters, and while I was lying there, with Goodrich and Keyes, a shrapnel burst among us, not hurting us in the least, but with the sweep of its bullets killing or wounding five men in our rear, one of whom was a singularly gallant young Harvard fellow, Stanley Hollister. An equally gallant young fellow from Yale, Theodore Miller, had already been mortally wounded. Hollister also died.

The Second Brigade lost more heavily than the First; but neither its brigade commander nor any of its regimental commanders were touched, while the commander of the First Brigade and two of its three regimental commanders had been killed or wounded.

In this fight our regiment had numbered 490 men, as, in addition to the killed or wounded of the first fight, some had had to go to the hospital for sickness and some had been left behind with the baggage, or were detailed on other duty. Eighty-nine were killed or wounded: the heaviest loss suffered by any regiment in the cavalry division. The Spaniards made a stiff fight, standing firm until we charged home. They fought much more stubbornly than at Las Guasimas. We ought to have expected this, for they have always done well in holding intrenchments. On this day they showed themselves to be brave foes, worthy of honor for their gallantry.

In the attack on the San Juan hills our forces numbered about 6,600. There were about 4,500 Spaniards against us. Our total loss in killed or wounded was 1,071. Of the cavalry division there were, all told, some 2,300 officers and men, of whom 375 were killed or wounded. In the division over a fourth of the officers were killed or wounded, their loss being relatively half as great again as that of the enlisted men – which was as it should be.

I think we suffered more heavily than the Spaniards did in killed or wounded (though we also captured some scores of prisoners). It would have been very extraordinary if the reverse was the case, for we did the charging; and to carry earthworks on foot with dismounted cavalry, when these earthworks are held by unbroken infantry armed with the best modern rifles, is a serious task.

Virgil

The Trojan Horse

The story of the Wooden Horse, by means of which the Greeks gained access to Troy, is told in Virgil's Aeneid.

THE Grecian leaders, now disheartened by the war, and baffled by the Fates, after a revolution of so many years, build a horse to the size of a mountain, and interweave its ribs with planks of fir. This they pretend to be an offering, in order to procure a safe return; which report spread. Hither having secretly conveyed a select band, chosen by lot, they shut them up into the dark sides, and fill its capacious caverns and womb with armed soldiers. In sight of Troy lies Tenedos, an island well known by fame, and flourishing while Priam's kingdom stood: now only a bay, and a station unfaithful for ships. Having made this island, they conceal themselves in that desolate shore. We imagined they were gone, and that they had set sail for Mycenae. In consequence of this, all Troy is released from its long distress: the gates are thrown open; with joy we issue forth, and view the Grecian camp, the deserted plains, and the abandoned shore. Some view with amazement that baleful offering of the virgin Minerva, and wonder at the stupendous bulk of the horse; and Thymoetes first advises that it be dragged within the walls and lodged in the tower, whether with treacherous design, or that the destiny of Troy now would have it so. But Capys, and all whose minds had wiser sentiments, strenuously urge either to throw into the sea the treacherous snare and suspected oblation of the Greeks; or by applying flames consume it to ashes; or to lay open and ransack the recesses of the hollow womb. The fickle populace is split into opposite inclinations. Upon this, Laocoön, accompanied with numerous troops, first before all, with ardour hastens down from the top of the citadel; and while yet a great way off cries out, "O, wretched countrymen, what desperate infatuation is this? Do you believe the enemy gone? Or think you any gifts of the Greeks can be free from deceit? Is Ulysses thus known to you? Either the Greeks lie concealed within this wood, or it is an engine framed against our walls, to overlook our houses, and to come down upon our city; or some mischievous design lurks beneath it. Trojans, put no faith in this horse. Whatever it be, I dread the Greeks, even when they bring gifts." Thus said, with valiant strength he hurled his massive spear against the sides and belly of the monster, where it swelled out with its jointed timbers; the weapon stood quivering, and the womb being shaken, the hollow caverns rang, and sent forth a groan. And had not the decrees of heaven been adverse, if our minds had not been infatuated, he had prevailed on us to mutilate with the sword this dark recess of the Greeks; and thou, Troy, should still have stood, and thou, lofty tower of Priam, now remained!

In the meantime, behold, Trojan shepherds, with loud acclamations, came dragging to the king a youth, whose hands were bound behind him; who, to them a mere stranger, had voluntarily thrown himself in the way, to promote this same design, and open Troy to the Greeks; a resolute soul, and prepared for either event, whether to execute his perfidious purpose, or submit to inevitable death. The Trojan youth pour tumultuously around from every quarter, from eagerness to see him, and they vie with one another in insulting the captive. Now learn the treachery of the Greeks, and from one crime take a specimen of the whole nation. For as he stood among the gazing crowds perplexed, defenceless, and threw his eyes around the Trojan bans, "Ah!" says he, "what land, what seas can now receive me? Or to what further extremity can I, a forlorn wretch, be reduced, for whom there is no shelter anywhere among the Greeks? And to complete my misery the Trojans too, incensed against me, sue for satisfaction with my blood." By which mournful accents our affections at once were moved towards him, and all our resentment suppressed.

At these tears we grant him his life, and pity him from our hearts. Priam himself first gives orders that the manacles and strait bonds be loosened from the man, then thus addresses him in the language of a friend: "Whoever you are, now henceforth forget the Greeks you have lost; ours you shall be: and give me an ingenuous reply to these questions: To what purpose raised they this stupendous bulk of a horse? Who was the contriver? Or what do they intend? What was the religious motive? Or what warlike engine is it?" he said. The other, practised in fraud and Grecian artifice, lifted up to heaven his hands, loosed from the bonds. "Troy can never be razed by the Grecian sword, unless they repent the omens at Argos, and carry back the goddess whom they had conveyed in their curved ships. And now, that they have sailed for their native Mycenae with the wind, they are providing themselves with arms; and, they will come upon you unexpected." For he declared that "if your hands should violate this offering sacred to Minerva, then signal ruin awaited Priam's empire and the Trojans. But, if by your hands it mounted into the city, that Asia, without further provocation given, would advance with a formidable war to the very walls, and our posterity be doomed to the same fate." By such treachery and artifice of perjured Sinon, the story was believed: and we, whom neither Diomede, nor Achilles, nor a siege of ten years, nor a thousand ships, had subdued, were ensnared by guile and constrained tears.

Meanwhile they urge with general voice to convey the statue to its proper seat, and implore the favour of the goddess. We make a breach in the walls, and lay open the bulwarks of the city. All keenly ply the work; and under the feet apply smooth-rolling wheels; stretch hempen ropes from the neck. The fatal machine passes over our walls, pregnant with arms. It advances, and with menacing aspect slides into the heart of the city. O country, O Ilium, the habitation of gods, and ye walls of Troy by war renowned! Four times it stopped in the very threshold of the gate, and four times the arms resounded in its womb: yet we, heedless, and blind with frantic zeal, urge on, and plant the baneful monster in the sacred citadel. Unhappy we, to whom that day was to be the last, adorn the temples of the gods throughout the city with festive boughs. Meanwhile, the heavens change, and night advances rapidly from the ocean, wrapping in her extended shade both earth and heaven, and the wiles of the Myrmidons. The Trojans, dispersed about the walls, were hushed: deep sleep fast binds them weary in his embraces. And now the Grecian host, in their equipped vessels, set out for Tenedos, making towards the well-known shore, by the friendly silence of the quiet moonshine, as soon as the royal galley stern had exhibited the signal fire; and Sinon, preserved by the will of the adverse gods, in a stolen hour unlocks the wooden prison to the Greeks shut up in its tomb: the horse, from his expanded caverns, pours them forth to the open air. They assault the city buried in sleep, and wine. The sentinels are beaten down; and with opened gates they receive all their friends, and join the conquering bands.

Meanwhile the city is filled with mingled scenes of woe; and though my father's house stood retired and enclosed with trees, louder and louder the sounds rise on the ear, and the horrid din of arms assails. I start from sleep and, by hasty steps, gain the highest battlement of the palace, and stand with erect ears: as when a flame is driven by the furious south winds on standing corn; or as a torrent impetuously bursting in a mountain-flood desolates the fields, desolates the rich crops of corn and the labours of the ox.

Then, indeed, the truth is confirmed and the treachery of the Greeks disclosed. Now Deiphosus' spacious house tumbles down, overpowered by the conflagration; now, next to him, Ucalegon blazes: the straits of Sigaeum shine far and wide with the flames. The shouts of men and clangour of trumpets arise. My arms I snatch in mad haste: nor is there in arms enough of reason: but all my soul burns to collect a troop for the war and rush into the citadel with my fellows: fury and rage hurry on my mind, and it occurs to me how glorious it is to die in arms.

The towering horse, planted in the midst of our streets, pours forth armed troops; and Sinon victorious, with insolent triumph scatters the flames. Others are pressing at our wide-opened gates, as many thousands as ever came from populous Mycenae: others with arms have blocked up the lanes to oppose our passage; the edged sword, with glittering point, stands unsheathed, ready for dealing death: hardly the foremost wardens of the gates make an effort to fight and resist in the blind encounter. By the impulse of the gods, I hurry away into flames and arms, whither the grim Fury, whither the din and shrieks that rend the skies, urge me on. Ripheus and Iphitus, mighty in arms, join me; Hypanis and Dymas come up with us by the light of the moon, and closely adhere to my side. Whom, close united, soon as I saw resolute to engage, to animate them the more I thus begin: "Youths, souls magnanimous in vain! If it is your determined purpose to follow me in this last attempt, you see what is the situation of our affairs. All the gods, by whom this empire stood, have deserted their shrines and altars to the enemy: you come to the relief of the city in flames: let us meet death, and rush into the thickest of our armed foes. The only safety for the vanquished is to throw away all hopes of safety." Thus the courage of each youth is kindled into fury. Then, like ravenous wolves in a gloomy fog, whom the fell rage of hunger hath driven forth, blind to danger, and whose whelps left behind long for their return with thirsting jaws; through arms, through enemies, we march up to imminent death, and advance through the middle of the city: sable Night hovers around us with her hollow shade.

Who can describe in words the havoc, who the death of that night? Or who can furnish tears equal to the disasters? Our ancient city, having borne sway for many years, falls to the ground: great numbers of sluggish carcasses are strewn up and down, both in the streets, in the houses, and the sacred thresholds of the gods. Nor do the Trojans alone pay the penalty with their blood: the vanquished too at times resume courage in their hearts, and the victorious Grecians fall: everywhere is cruel sorrow, everywhere terror and death in a thousand shapes.

We march on, mingling with the Greeks, but not with heaven on our side; and in many a skirmish we engage during the dark night: many of the Greeks we send down to Hades. Some fly to the ships, and hasten to the trusty shore; some through dishonest fear, scale once more the bulky horse, and lurk within the well-known womb.

Ye ashes of Troy, ye expiring flames of my country! Witness, that in your fall I shunned neither darts nor any deadly chances of the Greeks. Thence we are forced away, forthwith to Priam's palace called by the outcries. Here, indeed, we beheld a dreadful fight, as though this had been the only seat of the war, as though none had been dying in all the city besides; with such ungoverned fury we see Mars raging and the Greeks rushing forward to the palace, and the gates besieged by an advancing testudo. Scaling ladders are fixed against the walls, and by their steps they mount to the very door-posts, and protecting themselves by their left arms, oppose their bucklers to the darts, while with their right hands they grasp the battlements. On the other hand, the Trojans tear down the turrets and roofs of their houses; with these weapons, since they see the extremity, they seek to defend themselves now in their last death-struggle, and tumble down the gilded rafters; others with drawn swords beset the gates below; these they guard in a firm, compact body. I mount up to the roof of the highest battlement, whence the distressed Trojans were hurling unavailing darts. With our swords assailing all around a turret, situated on a precipice, and shooting up its towering top to the stars (whence we were wont to survey all Troy, the fleet of Greece, and all the Grecian camp), where the topmost story made the joints more apt to give way, we tear it from its deep foundation, and push it on our foes. Suddenly tumbling down, it brings thundering desolation with it, and falls with wide havoc on the Grecian troops. But others succeed: meanwhile, neither stones, nor any sort of missile weapons, cease to fly. Just before the vestibule, and at the outer gate, Pyrrhus exults, glittering in arms and gleamy brass. At the same time, all the youth from Scyros advance to the wall, and toss brands to the roof. Pyrrhus himself in the front, snatching up a battleaxe, beats through the stubborn gates, and labours to tear the brazen posts from the hinges; and now, having hewn away the bars, he dug through the firm boards, and made a large, wide-mouthed breach. The palace within is exposed to view, and the long galleries are discovered: the sacred recesses of Priam and the ancient kings are exposed to view; and they see armed men standing at the gate.

As for the inner palace, it is filled with mingled groans and doleful uproar, and the hollow rooms all throughout howl with female yells: their shrieks strike the golden stars. Then the trembling matrons roam through the spacious halls, and in embraces hug the door-posts, and cling to them with their lips. Pyrrhus presses on with all his father's violence: nor bolts, nor guards themselves, are able to sustain. The gate, by repeated battering blows, gives way, and the door-posts, torn from their hinges, tumble to the ground. The Greeks make their way by force, burst a passage, and, being admitted, butcher the first they meet, and fill the places all about with their troops. Those fifty bedchambers, those doors, that proudly shone with barbaric gold and spoils, were leveled to the ground: where the flames relent, the Greeks take their place.

Perhaps, too, you are curious to hear what was Priam's fate. As soon as he beheld the catastrophe of the taken city, and his palace gates broken down, and the enemy planted in the middle of his private apartments, the aged monarch, with unavailing aim, buckles on his shoulders (trembling with years) arms long disused, girds himself with his useless sword, and rushes into the thickest of the foes, resolute on death. And lo! Polites, one of Priam's sons, who had escaped from the sword of Pyrrhus, through darts, through foes, flies along the long galleries, and wounded traverses the waste halls. Pyrrhus, all afire, pursues him with the hostile weapon, is just grasping him with his hand, and presses on him with the spear. Soon as he at length got into the sight and presence of his parents, he dropped down, and poured out his life with a stream of blood. Upon this, Priam, though now held in the very midst of death, yet did not forbear, nor spared his tongue and passion; and, without any force, threw a feeble dart: which was instantly repelled by the hoarse brass, and hung on the highest boss of the buckler without any execution. Pyrrhus made answer and dragged him to the very altar, trembling and sliding in the streaming gore of his son: and with his left hand grasped his twisted hair, and with his right unsheathed his glittering sword, and plunged it into his side up to the hilt. Such was the end of Priam's fate: this was the final doom allotted to him, having before his eyes Troy consumed, and its towers laid in ruins; once the proud monarch over so many nations and countries of Asia: now his mighty trunk lies extended on the shore, the head torn from the shoulders, and a nameless corpse.

T. J. Waldron & James Gleason

The Frogmen

Manned torpedoes were first developed by the Italian Navy, who used them to considerable effect in attacks on the British fleet at Gibraltar and Alexandria in 1941. Three years later, the British decided to repay the compliment – with a two-man torpedo attack on La Spezia, the birthplace of the "human torpedo".

In February 1944, three crews of charioteers established a secret base at San Vito, at the entrance to Taranto harbour. One of the crews, consisting of Lt. M. R. Causer and Seaman Harry Smith, was chosen to attack the Italian 10,000-ton cruiser *Bolzano* which was lying in the heavily defended anchorage of La Spezia. The *Bolzano* was the last of the eight-inch gun cruisers with which Italy had entered the war, and she was at that time German controlled.

Causer and Smith made contact with the captain of an Italian Motor Torpedo Boat which had been based on La Spezia before coming over to the Allies when Italy surrendered. The plan was for the M.T.B. to drop them within two miles of the harbour entrance. The entrance to the harbour consisted of a mile-long channel with a breakwater and boom defences stretched across at the mouth of the harbour. Causer and Smith were to attack the cruiser – get ashore – lie low all day – and then make their way about a mile up the coast where they were told they would find a large rock about two hundred yards out to sea. They were to swim to the rock, and during the night a fast motor-boat would come in and take them off. That, however, is not the way things worked out.

The two-man torpedo was placed aboard the M.T.B., together with the diving equipment, and from Naples they headed for Corsica, which was held at that time by Allied troops. On

22nd June they sailed from Corsica in an Italian destroyer with the M.T.B. in attendance, and about twenty miles from the Italian coast they transferred to the M.T.B. for the last leg of the journey to La Spezia.

Shortly afterwards the M.T.B. stopped and the captain said that that was as far as he could possibly go. He had let them down, he promised to go within two miles of the harbour, but he evidently got frightened, and he dropped them at least seven miles from the harbour – which didn't help the original plan much. The torpedo was lowered over the side and Causer and Smith got aboard, and with a final wave to the crew of the M.T.B. headed for La Spezia at 22.30 hrs.

They travelled on the surface until they sighted the harbour entrance, when they took a bearing and dived to twenty feet and proceeded at that depth for about thirty minutes. When they surfaced again they were well within the harbour approaches, and heading for the breakwater and boom defences. Searchlights swept the sky and tracers from two shore-based guns were firing out to sea. They dived quickly and carried on towards the breakwater. They had not been going long before they heard the sound of an approaching engine. They slowed down and held their breath while a motor launch passed directly over them. After a while they gently broke surface again, and sighted the harbour entrance to the right of the breakwater; they dived once more and nosed their way towards the entrance.

Here they met their first obstacle, an anti-submarine net, stretched right across the entrance, and down to the sea-bed. They dived to thirty feet and started cutting their way through the net. Unfortunately that was not the only net. Altogether they encountered at least half a dozen more, set at different angles, but they hacked and struggled and wormed their way through one net after another until finally they had penetrated the harbour boom defences. They were inside La Spezia.

Slowly they surfaced, and there was the target lying in the centre of the harbour. Lt. Causer turned round in his seat, and he and Smith solemnly shook hands. Cutting through the nets must have taken several hours and they were well behind schedule; in fact it was almost daylight. The Italian captain had certainly let them down very badly by dropping them so far out to sea; it had taken much longer than they had anticipated to reach the harbour.

However, there was no turning back now, so they dived to

twenty feet and slowly approached the ship until they could see her dark outline on the surface. Quietly they came up beneath her great propellers and pulled themselves along until they were just forward of 'midships, underneath the bridge. They placed four magnets on the bottom of the ship, then released the warhead of the torpedo and lashed it to the magnets. Lt. Causer then set the time fuse for two hours to give them time to get clear; they mounted the torpedo and dropped down to thirty feet.

It was at this point that they discovered that the batteries that operated the torpedo were very low indeed, and that they would not be able to make the shore. The plan had been to sink the machine in deep water, get ashore, hide the diving gear and then lie low for a time. This latest discovery, however, caused a complete change of plan. They managed to get to the breakwater where they scuttled the torpedo. They then surfaced near some rocks, climbed out of the water and took off their diving gear. They had been in the water seven and a half hours. They were both all in, so they slept for about an hour and a half on the rocks. Then at 06.30 hrs. the explosion occurred. Two large water spouts shot up well above the level of the bridge, she shuddered, started to settle, and then rolled over on to her starboard side. She sank in a quarter of an hour, leaving her port side showing above the water. Thus the wheel had turned the full circle – as this ship of war was destroyed in the centre of the very harbour from which this form of warfare had started.

Meanwhile, Causer and Smith were still quite literally on the rocks. It was about half a mile to the shore, but it would have been impossible to swim it because of the activity in the harbour following the explosion so they stayed where they were and hoped for the best. Later in the day they were spotted by an Italian fisherman and his small son. After some parleying the fisherman agreed to get another boat for them. They had no option but to trust him, and sure enough he returned later with another rowing boat in tow. They got into the boat and mingled with the Italian fishermen in the harbour all day, and at night set out with the intention of rowing to Corsica, ninety miles away. They kept going all night, and the next day they made a landfall, which they thought was Corsica. After they beached the boat they asked a small boy if this was Corsica – only to learn that they were still in Italy. They had in fact travelled some twenty miles farther down the coast.

After wandering and hiding, with the notion of making vaguely in the direction of the Allied lines, they met a woman who put them in touch with some Italian guerrillas. The two Britishers joined the guerrillas and fought with them for about six weeks. The band that they had been fighting with was practically wiped out in a skirmish, so they decided to try and make our own lines and set out in civilian clothes with this intention. They carried their battledress in a sack as an additional precaution because they had both been warned that if they were picked up in civilian clothes they would be shot as spies. They carried on daily getting nearer and nearer to the front line, when by the greatest stroke of bad luck they were picked up by a German patrol in forbidden territory and taken in for interrogation. They were suspected of having been involved in the sinking of the *Bolzano*, but they both stoutly maintained that they were survivors of a big submarine.

Eventually Smith was sent to a prison camp in Bremen and later to Lubeck where he was put in solitary confinement, and otherwise badly used in an endeavour to get him to admit complicity in the *Bolzano* affair, but he stuck to his story until his ultimate release by the 11th Hussars.

W. Stanley Moss

The Kidnapping of General Kreipe

Major General Karl Kreipe was the German officer commanding in Crete. On 26 April 1944 Captain Stanley Moss and Major Patrick Leigh Fermor of the British SOE determined upon Kreipe's kidnap.

It was eight o'clock when we reached the T-junction. We had met a few pedestrians on the way, none of whom seemed perturbed at seeing our German uniforms, and we had exchanged greetings with them with appropriately Teutonic gruffness. When we reached the road we went straight to our respective posts and took cover. It was now just a question of lying low until we saw the warning torch-flash from Mitso, the buzzer-man. We were distressed to notice that the incline in the road was much steeper than we had been led to believe, for this meant that if the chauffeur used the foot-brake instead of the hand-brake when we stopped him there would be a chance of the car's running over the edge of the embankment as soon as he had been disposed of. However, it was too late at this stage to make any changes in our plan, so we just waited and hoped for the best.

There were five false alarms during the first hour of our watch. Two *Volkswagen*, two lorries, and one motor-cycle combination trundled past at various times, and in each of them, seated primly upright like tailors' dummies, the steel-helmeted figures of German soldiers were silhouetted against the night sky. It was a strange feeling to be crouching so close to them – almost within arm's reach of them – while they drove past with no idea that nine pairs of eyes were so fixedly watching them. It felt rather like going on patrol in action, when you find yourself very close to the enemy trenches, and can hear the sentries talking or quietly whistling, and can see them lighting cigarettes in their cupped hands.

It was already one hour past the General's routine time for making his return journey when we began to wonder if he could possibly have gone home in one of the vehicles which had already passed by. It was cold, and the canvas of our German garb did not serve to keep out the wind.

I remember Paddy's asking me the time. I looked at my watch and saw that the hands were pointing close to half-past nine. And at that moment Mitso's torch blinked.

"Here we go."

We scrambled out of the ditch on to the road. Paddy switched on his red lamp and I held up a traffic signal, and together we stood in the centre of the junction.

In a moment – far sooner than we had expected – the powerful headlamps of the General's car swept round the bend and we found ourselves floodlit. The chauffeur, on approaching the corner, slowed down.

Paddy shouted, "Halt!"

The car stopped. We walked forward rather slowly, and as we passed the beams of the headlamps we drew our ready-cocked pistols from behind our backs and let fall the life-preservers from our wrists.

As we came level with the doors of the car Paddy asked, "Ist das das General's Wagen?"

There came a muffled "Ja, ja" from inside.

Then everything happened very quickly. There was a rush from all sides. We tore open our respective doors, and our torches illuminated the interior of the car – the bewildered face of the General, the chauffeur's terrified eyes, the rear seats empty. With his right hand the chauffeur was reaching for his automatic, so I hit him across the head with my cosh. He fell forward, and George, who had come up behind me, heaved him out of the driving-seat and dumped him on the road. I jumped in behind the steering-wheel, and at the same moment saw Paddy and Manoli dragging the General out of the opposite door. The old man was struggling with fury, lashing out with his arms and legs. He obviously thought that he was going to be killed, and started shouting every curse under the sun at the top of his voice.

The engine of the car was still ticking over, the hand-brake was on, everything was perfect. To one side, in a pool of torchlight in the centre of the road, Paddy and Manoli were trying to quieten the General, who was still cursing and struggling. On the other side George and Andoni were trying to pull the chauffeur to his feet, but the man's head was pouring with blood, and I think he must have been unconscious, because every time they lifted him up he simply collapsed to the ground again.

This was the critical moment, for if any other traffic had come along the road we should have been caught sadly unawares. But now Paddy, Manoli, Nikko, and Stratis were carrying the General towards the car and bundling him into the back seat. After him clambered George, Manoli, and Stratis – one of the three holding a knife to the General's throat to stop him shouting, the other two with their Marlin guns poking out of either window. It must have been quite a squash.

Paddy jumped into the front seat beside me.

The General kept imploring, "Where is my hat? Where is my hat?"

The hat, of course, was on Paddy's head.

We were now ready to move. Suddenly everyone started kissing and congratulating everybody else; and Micky, having first embraced Paddy and me, started screaming at the General with all the pent-up hatred he held for the Germans. We had to push him away and tell him to shut up. Andoni, Grigori, Nikko, and Wallace Beery were standing at the roadside, propping up the chauffeur between them, and now they waved us good-bye and turned away and started off on their long trek to the rendezvous on Mount Ida.

We started.

The car was a beauty, a brand-new Opel, and we were delighted to see that the petrol-gauge showed the tanks to be full.

We had been travelling for less than a minute when we saw a succession of lights coming along the road towards us; a moment later we found ourselves driving past a motor convoy, and thanked our stars that it had not come this way a couple of minutes sooner. Most of the lorries were troop transports, all filled with soldiery, and this sight had the immediate effect of quietening George, Manoli, and Stratis, who had hitherto been shouting at one another and taking no notice of our attempts to keep them quiet.

When the convoy had passed Paddy told the General that the two of us were British officers and that we would treat him as an honourable prisoner of war. He seemed mightily relieved to hear this and immediately started to ask a series of questions, often not even waiting for a reply. But for some reason his chief concern still appeared to be the whereabouts of his hat – first it was the hat, then his medal. Paddy told him that he would soon be given it back, and to this the General said, "Danke, danke."

It was not long before we saw a red lamp flashing in the road before us, and we realized that we were approaching the first of the traffic-control posts through which we should have to pass. We were, of course, prepared for this eventuality, and our plan had contained alternative actions which we had hoped would suit any situation, because we knew that our route led us through the centre of Heraklion, and that in the course of our journey we should probably have to pass through about twenty control posts.

Until now everything had happened so quickly that we had felt no emotion other than elation at the primary success of our venture; but as we drew nearer and nearer to the swinging red lamp we experienced our first tense moment.

A German sentry was standing in the middle of the road. As we approached him, slowing down the while, he moved to one side, presumably thinking that we were going to stop. However, as soon as we drew level with him – still going very slowly, so as to give him an opportunity of seeing the General's pennants on the wings of the car – I began to accelerate again, and on we went. For several seconds after we had passed the sentry we were all apprehension, fully expecting to hear a rifle-shot in our wake; but a moment later we had rounded a bend in the road and knew that the danger was temporarily past. Our chief concern now was whether or not the guard at the post behind us would telephone ahead to the next one, and it was with our fingers crossed that we approached the red lamp of the second control post a few minutes later. But we need not have had any fears, for the sentry behaved in exactly the same manner as the first had done, and we drove on feeling rather pleased with ourselves.

In point of fact, during the course of our evening's drive we passed twenty-two control posts. In most cases the above-mentioned formula sufficed to get us through, but on five occasions we came to road-blocks – raisable one-bar barriers – which brought us to a standstill. Each time, however, the General's pennants did the trick like magic, and the sentries would either give a smart salute or present arms as the gate was lifted and we passed through. Only once did we find ourselves in what might have developed into a nasty situation. . . .

Paddy, sitting on my right and smoking a cigarette, looked quite imposing in the General's hat. The General asked him how long he would have to remain in his present undignified position, and in reply Paddy told him that if he were willing to give his parole that he would neither shout nor try to escape we should treat him, not as a prisoner, but, until we left the island, as one of ourselves. The General gave his parole immediately. We were rather surprised at this, because it seemed to us that anyone in his position might still entertain reasonable hopes of escape – a shout for help at any of the control posts might have saved him.

Baron de Marbot

Lisette

An incident from the Napoleonic battle of Eylau, 1807.

To enable you to understand my story, I must go back to the autumn of 1805, when the officers of the Grand Army, among their preparations for the battle of Austerlitz, were completing their outfits. I had two good horses, the third, for whom I was looking, my charger, was to be better still. It was a difficult thing to find, for though horses were far less dear than now, their price was pretty high, and I had not much money; but chance served me admirably. I met a learned German, Herr von Aister, whom I had known when he was a professor at Sorèze. He had become tutor to the children of a rich Swiss banker, M. Scherer, established at Paris in partnership with M. Finguerlin. He informed me that M. Finguerlin, a wealthy man, living in fine style, had a large stud, in the first rank of which figured a lovely mare, called Lisette, easy in her paces, as light as a deer, and so well broken that a child could lead her. But this mare, when she was ridden, had a terrible fault, and fortunately a rare one: she bit like a bulldog, and furiously attacked people whom she disliked, which decided M. Finguerlin to sell her. She was bought for Mme de Lauriston, whose husband, one of the Emperor's aides-de-camp, had written to her to get his campaigning outfit ready. When selling the mare, M. Finguerlin had forgotten to mention her fault, and that very evening a groom was found disembowelled at her feet. Mme de Lauriston, reasonably alarmed, brought an action to cancel the bargain; not only did she get her verdict, but, in order to prevent further disasters, the police ordered that a written statement should be placed in Lisette's stall to inform purchasers of her ferocity, and that any bargain with regard to her should be void unless the purchaser declared in writing that his attention had been called to the notice. You may suppose that with such a character as this the mare was not easy to dispose of, and thus Herr von Aister informed me that her owner had decided to let her go for what anyone would give. I offered 1,000 francs, and M. Finguerlin delivered Lisette to me, though she had cost him 5,000. This animal gave me a good deal of trouble for some months. It took four or five men to saddle her, and you could only bridle her by covering her eyes and fastening all four legs; but once you were on her back, you found her a really incomparable mount.

However, since while in my possession she had already bitten several people, and had not spared me, I was thinking of parting with her. But I had meanwhile engaged in my service Francis Woirland, a man who was afraid of nothing, and he, before going near Lisette, whose bad character had been mentioned to him, armed himself with a good hot roast leg of mutton. When the animal flew at him to bite him, he held out the mutton; she seized it in her teeth, and burning her gums, palate, and tongue, gave a scream, let the mutton drop, and from that moment was perfectly submissive to Woirland, and did not venture to attack him again. I employed the same method with a like result. Lisette became as docile as a dog, and allowed me and my servant to approach her freely. She even became a little more tractable towards the stablemen of the staff, whom she saw every day, but woe to the strangers who passed near her!

Such was the mare which I was riding at Eylau at the moment when the fragments of Augereau's army corps, shattered by a hail of musketry and cannon-balls, were trying to rally near the great cemetery. You will remember how the 14th of the line had remained alone on a hillock, which it could not quit except by the Emperor's order.

I found the 14th formed in square on the top of the hillock, but as the slope was very slight the enemy's cavalry had been able to deliver several charges. These had been vigorously repulsed, and the French regiment was surrounded by a circle of dead horses and dragoons, which formed a kind of rampart, making the position by this time almost inaccessible to cavalry; as I found, for in spite of the aid of our men, I had much difficulty in passing over this horrible entrenchment. At last I was in the square. Since Colonel Savary's death at the passage of the Wkra, the 14th had been commanded by a major. While I imparted to this officer, under a hail of balls, the order to quit his position and try to rejoin his corps, he pointed out to me that the enemy's artillery had been firing on the 14th for an hour, and had caused it such loss that the handful of soldiers which remained would inevitably be exterminated if they went down into the plain, and that, moreover, there would not be time to prepare to execute such a movement, since a Russian column was marching on him, and was not more than a hundred paces away. "I see no means of saving the regiment," said the major; "return to the Emperor, bid him farewell from the 14th of the line, which has faithfully executed his orders, and bear to him the eagle which he gave us, and which we can defend no longer: it would add too much to the pain of death to see it fall into the hands of the enemy." Then the major handed me his eagle. Saluted for the last time by the glorious fragment of the intrepid regiment with cries of "Vive l'Empereur!" they were going to die for him. It was the *Cæsar morituri te salutant* of Tacitus, but in this case the cry was uttered by heroes. The infantry eagles were very heavy, and their weight was increased by a stout oak pole on the top of which they were fixed. The length of the pole embarrassed me much, and as the stick without the eagle could not constitute a trophy for the enemy, I resolved with the major's consent to break it and only carry off the eagle. But at the moment when I was leaning forward from my saddle in order to get a better purchase to separate the eagle from the pole, one of the numerous cannon-balls which the Russians were sending at us went through the hinder peak of my hat, less than an inch from my head. The shock was all the more terrible since my hat, being fastened on by a strong leather strap under the chin, offered more resistance to the blow. I seemed to be blotted out of existence, but I did not fall from my horse; blood flowed from my nose, my ears, and even my eyes; nevertheless I still could hear and see, and I preserved all my intellectual faculties, although my limbs were paralysed to such an extent that I could not move a single finger.

Meanwhile the column of Russian infantry which we had just perceived was mounting the hill; they were grenadiers wearing mitre-shaped caps with metal ornaments. Soaked with spirits, and in vastly superior numbers, these men hurled themselves furiously on the feeble remains of the unfortunate 14th, whose soldiers had for several days been living only on potatoes and melted snow; that day they had not had time to prepare even this wretched meal. Still our brave Frenchmen made a valiant defence with their bayonets, and when the square had been broken, they held together in groups and sustained the unequal fight for a long time.

During this terrible struggle several of our men, in order not to be struck from behind, set their backs against my mare's flanks, she, contrary to her practice, remaining perfectly quiet. If I had been able to move I should have urged her forward to get away from this field of slaughter. But it was absolutely impossible for me to press my legs so as to make the animal I rode understand my wish. My position was the more frightful since, as I have said, I retained the power of sight and thought. Not only were they fighting all round me, which exposed me to bayonet-thrusts, but a Russian officer with a hideous countenance kept making efforts to run me through. As the crowd of combatants prevented him from reaching me, he pointed me out to the soldiers around him, and they, taking me for the commander of the French, as I was the only mounted man, kept firing at me over their comrades' heads, so that bullets were constantly whistling past my ear. One of them would certainly have taken away the small amount of life that was still in me had not a terrible incident led to my escape from the mêlée.

Among the Frenchmen who had got their flanks against my mare's near flank was a quartermaster-sergeant, whom I knew from having frequently seen him at the marshal's, making copies for him of the "morning states". This man, having been attacked and wounded by several of the enemy, fell under Lisette's belly, and was seizing my leg to pull himself up, when a Russian grenadier, too drunk to stand steady, wishing to finish him by a thrust in the breast, lost his balance, and the point of his bayonet went astray into my cloak, which at that moment was puffed out by the wind. Seeing that I did not fall, the Russian left the sergeant and aimed a great number of blows at me. These were at first fruitless, but one at last reached me, piercing my left arm, and I felt with a kind of horrible pleasure my blood flowing hot. The Russian grenadier with redoubled fury made another thrust at me, but, stumbling with the force which he put into it, drove his bayonet into my mare's thigh. Her ferocious instincts being restored by the pain, she sprang at the Russian, and at one mouthful tore off his nose, lips, eyebrows, and all the skin of his face, making of him a living death's-head, dripping with blood. Then hurling herself with fury among the combatants, kicking and biting, Lisette upset everything that she met on her road. The officer who had made so many attempts to strike me tried to hold her by the bridle; she seized him by his belly, and carrying him off with ease, she bore him out of the crush to the foot of the hillock, where, having torn out his entrails and mashed his body under her feet, she left him dying on the snow. Then, taking the road by which she had come, she made her way at full gallop towards the cemetery of Eylau.

Louis-Francois Lejeune

1812: The Retreat from Moscow

General Lejeune was the Chief of Staff to Marshal Davout in the Grand Armée that Napoleon led to Moscow – and, finding it deserted and afire, led away again.

On October 18 we received orders to leave Moscow, and to march by way of the Kalouga road on the 19th. Thus, after a month's delay at Moscow, which had been of no special advantage to us, during which our army had received few reinforcements and our troops had been worn out hunting for provisions, we left that city and sadly began our retreat towards France. We were fortunate in having beautiful autumn weather, and the first few days' march was peaceful enough, for we only had to drive off a few Cossacks who hovered on the flank of our columns. But, as on our advance, we were everywhere harassed by the Russian plan of burning everything on our approach, and we could do nothing to prevent it. About ten leagues from Moscow the first corps halted at the base of a fine castle, the foundations and first floor of which were of hewn stone. I had several orders to write, and I went up a grand staircase into a suite of rooms which seemed to have been but recently deserted, for they still contained a piano, a harp, and a good many chairs, on which lay a guitar, several violins, some music, drawings, embroideries, and lady's unfinished needlework. I had scarcely been writing ten minutes, when we all noticed a smell of smoke. This smoke quickly filled the place, becoming so dense that we were obliged to give up work to try and find out whence it came. It seemed to issue chiefly from the wooden framework of one door. I had it broken in, and thick smoke at once poured through the aperture. I then went down into the cellars to see if the fire originated there, but I could discover nothing. I tried having a few buckets of water flung into the opening we had made, but even as my orders were obeyed such masses of flames rushed out upon us that we had only just time to collect our papers and escape. We had scarcely got downstairs when we heard the windows breaking with a crash, and as we looked back on our way to join the bivouac of our corps we saw volumes of flames issuing from all the windows of the castle, which fell at last, bearing witness by its destruction to the patriotic fury with which the Russians, torch in hand, were determined to pursue us.

Of course other fires occurred accidentally, with which the Russians had nothing to do. The little grain or flour found by our soldiers was made into cakes and put in the ovens with which all the peasants' huts were provided. Scarcely was one batch of cakes done before other troops came up, and the oven was heated again till the chimney would suddenly catch fire. This was how most of the fires which were to light up our passage from Moscow to the Niemen came about.

My courage almost fails me when I try to relate the horrors of those awful days and nights of suffering. But in spite of that, I shall put down here all that I find in my notes, for I think that the lessons taught by the past should be brought forcibly before the eyes of those whose genius leads to their being called to command armies.

Kutusoff[1] justly felt that the best way to make war on us was to cut off our communications, so as to isolate us in the midst of a hostile population to whom our loss would be gain. He therefore took up a position at Vinkowo commanding the Kalouga road, by which he thought it probable that we should retire. The success he had achieved on the 18th confirmed his belief that the course he had adopted was the best, and the aim of his later manœuvres was to bar our passage.

Under these circumstances it was important that we should push on as rapidly as possible during the first days of our retreat, so as to gain a couple of days' march on the enemy and get possession, without fighting, of the principal passes. But, alas! this was just what we did not do. Although much of the impedimenta of the French army had been sent on some days before, we were still encumbered with a great number of

1 Field Marshal Mikhail Golenischev-Kutusov (1745–1813), appointed commander of the Russian armies in the West in late summer 1812. He died of exhaustion following his long campaign against the French.

wagons and carriages laden with provisions for the prisoners, and with the booty we had taken, which included warm garments to protect us from the cold we should have to encounter. The amount of baggage was really enormous, and to give some idea of it I will just mention what I, an officer, who realised as much as any one the importance of getting rid of encumbrances, was trying to take with me. I still had: 1. Five saddle horses; 2. a barouche, drawn by three horses, containing my personal effects and the furs in which I meant to wrap myself when bivouacking in the open; 3. the wagon, drawn by four horses, in which were all the papers of the staff, the maps, and the cooking utensils for the officers and their servants; 4. three smaller wagons, each drawn by three little Russian horses, in which rode our servants and the cook, under whose care were the stores of oats, a few precious trusses of hay, with the sugar, coffee, and flour belonging to the staff; 5. my secretary's horse; 6. the three horses I had lent to my sister, which had gone on in front, making altogether six carriages and twenty-five horses, to take along little more than bare necessaries. The traces of the carriages were constantly breaking, the march was retarded whilst they were mended, there were perpetual blocks in the sand, the marshes, or in the passes, and it often took our troops twelve hours to do a distance which a single carriage could have accomplished in two.[2]

The Emperor, who was very much concerned at these delays, ordered that all the carriages not absolutely necessary for the transport of the few provisions we had with us were to be burnt and the horses used to help drag the artillery. So many were, however, interested in eluding this stern but wise sentence, that it was very insufficiently carried out. To set an example the Emperor had one of his own carriages burnt, but no one felt drawn towards imitating him, and the army, which still numbered between 105,000 and 106,000 combatants, and had 500 pieces of cannon, took six days to cover some thirty leagues. Further precious time was lost in getting across country by difficult roads, from the main Kalouga route, which was very bad, to a better one, and the Emperor leaving only Murat's cavalry and Marshal Ney's infantry on the old road to cover us from the attacks of the army under Kutusoff.

The Viceroy's corps marched at the head of the column on the

2 Marbot says that the army in retreat was followed by 40,000 vehicles.

new road, whilst the Delzons division, as advanced guard, occupied Malo-Jaroslavitz,[3] the passage through which was extremely difficult.

Malo-Jaroslavitz was a little town of wooden houses, with tortuous streets, built on the steep sides of a lofty hill, at the base of which wound the little river Luya in a deep valley it had hollowed out for itself. A narrow bridge spanned the river below the only road by which the town could be reached from Moscow, and this road was here bounded on either side by impassable ravines, down which flowed the rapid torrents of such frequent occurrence in Russia during the rainy season. On the evening of the 23rd the town was occupied without resistance by the first of our battalions to arrive, the inhabitants having all fled at their approach.

That same evening the Emperor halted at the post-house of Malo-Jaroslavitz, a mere peasant's hut, where he passed the night after having sent out officers bearing his orders to the corps écheloned on the road from Moscow to Smolensk, telling them to meet him at the latter town.

On October 24 the Emperor was riding with the first corps and his Guard, as he thought, in perfect security, when a considerable body of cavalry appeared on the right, which we all took at first to be Murat's troops. We were not long left in our error. It appeared that a certain Platoff,[4] a celebrated Cossack hetman or general, had promised Kutusoff to carry off Napoleon, and now with several thousands of his men he suddenly flung himself upon that part of the French army which he fancied included the Imperial staff. He had guessed rightly, and in the twinkling of an eye Napoleon was surrounded by Cossacks and compelled to draw his sword in his own defence. Fortunately, however, his escort was made up of men devoted to his person, and they pressed round him, breaking the shock of the barbarian charge. General Rapp, as he was engaged in trying to get the Emperor away, was overthrown in the mêlée, whilst his horse was pierced by a lance. Several officers near the Emperor were wounded. The mounted grenadiers and chasseurs of the Guard, however, recovering from the momentary surprise caused by the bold attack and wild cries of the hordes of

3 Maloyaroslavetz, on the River Lusha.

4 General Matvei Ivanovich Platov (1751–1818), Hetman of the Don Cossacks, whom he led into Paris in 1814.

Cossacks, dashed into their midst and put them to rout. In this struggle Emmanuel Lecouteulx, one of Prince Berthier's[5] aides-de-camp, having broken his sword in the body of a Cossack, seized his lance and brandishing it above his head pursued his other enemies with it. A green furred pelisse, rather like those often worn by Russian officers, hid his uniform, and a French grenadier, taking him for a Cossack, plunged his long sabre into his shoulder, the point coming out through his breast. We all thought this terrible wound would lose us a favourite comrade, but God preserved him, and he is still alive. After being thus quickly dispersed, the Cossacks, leaving many of their numbers behind, dashed away, but they were stopped in their flight by a Dutch battalion, which flung itself upon them and greeted them with volleys of musketry.

Soon after this skirmish news was brought to the Emperor that our advanced guard had been vigorously attacked at Malo-Jaroslavitz, and the army quickened its march to go to the rescue. General Kutusoff, informed of our change of route, had at once sent a body of infantry and artillery more than 60,000 strong, under command of General Doctoroff, with orders to take possession of Malo-Jaroslavitz. These troops easily turned out the few French battalions under General Delzons, and occupied the town in their place. General Delzons, it is true, drove the Russians back to the centre of the town, but a ball fractured his skull, killing him and his brother, who was beside him, on the spot. The French began to give way, and the Russians recovered the ground they had lost. General Guilleminot was sent to replace Delzons, and Prince Eugène[6] supported him with the Broussier division. Again and again Guilleminot drove the Russians beyond the principal square, but fresh efforts on the part of the enemy forced him in his turn to retreat. The formidable Russian artillery placed on the heights overlooking the town and in its gardens poured a murderous fire down upon the road on which the French were coming up, whilst we were unable to reply with an effective fire

5 Marshal Louis-Alexandre Berthier, prince de Neuchatel et de Wagram (1753–1815), chief-of-staff of the *Grande Armée en Russie*. After refusing to rejoin Napoleon in 1815, he was overcome by remorse and committed suicide by defenestration.

6 i.e. Viceroy Eugene de Beauharnais (1781–1824), son of Josephine de Beauharnais and stepson of Napoleon. He assumed command of the retreat towards its close.

from below, as we could only get our guns into position in the meadows by the Luya. Everything had therefore to be done by us with the bayonet in a space so limited that any flank manoeuvres were impossible. The enemy had all the advantages alike of the ground and of superiority of numbers. During the thick of the struggle, Prince Eugène had a second bridge flung across the river beside the first, so as to facilitate the passage of his troops.

Ten times at least we drove the Russians back, only in our turn to lose the ground we had gained; but at last the united efforts of Guilleminot and Broussier, supplemented later by the gallant charge of the Italian grenadiers under Pino, compelled the enemy to retire, leaving us masters of the town, which was now in flames. Our artillery was at last able to scale the hill so as to debouch in the plain, and dashing through the burning streets, crushing the wounded, the burnt, and the dead as they went, they succeeded in getting through the town and taking up their position on the hills, to pour down in their turn a hot cannonade upon the Russian troops posted within range across and behind the road to Kalouga.

The first corps seconded this operation, and spent the evening actually amongst the bodies of the dead all kneaded up by the wheels of the guns. A dark night fortunately shrouded the horrors from our sight, and we were able in the end to take up our position on the plain beyond the mangled remains of our comrades.

In this battle, in which our numbers were but one to four of our adversaries, we lost many men, and we were threatened with a similar struggle at every pass.

The Emperor went over the scene of the awful combat, overwhelmed Prince Eugène and his generals with praise, and then withdrew to a distant hut, where he passed the night. I was told afterwards that he held a council of war with several of the Marshals and King Murat, and that, after having the maps spread out and discussing them, he seemed to be for some time plunged in the greatest uncertainty. He finally dismissed every one at midnight without having come to any decision. This must have been indeed a cruel night for the great man, who now saw his star beginning to set, his power crumbling away, and who must already have begun to wonder if he could ever re-establish it, or even if he would get back to France.

On the morning of the 25th, Marshal Davout, Colonel Kobilinski, and I went the round of our outposts and saw with regret that the Russian army was drawn up in good order not far off, completely blocking the road to Kalouga, which we hoped to take. We made our dispositions for forcing a passage, and as we stood in a close group, bending over our maps, we offered an excellent mark for a Russian artilleryman. A ball from a twelve-pounder passed between the Marshal and me, and carried away one of Colonel Kobilinski's legs. The unfortunate officer fell against me, and we thought he was killed, but he recovered miraculously, and I shall speak of him again later.

The Marshal and I, with hearts torn by this catastrophe, impatiently awaited the signal to advance, when a very different order filled us with surprise and dismay.

I must explain that Kalouga, whither we thought we were bound, was a town of great importance to the Russians, as it was their emporium of provisions and weapons. Built as it was beside a river divided into several branches, it could easily be fortified, and afforded an admirably defended position for the Russian army, which had arrived before us. The Emperor doubted whether we were strong enough to force a passage through it, and he had lost so much time hesitating that he was now reduced to the necessity of ordering the army to abandon the Kalouga road for the one leading to Mojaisk, by which we had come. This was to fling us back upon the desert without provisions, to make us tramp once more over the ashes we had left behind us on our way to Moscow; in a word, it was to deprive us of all hope of finding a scrap of food. Needless to say that this decision afflicted us all most cruelly. The Emperor with his Guard went first, followed by the various corps of the army with all their terrible encumbrances, whilst henceforth Marshal Davout's corps, which of course was also my own, formed the rearguard.

It became momentarily more difficult to reassure our soldiers on the subject of this loss of time and retrograde movement. On the second day of our retreat for the Mojaisk road, that is to say on October 26, a fine cold rain set in, which damped every one's spirits yet more, and greatly increased the difficulties of the march. Once more we saw a fine castle, which looked as if it would provide us with a comfortable shelter, but we had no sooner entered it than fire broke out, and that before our people had lit a match. We found the incendiary apparatus, which had been left in position by the owner, but too late to extinguish the flames. Our troops were already beginning to suffer from dysentery through insufficient and badly cooked food, a few cakes and a little poor soup being all they had even now. The sick who were unable to march with the rest were abandoned on the road. Meanwhile Marshal Mortier[7] rejoined us at Verea with the two divisions of the Young Guard; he had accomplished the melancholy task assigned to him of blowing up the Kremlin, against which his noble soul had revolted.[8] Before leaving Moscow some of Mortier's troops took prisoner one of our bitterest enemies, the Russian Lieutenant-General Vintzingerode, and he was being taken to the Emperor, when he had the good fortune to make his escape.

On the 27th the advanced guard of the army re-entered Mojaisk, still encumbered with the wounded left behind after the battle of Borodino, and on the 28th the rearguard arrived. How painful and touching was the meeting with our unfortunate wounded, to whom we now returned with none of the comforts or the cheering news which they expected us to bring them! All we could offer them was an exhortation to resignation; we dared not tell them that we were about to abandon them once more, this time finally, and we were ourselves slowly beginning to face the fact that their terrible lot would most probably soon be our own. Wherever we passed, every refuge still left standing was crowded with wounded, and at Kolinskoy alone there were more than 2,000. Hitherto we had only been pursued by a few Cossacks, but every day their numbers increased, and they became more aggressive. Just before we left Kolinskoy on the 30th, wishing to reconnoitre the enemy on the plain, I was walking along the terrace of a convent, when I suddenly found myself in the presence of about a hundred Cossacks, who were like myself approaching to reconnoitre. When they caught sight of me they at first took to flight; but seeing that I was alone, they returned, and I had only just time to mount and gallop off to rejoin our troops, who had started and were already some distance off. Here and there we passed carriages left on the road because the starved horses, exhausted with fatigue, had fallen down. The few which could be made to get up again were at once harnessed to the wagons containing some of the

7 Marshal Adolphe Mortier, duc de Trevise (1768–1835).
8 Mortier, in fact, refused to demolish the Kremlin.

wounded, but they all died after dragging their new burdens for a few steps only. Then the wounded were in their turn abandoned, and as we rode away we turned aside our heads that we might not see their despairing gestures, whilst our hearts were torn by their terrible cries, to which we tried in vain to shut our ears. If our own condition was pitiable, how much more so was theirs with nothing before them but death from starvation, from cold, or from the weapons of the Russians! The 30th was a sad and terribly long day, for we had to march nearly all night in intense cold, the severest we had yet had to encounter, it being important that we should arrive at Giatz before the enemy, who were pushing on rapidly by cross roads in the hope of getting there first.

The first corps arrived at Giatz on the 31st, and a few hours afterwards the Russians appeared in great force. The next day, November 1, they tried to force a passage through our troops, and having failed they had to be content with hotly cannonading one of our big convoys which had been considerably delayed, and was defiling near the entrance to the town just in front of the enemy's guns. The balls wrought terrible havoc in this convoy of ours, and amongst the carriages was that in which I had sent my sister on in advance. I was fortunate enough to be able to save her. The coachman assured me that the three horses were still fresh and in first-rate condition, so I said to my sister, "Dare you face the guns?" She replied trembling, "I will do what you tell me." I at once turned to the coachman with the words, "Cross that meadow at a gallop; the balls will go over your head, and you will succeed in getting in front of the rest of the convoy. You will then be able to push on without stopping." He followed my advice and with the best results, for my sister got off unhurt. The convoy consisted of some hundreds of badly harnessed carriages, containing many wounded, with the wives and children of several French merchants of Moscow, who were flying the country after having been robbed of their all by the Russians. The company of the Théâtre Français of Moscow had also joined this party, the unlucky actors little dreaming of the terrible tragedy in which they were to play their part through placing themselves under our protection, so soon, alas! to avail them nothing.

On November 2 snow began to fall, and there were already eight or nine degrees of frost. The various divisions of the first corps took it in turns to act as rearguard, and on that day it was the turn of the Gérard division, with which we had passed the night in a wood beneath the snowflakes. The effects of the great cold were already disastrous; many men were so benumbed at the moment for departure as to be unable to rise, and we were obliged to abandon them.

We reached Viasma on November 3 at the same time as the Russians, whose advanced guard was checked by Marshal Ney's troops drawn up in front of them. It was evident that a battle must take place here, and every preparation was made on both sides. The French troops in a position to take part in it numbered about 30,000 or 40,000, whilst the Russians had two corps consisting together of more than 60,000 men. Marshal Davout's and the Viceroy's corps, with the Poles under Prince Poniatowski,[9] were successively engaged, and for a long time exposed to an overwhelming fire from a strong body of artillery with horses better harnessed and in far better condition than ours, which were too worn out for manoeuvring. The first corps and that under Prince Eugène became separated from each other twice, and were both for a time in very critical positions. Fortunately Marshal Ney was able to send a regiment to the rear of the Russian army, which threw it into confusion, and Generals Kutusoff, Miloradowich, Platoff, and Suvoroff, who had hoped to make us lay down our arms, stopped the pursuit, though fifty pieces of cannon still poured out their fire upon our luckless convoys, which were defiling past during the battle. Many men were lost on both sides here as elsewhere, for our soldiers were still undaunted, and nearly every shot from us told in the Russian ranks, which were more numerous and more closely serried than ours.

Marshal Ney, whose turn it was to act as rearguard, now protected the Viasma pass, and the French army marched towards Dorogobouj. After having passed the night of the 3rd and the day of the 4th on the road, we halted in the evening in a pine forest on the borders of a frozen lake not far from the Castle of Czarkovo, where the Emperor had been for two days. On the 5th the first corps took up its position at Semlevo, so as to let that of Marshal Ney pass on, it being our turn now to be rearguard. The Cossacks harassed us greatly, and many of our

9 Prince Poniatowski, nephew of the King of Poland, commander of the Polish and Saxon troops which formed V Corps. He was drowned in the withdrawal from Leipzig in October 1813.

stragglers, whose numbers increased every day, fell into their hands. We now also made out on our flanks numerous columns of Russian cavalry and artillery, which were trying to pass us so as to await us at the entrance to the pass on this side of Dorogobouj. Marshal Ney foresaw this danger, and instead of going through the pass he halted near it for us to come up. Thanks to his forethought and support, we only suffered from a slack cannonade and reached Dorogobouj safely.

Leaving that place on the 6th we made a long march, and at nightfall we camped in a large wood, where General Jouffroy had been obliged to halt with the badly harnessed and damaged artillery under his care. We spent the night in packing the wagons which were still in a fit state to proceed, blowing up those we had to abandon, and burning the gun-carriages we could not take with us. These explosions, which were now of very frequent occurrence, were signals of our misfortunes, and affected us much as the tolling of the bell at her child's funeral would some bereaved mother.

General Jouffroy had had a tent pitched, and invited me to share its shelter and his supper. A supper! Good heavens! What a luxurious treat in the midst of our misery!

I had a new experience at that supper. Hitherto a few cows had still remained to the staff, and I had not been reduced to eating horseflesh. But now the General had nothing to offer me but a repast of horseflesh so highly spiced that in spite of its toughness and the coarse veins, which resisted the efforts of the sharpest teeth to masticate them, it really tasted not unlike what the French call *boœf à la mode*. Generally horseflesh is so black, and its gravy is so yellow and insipid, so very like liquid sulphur, that it looks most repulsive, but we quite enjoyed our meal, washed down as it was by a flask of good wine which had belonged to some great man of Moscow.

For a long time the only meat our soldiers ever tasted had been horseflesh, and the poor fellows were so brutalised by misery and famine that they often did not wait till an animal was dead to cut it up and carry off the fleshy portions. When a horse stumbled and fell, no one tried to help it up, but numbers of soldiers at once flung themselves upon it, and cut open its side to get at the liver, which is the least repulsive part. They would not even put it out of its misery first, and I have actually seen them angry at the poor beast's last struggles to escape its butchers, and heard them cry, "Keep quiet, will you, you rogue?"

The numbers of the stragglers increased in a perfectly appalling way; they stopped in crowds to roast a few shreds of horseflesh, and the French, who must always have their joke whatever their misery, called the tattered wretches the *fricoteurs* or revellers.

During the night of the 6th and the day of the 7th, a heavy fall of snow drifting before a strong wind rendered our march extremely arduous. We were often unable to see two paces before us, but all the time the balls from the enemy were ploughing up the ground, and every now and then a few victims fell. No one had the heart to stop to help those who were struck, for the most selfish egotism crushed all kindly feeling in almost every breast. It was in a state of bodily and mental torpor that we reached Pnevo on one of the tributaries of the Dnieper, a very difficult river to cross. To protect its passage during our occupation of the country, we had had a big log hut built surrounded by a weak earthwork. This little redoubt was the only shelter which had not been burnt on the long road we were traversing, and the first corps halted there to pass the night. It was built in the same way as the huts of the peasants, with big squared trunks of trees laid horizontally on each other, forming walls almost impervious to balls, but not more than fifty people could get inside, so that the rest of the troops had to camp around it. The heavy snow and the bitter wind prevented us from going to get fuel, for there were no trees near, and every one suffered terribly from cold during the night.

In our halts we always faced the north. The Viceroy, who was marching on our left, met with the greatest difficulties, for he had counted on finding a bridge over the Wop, but this bridge was broken, so that he had to cross by the ford, and the water was all frozen over. His artillery and baggage stuck fast in the mud on the banks, and he was compelled to abandon them.

General Rapp and several other officers came to share our small quarters, where we were all very closely packed together. At nine o'clock the next morning when we left our shelter we found the ground near the wretched little redoubt encumbered with poor fellows, who, after having with infinite trouble managed to light fires, had been overcome with the cold, and burnt by flying sparks though covered with snow. Many of them were never to rise again from the spot on which they had fallen. Before we left we had the log hut burnt down.

The coating of ice on the roads made them so slippery that men and horses could scarcely keep their feet. My horse fell with me, and I was so much hurt that I could not remount, and I went to share Marshal Davout's *wurst*[10] or ambulance wagon, drawn by a very strong pair of cobs, which galloped along on the ice as easily as others would on turf.

Having burnt the bridge behind us, we imagined ourselves to be in security for the rest of the day. But when we halted for the first time about noon, we heard a brisk firing a short distance off, which evidently came from the twelve-pounders of our own park of artillery. This made the Marshal both uneasy and angry, and he sent for the officer in command of the artillery. He came hurrying up with a smile on his face, as if he were the bearer of good news. Davout, however, frowning at him from his *wurst*, accosted him roughly with the words, "So it's you, you scoundrel, who have dared to fire my reserve guns without orders from me!" Greatly surprised at this address, the officer had the presence of mind to pretend not to know to whom Davout was speaking, and after looking about him, he said as he set spurs to his horse to return to his post, "Surely that language cannot be addressed to me!" A few minutes later we learnt that some 1,200 Cossacks had flung themselves upon the big park of artillery, but when the commanding officer halted he had prudently prepared his guns for action and formed squares to guard against a cavalry charge, so that when that charge was made a volley of grapeshot from thirty guns overthrew one half of the assailants and put the other half to flight. I now once more entreated the Marshal, as I had so often done before, to choose another chief of the staff, pointing out to him that half our aides-de-camp and commissaries were already killed or taken prisoners, and that I really could not do all the work he required alone. He, however, begged me to remain, with a politeness which was so truly remarkable from him, that General Haxo maliciously asked me, "Whatever have you done to the Marshal? He must be very fond of you, for I never saw him pet any one as he does you."

On this same eventful 9th of November we re-entered Smolensk, where the Emperor received news of the Malet and

10 The *wurst* was an open ambulance wagon, now no longer in use.

Lahorie conspiracy at Paris,[11] and of the check received by the corps which he had ordered to debouch on his flanks. The tidings from Spain were not of a kind to afford him any consolation, for there was no unity of purpose or of action amongst the French Generals there, a fact by which the enemy was not slow to profit. The Emperor, fearing lest discouragement should spread through the ranks of our retreating army, pretended to be quite unmoved by all this distressing news. He wanted to appear superior to every adversity and ready to face calmly every event, however untoward, but his assumed indifference was misinterpreted and had a bad effect.

We no longer had a smithy for rough-shoeing our horses, so that they nearly all fell and were too weak to get up again. Our cavalry was thus completely destroyed, and the dismounted men even flung away their weapons, which their fingers were too frozen to hold. Some 300 officers, who had lost all their men, then proposed forming themselves into a kind of picked corps, ready to fight together on every emergency; but with them, as with the common soldiers, strength and discipline soon gave way, and what might have been a noble band, bound together by misfortune, fell to pieces in a few days without having rendered the slightest service to any one.

We camped for the night of the 10th on the banks of the Dnieper, beside the bridge where General Gudin had been killed. Our bivouac fires were soon surrounded by those of the numerous stragglers who had met here. Their appearance would have torn our hearts if we had not already been reduced to the level of the brutes, without the power of feeling compassion. Many of the poor wretches, who were all without weapons, were wearing silk pelisses trimmed with fur, or women's clothes of all manner of colours, which they had snatched from the flames of Moscow or taken from carriages abandoned by the way. These garments, which were fuller and looser than those of men, were a better protection from the cold. Some also wore the clothes of their comrades who had died on the road.

Numbed with cold and famishing with hunger, those who had

11 This conspiracy all but succeeded in overthrowing the Imperial Government; Malet, who had escaped from prison, where he was confined for participation in the plot of 1801; having circulated the news of Napoleon's death, and forged a decree of the Senate. He was, however, taken prisoner by Laborde, and shot, with Lahorie and other traitors to the Emperor, on 29 October 1812.

been unable to make a fire would creep up to their more fortunate comrades and plead for a little share in the warmth, but no one dreamt of sacrificing any of the hardly won heat for the sake of another. The new arrivals would remain standing behind those seated for a little while, and then, too weak to support themselves longer, they would stagger and fall. Some would sink on to their knees, others into a sitting posture, and this was always the beginning of the end. The next moment they would stretch out their weary limbs, raise their dim and faded eyes to heaven, and as froth issued from their mouths their lips would quiver with a happy smile, as if some divine consolation had soothed their dying agony. Often before the last breath was drawn, and even as the failing limbs stretched themselves out with an appearance of heavenly calm, some other poor wretch, who had been standing by, would seat himself upon the still heaving breast of his dying comrade, to remain resting upon the corpse with his living weight until, generally very little later, his own turn should come, and he also, finding himself too weak to rise, should yield up his breath. The horror of it all was but slightly shrouded by the falling snow, and we had to witness this kind of thing for yet another thirty days!

The first corps entered Smolensk on the 11th, and remained there till the 16th. The interval was employed in distributing to the troops the few provisions and clothes which had been collected in the storehouses by order of the Emperor,[12] and in seeing off for Wilna all the convoys which could still be supplied with horses.

The Imperial Guard, which was always held in reserve, had fought very little and lost fewer men than any of the other corps. The Emperor still owned in that Guard a force of from 3,000 to 4,000 men in good fighting condition, but these troops, though the discipline to which they had to submit was much less severe than that enforced in the rest of the army, really suffered quite as much as we did. The Emperor had their affection for him very much at heart, and in the friendly familiar way which he knew would please them he used sometimes to go amongst them, and pulling the long moustaches, all stiff with ice, of one or another, he would say, "Ah, old *Grognards*,[13] you may count

12 Large quantities of food and clothing had been brought together at Smolensk, but there was some mistake about their distribution, and many men got too much, whilst others received nothing.

13 The *Grognards* was the popular name given to the Old Guard of Napoleon I.

on me as I count on my Guard to fulfil the high destiny to which they are called." These few words would at once restore the confidence of the brave fellows in their chief, and to the end of the journey the Emperor was always surrounded by them.

We had still more than 120 leagues to cross between Smolensk and the Niemen. There were already from 12 to 15 degrees of frost, and the cold was still increasing. The roads grew worse every day, and there was too little of everything at Smolensk for the four days' halt to have done much to recruit the exhausted strength of the troops, or to restore anything like order in the disorganised army. My chief, the Prince of Eckmühl (Marshal Davout), who, as the Emperor justly remarked, was a man of iron constitution,[14] was very exacting, and expected the Staff accounts to be written up every day just as in times of peace. Now all my assistants but one had disappeared, and I therefore again tendered my resignation. Just as we were leaving Smolensk, the Emperor consented to my leaving Davout, and named Charpentier, a general of division just removed from the Government of Smolensk, to take my place. The general, however, being not at all anxious to take up a task of which he knew the difficulties, evaded appearing at his post for ten or twelve days, so that I had to go on doing the work of the Chief of the Staff without the title or pay.

Before he left Smolensk the Emperor ordered Marshal Ney to remain there until the 17th, when he was to blow up the fortifications. He also told him that his corps was to act as rearguard after the departure of the first corps, which was to precede him by one day and await him at the Krasnoe ravine. This ravine, which was a very difficult pass, had been encumbered for nine days with carriages, many of which were being burnt to clear the way.

The Viceroy's corps, now reduced to 1,200 or 1,500 men, which was marching in advance of ours, had been greatly harassed ever since leaving Smolensk by some 12,000 or 14,000 Russians, with a strong force of artillery mounted on sledges. The Emperor and his Guard had waited at the entrance to the terrible defile for the Viceroy to come up with his corps and protect his passage through it, and he now determined not to enter it until the arrival of the first corps also, which had not been able to leave Smolensk until two

14 Davout (1770–1823) was one of Napoleon's most cherished Marshals; the Emperor also wrote of his "distinguished bravery and firmness of character, the first qualities in a warrior."

o'clock in the morning of Monday, the 16th. During this halt Napoleon learnt that the enemy was advancing in force upon Orcha to intercept the passage of the Dnieper, and had massed a large number of troops in the village of Kourkovo, not far from us. The Young Guard, commanded by General Roguet, whose gallant audacity was well known to the Emperor, had joined him during the day, and Napoleon now sent him to create a diversion in the night by attacking the enemy's corps which was causing us so much uneasiness.

The first corps, which had been the 4th under the Viceroy, was terribly harassed all through the march on the 16th by numerous Cossacks with artillery. When darkness fell the attacks lessened, and we availed ourselves of the reprieve by marching all night towards the Krasnoe pass. With the first gleams of light on the 17th, however, we found ourselves threatened by great masses of Russian infantry and cavalry struggling to surround us and make us lay down our arms; and though they did not venture actually to attack us, the fire from their guns wrought great havoc amongst us. Again and again our little army, reduced to 4,000 men bearing arms, but hampered by numerous stragglers, halted to face the enemy and await Marshal Ney, who was to cover our retreat. On this occasion I had a fresh opportunity of admiring the courage and *sang-froid* of General Compans. Severely wounded in the shoulder, and suffering greatly, he was compelled, like most of us, to march on foot. This, however, did not prevent him from facing the enemy with a smiling face and as unruffled a calm as if he were walking about in his own garden at home. The sight of his happy face and composed demeanour had the best results on his soldiers, giving them a sense of security, and leading them to imitate their general's stoicism.

Our position at Krasnoe was, however, anything but pleasant. Surrounded by enemies ten times as numerous as ourselves, we could not imagine how it was that Marshal Ney, whom we supposed to be just behind us, had not managed to beat off at least some of them. We fought steadily, hoping every moment to see him appear. But the enemy's cannonade became hotter and hotter, making terrible gaps in our ranks, and the snow, which had been falling heavily ever since the evening before, added to our difficulties, rendering our situation all but desperate. The Emperor, who was becoming very anxious about our fate, generously turned back and came to meet us, cutting a passage through our assailants at the head of the Old Guard, and meeting Marshal Davout's advanced guard beyond Krasnoe.

Meanwhile nothing had been heard of Ney and the rearguard with him, but it turned out afterwards that on leaving Smolensk the Marshal, with the few troops still remaining to him, had been immediately pursued by thousands of the enemy, who poured such a hot fire into his already diminished ranks from every side, that after three days' continuous struggle he was compelled to abandon the attempt to cut through the enemy's forces, and to deviate from the main road to Krasnoe, where we were so anxiously awaiting him. When darkness was beginning he found himself far away from us, with the Dnieper between him and safety. He took up a position parallel with the river, and allowed his troops to light their bivouac fires. Kutusoff, who had followed Ney, now looked upon him as his certain prey, for he could see no way of escape for him, and sent an officer with a flag of truce to summon him to surrender.

The envoy, who performed his mission with the greatest politeness, was received with assumed courtesy, and detained on various pretexts whilst the Marshal was having the depth of the river sounded, and the strength of the ice on it tested. He was told that several men had gone over to the other bank and returned safely. He then ordered the throwing of fresh fuel on the fires, as if he had decided to remain where he was, and telling the envoy that he would have to accompany him, he gave the signal for crossing the river, instructing his subordinates to make the men go over in single file, and to keep well away from each other. Everything – artillery, baggage, even wounded – which would have hindered the safe crossing or broken the ice, was abandoned on the banks.[15] The transit was accomplished without accident, and at daybreak on the 18th the Marshal was several leagues from the further bank, but he was now attacked by a considerable body of Cossacks under Platoff, but he managed to fight his way through them,[16] though he had none

15 This crossing of the Dnieper was one of the most brilliant feats achieved by the French in this or any campaign. The story of its accomplishment is variously told by eye-witnesses, but all agree that but for its successful performance the whole of the rear-guard would have been cut to pieces or taken prisoner by the Russians.

16 Marbot says that Platoff was in a drunken sleep when the French came up, and that, discipline in the Russian army being very strict, no one ventured to wake him or to stand to arms without his orders, and that it was to this circumstance that the Marshal owed the final escape of his little body of men.

but infantry with him, and after three days' march along the winding banks of the river, his rear harassed perpetually by Bashkirs and Tartars, who picked off and ill treated all stragglers, Ney at last rejoined the Emperor at Orcha.

In the struggle at Krasnoe, which lasted the whole day and in which we were exposed to a terrible artillery fire, my servant was wounded by grapeshot, and the two saddle horses he was leading were killed, whilst that on which he was mounted was very badly hurt. I thought the poor animal would certainly die of his awful wound, but, strange to say, he was the only one of all my horses to live to reach the Vistula. He was, in fact, quite well again when he was taken by the enemy at the gates of Thorn, and my poor servant was killed. With the horses I lost the furs they were carrying for my use, and nothing was left to me to protect me from the cold but a silk waterproof cloak, which turned out much more useful than I could have imagined, for it kept out the cold, and prevented my own animal heat, little as it was, from escaping.

As the day wore on at Krasnoe, and Ney did not appear, the anxiety of the Emperor and the army became more and more intense. Napoleon, fearing that his own retreat to Orcha would be cut off, dared not linger longer at Krasnoe, and he and his Guard left us an hour before nightfall, ordering us to wait for Marshal Ney. When the Emperor abandoned Krasnoe, the little town was full of those who had been wounded during the day. Nothing could have been more heart-rending than the sight of all the rooms in every house crowded with fine young fellows, their ages ranging from twenty to twenty-five years, who had but recently joined the army and had been under fire for the first time that day, but who within one short hour were to be left to their fate. Some few, who were able to march after their wounds had been dressed, were eager to be off again, but all the rest to the number of about 3,000 were left without surgeons or any necessaries.

The whole of the Guard was already gone, our much reduced first corps could no longer defend the heights beyond Krasnoe, and General Compans, who had remained till the very last, went down towards the town and crossed the ravine as night fell. He had scarcely done so, when, anxious to find out what the enemy were doing, I managed to creep along behind a hedge at the borders of the pass. The ravine was not more than thirty paces wide, and I very soon found myself almost face to face with a body of Russian artillery, which was being hastily put in position so as to riddle us with grapeshot. Beyond the ten or twelve guns of this battery I could see several considerable infantry corps advancing in line in our direction, leaving me no longer in any doubt of our being completely cut off from Marshal Ney. I hurried back with this distressing news to Marshal Davout, and recognising the hopelessness of waiting for Ney any longer, he did at once all that was left to us under the circumstances, by having a few guns placed in position to prevent the enemy from crossing the ravine, whilst the infantry was ordered to withdraw towards Lidoni, where we arrived a little before day-break, our retreat having been facilitated by a very dark night. The Russians, thinking that we were still in force at Krasnoe, did not enter it till the next morning.

The army was still deeply grieved at the supposed loss of Marshal Ney, for we were all certain that if he were alive he had been taken prisoner. The thought of his fate caused general discouragement, and all we hoped was to escape captivity ourselves. The numbers of the stragglers, ever on the increase, had now become immense; at every pass or difficult bit of road there was a block of wagons and carriages, and many vehicles broke through the ice on the marshes and remained embedded in the mud beneath. This was how I myself in our march from Lidoni to Koziani lost my baggage wagon and a barouche, which were still properly harnessed, for both of them with their drivers and horses were swallowed up by the mud. Hundreds of others met with similar misfortunes, and as I was going through Dombrowa the day after my loss, I came upon a carriage belonging to a M. de Servan, in which sat my sister. She had lost her carriage in the same marshes as I had mine, and de Servan, who had been more fortunate than either of us, had been good enough to take her on with him. A few hours later, just at the entrance to Orcha, de Servan's carriage was smashed by a cannon-ball. M. Levasseur, however, whose carriage got through safely, was good enough to allow me to transfer my sister to it, and showed her every possible attention.

On the evening of the 20th, when the moon was shining brightly, our advanced posts on the right bank of the Dnieper saw an officer approaching, whom they at first mistook for a Russian. They soon saw, however, that he was French, and on questioning him they learnt to their great joy that Marshal Ney, who had miraculously escaped from the clutches of Kutusoff, was but a league away from us. It would be difficult to describe our delight at receiving this news, which did much to restore the tone of the army, so lowered by discouragement. The Viceroy and Marshal Mortier hurried off to meet Ney, and the next day he was welcomed by the Emperor, who received him with the greatest enthusiasm, greeting him with the words, "I would have given everything rather than lose you."[17]

The first corps continued to cover the retreat, and we were aroused before daybreak after our arrival at Orcha by the coming up of the Russians in great force, who hoped to shut us into that town. Like every other place we passed through in our retreat, Orcha was encumbered with carriages crowded, as were the houses, with sick and wounded. A young cousin of mine, Alexander Lejeune, had been left at Orcha as manager of a hospital. I saw him as I went through, and he stuffed my pockets full of sugar and coffee ready roasted and ground. I urged him to fly whilst he could, and advised him to start in advance of us. He said he would just go and fetch a cloak and his money, but I never saw him again. He was probably delayed, and perished in the crowd after we left. If blood shed in the service of one's country is a patent of nobility, our family escutcheon ought to receive ten or twelve new chevrons of honour! For many of my nearest relations were wounded or killed in the Emperor's service, including five first cousins, namely, one Gérard, killed in Egypt; one Vignaux, killed in Spain; one Lejeune, killed in Russia; my brother, wounded at Friedland; the husband of my eldest sister, Baron Plique (General), wounded several times, who died whilst on active service; the brave and witty General Clary, brother of my wife, also wounded several times; and her other brother, who died at the age of twenty-two, as colonel of a regiment he had himself got together in Spain, and who was mourned as a son by his uncle, King Joseph.

I feel very proud at leaving such memories as these behind for my son, and have already too long delayed recording them in my Memoirs.

At Orcha the Dnieper is very wide, and so rapid that the ice was not firm enough to allow of our crossing it easily, although there were twenty-five degrees of frost. The two bridges, which were all we had been able to construct, were very narrow and far from strong, whilst the approaches to them were so slippery as

17 Henceforth Ney was nicknamed "the bravest of the brave."

to be very dangerous. We had had a great many carriages burnt in the streets of Orcha, but we were still terribly encumbered with them; the enemy harassed our rear perpetually, and had already gained a position from which they could cannonade our bridges. As soon as our troops were across we were compelled to set fire to these bridges, leaving behind us all who were without arms, or were for any reason unable to follow our rapid march. It was a terrible moment for us when we had thus to abandon so many of our wounded.

We passed the night of Sunday the 22nd with the rear-guard in a little wood on the road, and arrived in the evening of the 23rd at Kokonow, where we learnt, alas! that General Tchichakoff[18] had taken possession with a large force of Borisow,[19] so that we were cut off from that way of retreat. This news was distressing enough, but at three o'clock the next morning an event occurred which in its horror surpassed almost anything which had yet befallen us.

Opposite the house occupied by Marshal Davout and his officers, and not more than a couple of paces off, was a huge barn with four large doors, in which some five or six hundred persons, including officers, soldiers, stragglers, &c., had taken refuge as affording some shelter from the cold. Thirty or forty fires had been lighted, and the inmates of the barn, broken up into various groups, were all sleeping heavily in the warm air, which afforded such a contrast to the bitter cold of their usual bivouac, when the thatched roof caught fire, and in an instant the whole place was in flames. Suddenly, with a dull crash, the burning roof fell upon the sleepers, setting fire to the straw in which they lay, and to their clothes. Some few, who were near the doors, were able to escape; and with their clothes all singed they rushed to us screaming for help for their comrades. We were at the doors in a very few seconds; but what a terrible sight met our eyes! Masses of flames many yards thick rushed out from the doors to a distance of several yards, leaving only the narrow passage of exit some six feet high beneath a vault of fire which, fanned by the wind, spread with immense rapidity.

We could not, any of us, get near the poor creatures, whom we could see struggling wildly or flinging themselves face down-

18 Admiral Pavel Tshitsagov (1767–1849); his naval title was purely honorific and he was a land commander throughout his long career.
19 Borisov.

wards on the ground so as to suffer a little less. We hastily tied ropes, our handkerchiefs, anything we could get hold of, together, to fling to them, so as to be able to drag some of them out; but fresh shrieks soon stopped our efforts, for as we pulled they fell upon and were stabbed by each other's bayonets. Captain d'Houdetot got nearer to them than any of the rest of us were able to, but his clothes caught fire and he had to draw back. The 500 or 600 victims made several last despairing efforts to rise, but their strength was soon all gone, and presently the building fell, in upon them, their muskets became heated, the charges in them exploded, and their reports were the only funeral salute fired over the corpses of all the brave fellows. Very few escaped from the terrible conflagration, and those few had to tear off all their clothes. I saw one poor child of twelve or fourteen years old going about stark naked, but none of us could give him anything to put on, for we had lost our carriages, our horses, everything. There were now 13 degrees of frost, but we had to harden our hearts against the sufferers, for to help them was beyond our power.

On November 24 we passed the night at Tokotschin. The evening before Marshal Victor, pursued by Count von Wittgenstein, had joined the Emperor here, and would now protect his retreat. The Marshall had only just arrived from Germany, and had still 5,000 fresh troops in good order, whilst ours were thoroughly disorganised.

A singular episode occurred on the afternoon of the 24th, which had opened so tragically. I will just mention it here to give an idea of the vicissitudes we went through in our terrible retreat. Like the rest of us, I suffered very much from hunger, and for several days I had had nothing to eat but a little biscuit, whilst my only beverage was an occasional draught of cold coffee, which, however, kept me going somehow. I was marching sadly along, pondering on our woes, when an officer whom I scarcely knew by sight ran up to me, and with a pleasant smile asked me to do him a favour. "My position," I answered laughing, "is not such as to enable me to serve any one. But what do you want?" His reply was to hand me a parcel carefully done up in paper, and about the size of my two fists, which he begged me to accept. "But tell me what is in it," I said. "I entreat you not to refuse it." "But at least say what it is," I urged, trying to push it away with my right hand; but he closed my fingers over it and ran off. A good deal puzzled by suddenly receiving a present from a stranger, and quite at a loss to imagine his motive, I smelt the packet to begin with, and the result encouraged me to open it, when lo! and behold! a delicious odour of truffles greeted my nostrils, and I found myself the happy possessor of a quarter of a *pâté de foie gras* from Toulouse or Strasburg. I never saw the officer again, but I think my fervent expressions of gratitude must have found an echo in his heart. May he have escaped the fate which overtook so many of us, and from which his timely gift preserved me for a few days!

Another bit of good fortune marked this same day. As I have already said, I had lost all my furs and winter clothes, and in these deserted districts money was of no avail to buy new ones. I was feeling the want of them dreadfully, when I came across Colonel L. shut up in his carriage, and quite ill from the excessive precautions he was taking against the cold. "What do you want with all those furs?" I asked him. "You will be suffocated in them. Give me one." To which he replied, "Not for all the gold in the world!" "Bah!" I cried, "you will give me that bearskin, which really is in your way, and here are fifty gold napoleons for it." "Go to the devil! go to the devil with your napoleons! you bother me! – but there, General, I can't refuse you anything." He took the napoleons, and I hastily seized the bearskin, for fear he should think better of it. I went off with my treasure with indescribable joy, but the unlucky owner of so many sables and other furs was frozen to death a few days later.

The first corps passed the night of the 24th at Toloczin, and at daybreak the next morning the Russian firing recommenced, and we were pursued during the whole day by them, their balls mowing down our ranks. It was throughout our disastrous retreat the custom of the enemy to harass us all day, and when night fell to withdraw to distant villages, where they had a good rest and plenty of food, neither of which we were able to obtain, returning the next morning stronger than ever to attack us again with fresh vigour, whilst we were ever growing fewer and weaker.

On the 25th we passed the night in a wood close to a burnt village. The snow was very deep, adding greatly to our discomfort, and we made great piles of faggots on which to rest, all of us turning our feet towards one big fire. My bearskin was nearly fourteen feet square, and I let General Haxo roll himself into it with me. We were very warm and comfortable inside, and as we fell asleep we mentally blessed the man who had sold it to me. The next morning, the 26th, we started again before daybreak as usual, not daring to count those we had to leave behind.

On the same day we went through Borv, and the first corps halted for the night at Kroupski. A newly formed brigade of light Polish cavalry had just arrived in this village, and were heating the ovens in the cottages. An inn with stabling for twenty horses was assigned to Marshal Davout. In putting the horses which had followed – for, as I have said, we all went on foot now – in the stable, we found three children in a manger, one about a year old, the other two apparently only just born. They were very poorly dressed, and were so numbed with the cold that they were not even crying. I made my men seek their parents for an hour, but they could not be found; all the inhabitants had fled, and the three poor little things were left to our tender mercies. I begged the Marshal's cook to give them a little broth if he succeeded in making any, and thought no more about them. Presently, however, the warmth of the horses' breath woke the little creatures up, and their plaintive cries resounded for a long time in the rooms in which we were all crowded together. Our desire to do something to help them kept us awake for a long time, but at last we were overcome with sleep. At two o'clock in the morning we were roused by the news that the village was on fire; the overheating of the ovens had led to flames breaking out nearly everywhere. Our house, standing somewhat apart, was the only one to escape, and our three children were still crying. At daybreak, however, when we were starting, I could hear them no longer, and I asked the cook what he had done for them. He had, of course, suffered as much as we had, and he answered, with the satisfied air of a man who has done a good action, "Their crying so tore my heart that I could not close an eye. I had no food to give them, so I took a hatchet, broke the ice in the horse trough, and drowned them, to put them out of their misery!" Thus does misfortune harden the heart of man!

During the retreat many of the French were drowned. The wells in the village were all open and level with the ground, so that when troops arrived in the dark several men often fell into them, and rarely did any of their comrades try to save them. I saw more than ten wells, none of them very deep, on the surface of which the dead bodies of such victims were floating. Overcome with misery, other poor fellows committed suicide, and we often heard the discharge of a musket close by, telling of the end of some unfortunate wretch. On the other hand, some of the men who were simply covered with wounds kept up their courage and marched steadily on. One day, weary of walking, I sat down to rest on the trunk of a tree beside a fine young artilleryman who had just been wounded. Two doctors happened to pass us, and I called out to them to come and look at the wound. They did so, and at the first glance exclaimed, "The arm must be amputated!" I asked the soldier if he felt he could bear it. "Anything you like," he answered stoutly. "But," said the doctors, "there are only two of us to do it; so you, General, will be good enough to help us perform the operation." Seeing that I was anything but pleased at the idea, they hastened to add that it would be enough if I just let the artilleryman lean against me. "Sit back to back with him, and you will see nothing of the operation." I agreed, and placed myself in the required position. I think the operation seemed longer to me than it did to the patient. The doctors opened their cases of instruments; the artilleryman did not even heave a single sigh. I heard the slight noise made by the saw as it cut through the bone, and in a few seconds, or rather minutes, they said to me, "It is over! it is a pity we have not a little wine to give him, to help him to rally." I happened still to have half a bottle of Malaga with me, which I was hoarding up, only taking a drop at a time, but I gave it to the poor man, who was very pale, though he said nothing. His eyes brightened up, and he swallowed all my wine at a single gulp. Then, on returning the empty bottle with the words, "It is still a long way to Carcassonne," he walked off with a firm step at a pace I found it difficult to emulate.

Marshal Oudinot, who had recovered from his wound,[20] was now sent forward to Borisoff to try and take possession of the bridge over the Beresina, which had already been for several days in the hands of the Russian forces under Tchichakoff. This general had only just come from Moldavia, and on seeing the boldness with which Oudinot's troops advanced he took it for granted that the whole of the French army was approaching, and thinking his own position with the river behind him a very disadvantageous one, he wished to avoid a regular battle. He

20 The redoubtable Oudinot had been severely wounded at the battle of Polotsk, 17 August. He would be wounded again at Berezina, a bullet lodging inside his abdomen; he refused restraint, instead merely bit on a napkin while surgeons probed for the object.

therefore only made sufficient defence to cover the retreat of his army, and retired beyond the Beresina. Marshal Oudinot attacked Borisoff with his usual vigour, and entering it took 500 or 600 prisoners and all the baggage belonging to the Russian army. Tchichakoff had, however, burnt the bridge over the Beresina after crossing it, so that this victory gained us nothing.

The Emperor, who had no means of forcing the passage of the Beresina with an army of some 40,000 Russians opposing him, endeavoured to find a favourable point for throwing bridges across, and at the same time evading Wittgenstein, whom Marshal Victor was with infinite difficulty holding at bay, and Kutusoff, who was pursuing us. He was told that there was a ford at the village of Studzianka,[21] which he could reach by ascending the left bank of the river, but though the water was at the most four or five feet deep the approaches were very marshy and would be difficult for our carriages and artillery. The river, which was very muddy, was covered with ice, but it broke beneath those who tried to walk across it.

The difficulties on the other side, if we succeeded in reaching it, would be even greater, for heights commanding the banks were occupied by a Russian division, and the approach to these heights was a marshy tract without any firm road whatever. The road from Borisoff to Molodetschno by way of Zembino, the only one we could hope to reach, was a very narrow causeway, with many bridges raised to a good height above the marsh, much of which was quite under water. If any one of these little bridges should break, the march of the whole army would be arrested; but the Emperor had really no choice, and was compelled to resign himself to attempting the passage at Studzianka.

The engineers, pontonniers, and artillerymen therefore set to work at once, all the wood found in the village, even that of which the houses were built, being quickly converted into trestles, beams, planks, &c., and on the evening of the 26th, all appearing ready for the throwing across of the bridge, an attempt was made to place it in position. But the bed of the river was so muddy that the supports sank too deeply in it. It was, moreover, wider than had been supposed, and all the work had to be done over again. Two bridges instead of one were now

21 Studienka.

made, and the army began its march for Studzianka. On the 27th the first corps, now forming the rear guard, passed through Borisoff, and arrived at night at the ford chosen, where there was already a terrible block of carriages, those belonging to the corps of Marshal Oudinot and Marshal Victor, who had but recently rejoined us, being added to the others which had escaped from previous accidents, and whose owners had evaded the orders for burning them.

When we arrived at Studzianka about nine o'clock in the evening, the Emperor had already sent over in small rafts several hundred skirmishers to protect the bridges and those making them, whilst the corps of Marshals Ney and Oudinot with 500 or 600 cuirassiers of the Guard had crossed the river and taken up a position on the right in a wood beyond Studzianka. We passed the night in trying to bring something like order into our arrangements for crossing, sending the ammunition wagons first, and repairing the bridges where they had given way under the weight of the artillery. It was a very dark night, and many French, Dutch, Spanish, and Saxon soldiers fell into the wells of the village and were drowned. Their cries of distress reached us, but we had no ropes or ladders with which to rescue them, and they were left to their fate.

At daybreak the crossing of the river by the bridges went on without too much confusion, and I was able to go backwards and forwards several times, seeing to the safety of all that was of the greatest importance for the army; but at about eight o'clock in the morning, when the light revealed the immense crowds which had still to be got over, every one began to hasten to the bridges at once, and everything was soon thrown into the greatest disorder. Things became even worse when an hour later a combined attack was made on us by all the Russian forces, and we found ourselves between two fires. Truly our misfortunes had now reached their height.

Marshall Victor, who had taken up a position on the heights above Studzianka, was trying to beat off Wittgenstein, who had attacked him about ten o'clock with a large force of artillery, and although he had but very few troops with him, he managed to keep the Russians at a distance, but their balls, falling amongst the masses of carriages blocking the approaches to the bridges, flung their occupants and drivers into the most indescribable disorder, killing many and smashing up the vehicles. Some balls even rolled on to the bridges.[22]

On the right bank meanwhile Tchichakoff was attacking the French all along the line with some 25,000 or 30,000 Russians, whilst Marshals Ney and Oudinot had to oppose them only 9,000 or 10,000 men, with what was left of the Imperial Guard behind them as a reserve. Their front was but half a league in length, and the ground was very much broken up by woods. The Russians came to the fight well fed and warmed up by plenty of brandy; the French were debilitated by privations, and had moreover a cutting wind driving the snow in their faces. But with the enemy before them, they seemed to regain all their old energy, and Tchichakoff tried in vain to break their ranks, though he flung upon them in succession all the forces under his command. Marshal Oudinot, always in the front amongst the skirmishers, was wounded at the beginning of the action, and Marshal Ney took the command. Seizing a favourable moment he ordered General Doumerc, who had just brought up some 500 cuirassiers, to make a charge. This threw a Russian column into disorder, and won the French 1,500 prisoners. It was during this brilliant charge that a young officer, whom I loved for his many engaging qualities, met his death. Alfred de Noailles, only son of the Duc de Noailles, was struck in the heart by a ball,[23] and his face and body were so disfigured by being trampled beneath the feet of the horses, that he was only recognised by his height and by the mark on his fine white linen.

It was a melancholy consolation to his mourning widow and family to find his portrait in my album, in which I had collected likenesses of many young officers whom I numbered amongst my friends, and all of whom had been cut off in the flower of their age, before they had had time to fulfil the lofty destiny to which their noble names and exalted courage would have called them.

Towards three o'clock in the afternoon, when it was already beginning to get dark, for night falls very early in the winter in

22 Marbot relates that Marshal Victor's rearguard took the wrong road on its way to Studzianka, "and walked straight into the middle of Wittgenstein's army. The division was quickly surrounded and compelled to lay down its arms."

23 Marbot says that de Noailles escaped in the actual charge, but was killed by Cossacks after the engagement, he having ventured too far "to see what the enemy were doing."

these latitudes, Tchichakoff drew back, and we soon saw the fires of his bivouac, marking the position he had taken up about a league away from us.

Whilst all this was going on, the most awful scenes were being enacted at the entrance to the bridges on the right bank of the Beresina, and we could do absolutely nothing to prevent them.[24] Wittgenstein's artillery poured shells upon the struggling crowds, beneath whose weight the bridges were bending till they were under water. Those who could swim flung themselves into the river, trusting to their skill to save them, but they were overcome by the cold, and hardly any reached the further bank. On either side the hapless fugitives pressed on, driving others into the water, many clutching at the ropes of the bridges in the hope of being able to climb on to them. In the awful struggle none who fell ever rose again, for every one was immediately crushed to death by those behind, whilst all the while shells and balls rained upon the helpless masses. I was blessing God that my sister had escaped this terrible catastrophe, and had crossed some time before, when, to my horror, I saw M. Levasseur carrying her in his arms and endeavouring to make his way up to me. He had managed to extricate her from the crowd, and now brought her to me. "In what an awful moment do we meet again!" I exclaimed; "and what in the world can I do with you in your exhausted condition, now that you have found me? But courage," I added. "I got General Vasserot safely over in his carriage; I will find him, and put you under his care." This I managed to do, and two hours later my sister was kindly received by the General, to whom she said, "Oh, General, you have saved me; now I will take care of you."

The Beresina disaster was the Pultava with which the Russians had threatened us; it was not our only defeat or the last, but it was by far the most bloody of any which befell us. It involved the loss of the greater part of Marshall Victor's corps, which perished in defending our passage; and the loss of the whole of the Partouneaux[25] division, which had to surrender. In a word, it cost the French and their allies some 20,000 or 30,000

24 Many eye-witnesses of this awful disaster relate that the bridges were left almost empty on the night of the 27th, before the Russians came up, when all the non-combatants on the French side might quite easily have crossed.

25 This was Marshal Victor's rearguard, which mistook the road to the river.

men, killed, wounded, drowned, or taken prisoners. General Eblé, charged with the painful duty of burning the bridges after Marshal Victor's corps had passed over, had the greatest difficulty in cutting his way to them, and many of our own people were piteously struck down by the hatchets of his men before they were able to perform the task assigned to them. When at last the flames arose and the last hope of safety was cut off from those left on the other side, terrible were the cries of anguish which rent the air as thousands of poor wretches flung themselves into the water in a last despairing effort to escape. The ice broke beneath them; all was over, and the Cossacks swept down on the quarry, finding an immense amount of booty abandoned on the banks.[26]

On the evening of this terrible November 28 we halted at Zembino, a little town which had already been pillaged by our predecessors.[27] Marshal Davout and I took up our quarters in a little house crowded with others, which was heated by a stove. By dint of very close packing we managed to be able to lie down on the ground, and most of us slept profoundly till the time came to start again, which was before daybreak. I had been roused a few minutes before the clock struck the hour for departure by hearing stifled sobs, and by the dying light of a lamp I now made out the form of a tall and beautiful woman leaning against the stove, her face hidden by her hands, whilst the tears trickled through her fingers. It was a long time since I had seen any human creatures who had not lost all pretensions to good looks through their privations, and I was struck by the graceful attitude of this weeping figure, with the masses of light hair shading her ideal features. She reminded me of Canova's "Muse leaning on a Sepulchral Urn, and lost in Meditation." Whilst every one, wrapped in selfish egotism, left the room without taking any notice of the lady in distress, I approached her and asked her in a gentle voice what she was weeping about.

26 General Eblé delayed the burning of the bridges till the last possible moment, but the Russians were advancing to fall upon the rear of the fugitives, and, had they been able to use the bridges, the French loss would have been even greater than it was.

27 Lejeune passes very lightly over the march from the Beresina to Zembino, but it was a terribly arduous one, owing to the fact that the marshes, generally frozen over at this time of year, were still quite soft, and had the enemy pursued vigorously scarcely a man would have escaped to tell the tale.

She turned to me, revealing her beautiful face, wet with tears, and pointing to a pretty child asleep at her feet, she said, "I am the wife of M. Lavaux, a Frenchman, who had a library at Moscow. The Governor Rostopschin has sent him to Siberia, and I took refuge with my boy in the French army. The Duc de Plaisance and two other generals let us share their carriages till they were destroyed, and I have carried my child from the Beresina here, but my strength is exhausted. I can go no further, and I am in despair." "Could you keep your seat on horseback?" I asked at once. "I could try," she replied. "Well, do not lose courage; let us make haste. I will take your boy and place him on my sister's knee; she is in General Vasserot's carriage, and I will put you on a horse which a faithful servant shall lead. You will thus be able to follow your boy." A smile of hope lit up her expressive features. I fetched a wolf's skin, which was on the horse I meant to give her, and wrapped it about her to protect her from the intense cold which had now set in, took off several silk handkerchiefs I had about me, and tied them together to make sashes to fasten her on to her steed. I then placed her on her horse, put her under the care of one of my mounted servants, and they started together. I never saw lady, servant, or horses again; but Vasserot and my sister took care of the child, and gave him back to his mother, who came to claim him in the evening. I shall refer again to what I was able to learn of the adventurous career of this lady, who two years later was found by the Emperor Alexander I teaching the Demoiselles of the Légion d'Honneur at St Denis.

Beyond Zembino we had to cross a number of little bridges which the enemy had neglected to burn, and we felt that God had not entirely deserted us when He left us this means of getting over the marshes. We had not a scrap of food to give the 2,000 or 3,000 prisoners we were taking with us, and I purposely shut my eyes when they availed themselves of every chance of escape in the woods through which we passed. I could not bring myself to enforce their remaining with us by the cruel measures which alone could have availed, and I knew well enough that at any moment our fate might be worse than theirs.

Sunday, the 29th, was occupied by a dreary march to Kamen, which we reached about midnight. Our men, as tired out as ourselves, and longing for sleep, took a few bits of meat from the one wagon we still retained, in which tobacco and everything else were mixed together helter skelter. They did not notice in the darkness that some tobacco was sticking to the meat, and put it all into the pot on the fire together. At four o'clock in the morning, just before we started, the soup was given out, but it tasted most horribly of tobacco, and nobody but myself would take any of it. I was so hungry that I was not so prudent as the others, and I swallowed the whole of my portion. I had not marched far before a terrible headache came on; I felt sick, and soon began to vomit. I fainted away, and it was easy to see that I was poisoned. The news spread; even the Emperor heard of it, and in his despatches for Paris of that day he mentioned the matter, so that every one there thought I was dead. When we halted during the day, General Haxo and others, who had still a little humanity left, made me some tea, and drinking it saved my life. I remained with Marshal Davout in his *wurst*, and we arrived at Kotovitchi in the evening, where we put up at the house of the priest, a good old man, who spoke French very well, and who had declined to leave with the rest of the inhabitants because, though he had nothing with which to supply our bodily needs, he hoped to be able to minister to our spiritual necessities. Under his affectionate care I completely recovered, and when we set off again at four o'clock the next morning we were full of real gratitude to him.

During the whole of December 1 we were marching through dense forests, in which at every turn we came to difficult passes. We lost nearly all our prisoners here.

On December 2 we crossed the Ilia before daybreak, and entered yet other vast forests with no well-defined roads, and the snow added to the difficulties of our march, so that it was late before we got to Molodetschno. Whilst arranging for the camping of our troops in the dark, I fell into a swamp, and was only with great difficulty extricated. The cold was so intense that the mud froze about me immediately, so that it was hard work to get me out. On the very same day and at the same hour seven years before I had been seated on the snow beneath a tree, but it was after the battle of Austerlitz, and I was in a very happy frame of mind. The Emperor arrived at Molodetschno the same day, but instead of celebrating the anniversary of the greatest victory of his life, he had to dictate that terrible twenty-ninth bulletin describing succinctly the disasters his army had met with, though he disguised their true extent.

At four o'clock on the morning of the 3rd we started once more, without daring to count those who were unable to rise. Our route was strewn with the dead; and the wheels of the carriages, which were scarcely able to turn, went over the ice-covered corpses, often dragging them along for a little distance.

Haxo and I walked arm in arm, so as to save each other from slipping, and a soldier and an officer were walking one on either side of us. Presently the soldier drew a hunk of black Russian bread about the size of a fist out of his pocket, and began to gnaw at it greedily. The officer, surprised to see such a thing as bread, offered the grenadier a five-franc piece for it. "No, no!" said the man, tearing at his bread like a lion jealous of his prey. "Oh, do sell it to me," pleaded the officer; "here are ten francs." "No, no, no, no!" and the bread rapidly disappeared, till quite half was gone. "I am dying! I entreat you to save my life! Here are twenty francs!" Then with a savage look the grenadier bit off one more big mouthful, and, handing what was left to the officer, took the twenty francs, evidently feeling that he had made anything but a good bargain.

We were all covered with ice. Our breath, looking like thick smoke, froze as it left our mouths, and hung in icicles from our hair, eyebrows, moustaches, and beards, sometimes quite blinding us. Once Haxo, in breaking off the icicles which were bothering me, noticed that my cheeks and nose were discoloured. They looked like wax, and he informed me that they were frozen. He was right, for all sensation was gone from them. He at once began to rub them hard with snow, and a couple of minutes' friction restored circulation, but the pain was terrible, and it needed all my resolution not to resist having the rubbing continued. Colonel Emi, of the engineers, was frozen in exactly the same way a few minutes later, and in his despair he flung himself down and rolled about on the ground. We did not want to abandon him to his fate, but we had to strike him again and again before we could make him get up. Dysentery also worked terrible ravages amongst us, and its victims, with their dry and livid skin and emaciated limbs, looked like living skeletons. The poor creatures had had nothing to eat but a little crushed corn made into a kind of mash, for they had no means of grinding or of cooking it properly, and this indigestible food passed through the intestines without nourishing the body. Truly the unhappy wretches, many of them stark naked, presented, as they fell out by the way, a picture of death in its most revolting aspect.

Providence, however, had still a few moments in reserve for some of us, in which we found consolation for our woes, and gathered up fresh strength for the further trials awaiting us.

This was the case on December 4, as I will now relate. We had started before daybreak to escape a cannonade from some Cossacks, and we were already some distance from our bivouac when a second troop of Cossacks, bolder and more numerous than the first, flung itself across our path, and carried off two carriages belonging to the Commissary-General. Fortunately he was on foot, and managed to escape. A few miles beyond our party the same horde of Tartars drew up at the entrance to a ravine through which a body of some 300 or 400 Polish cavalry was endeavouring to pass so as to rejoin us. The Cossacks seemed likely to completely crush the Poles, when the noise of the firing attracted our attention, and we realised the danger of the brave fellows. General Gérard, with his usual chivalry, at once offered his services to Marshal Davout, and asked for volunteers to go to the aid of our allies. Though his men were worn out with fatigue, they were still full of confidence in him, and they one and all shouted, "I am ready! I am ready!" General Gérard dashed across the plain at their head, and when the Cossacks saw the little body of infantry approaching, they feared they were about to be caught between two fires and galloped off. The Poles thus rescued soon joined us, and a bit of really good fortune rewarded us all for our mutual help.

Some carriages belonging to a convoy from Germany had succeeded in reaching Markovo, a little village we were just about to enter. These carriages were packed full of fresh provisions of many different kinds, and the delight of our brave soldiers may be imagined when they found awaiting them a good meal of bread and cheese and butter, with plenty of wine to wash them down. What a feast it seemed after forty days of such scanty and miserable diet as theirs had been! We of the first corps shared in this rare good fortune.

General Guilleminot with his division had been the first to arrive at Markovo, and he had taken care that the precious carriages should not be pillaged. He was at the window of a little château when we were passing, and he called to us to join him. After having taken the necessary precaution of rubbing our faces with snow, but for which we should certainly have lost some of our features, we went into a warm room, where a very unexpected sight awaited us. Tea services of beautiful china were set out on handsome mahogany tables, whilst here and there were great piles of white bread and hampers of Brittany butter. At the sight of this wonderful spread, after our many weeks of privations, our eyes brightened and our nostrils became expanded like those of some Arab steed at the sound of the trumpet. Needless to say how eagerly and gladly we accepted the invitation to share in this delightful breakfast. We each did the part not of four, but of ten – our appetites were simply insatiable. Never did any breakfast party do greater justice to the fare provided than we did to the great bowls of tea poured out, and the thick slices of bread and butter cut for us by our host. It was hard work to tear ourselves away from this warm room with all its comforts to go and camp beneath the cold light of the stars near Smorgoni, where there were twenty-five degrees of frost.

The name of Smorgoni roused our curiosity, for we knew that the inhabitants of that village, situated as it was in the heart of a vast forest, devoted themselves to the chase of bears, selling the furs of the older animals, and training the young ones as gymnastic performers, often taking them the round of Europe to show off their tricks. The people of Smorgoni had not expected us, and took flight at our approach, carrying their furs and young bears with them, but for all that we expected to find the village interesting.

It was at Smorgoni on December 5 that the Emperor, yielding to the earnest entreaties of his most faithful servants, decided to leave the army and return to France, where his presence was most urgently needed. Before leaving, he signed the order for the promotion and reward of many officers and generals, which had been drawn up by Major-General Prince Berthier. He called his marshals together, frankly expressed to them his great regret at having lingered too long at Moscow, and announced to them his approaching departure, appointing King Murat of Naples to the command of the army.

It was eleven o'clock at night, and there were twenty-five degrees of frost when the Emperor left Smorgoni, accompanied by the Dukes of Vicenza and Friuli (Marshal Duroc) and the Count of Lobau (Marshal Mouton), and made his way to Osmiana, miraculously escaping from the 1,200 Cossacks whom he had to pass, and who would certainly have taken him prisoner if they had known he was so near them with an escort of scarcely 100 men. A little before dark these same Cossacks had been beaten by General Loison, and driven out of Osmiana, where they had hoped to arrest our retreat. Whilst waiting for daylight the enemy were sleeping a little distance from the road, and the

Emperor passed them unnoticed. Napoleon's departure threw the whole army into the greatest discouragement.

General Charpentier still declined to take my place, and I was compelled as before to perform the duties of Chief of the Staff. Fortunately, Marshal Davout now seemed to understand my position better, and was no longer so exacting. This made me willing to remain with him a few days longer.

On December 6 we passed through the little village of Pletchinzy just as a very interesting scene was taking place in it. Marshal Oudinot and General Pino, both wounded, had passed the night there with twenty-five or thirty officers and men belonging to their suite. A Cossack officer had heard of their presence, and thinking to take a great prize, he with some 200 men had surrounded the house in which they were. Speaking in good French, he politely summoned them to surrender. "We never surrender," was the reply, and a few well-aimed shots struck down some of the Cossacks. The hovel, for it was little more, was now regularly besieged, the French firing at close quarters into the ranks of the assailants, which they thinned considerably. Marshal Oudinot himself, though suffering greatly from a ball in the loins and unable to rise from the pallet on which he lay, made some holes in the walls between the planks, and firing through them picked off a good many Cossacks, for he never once missed his aim. Meanwhile, however, the enemy received reinforcements, and a gun was brought up to their aid. Four balls had already made a breach in the hut, but no one had been hurt. The French, after the manner of the Spanish, at once turned the opening to account by firing through it at their besiegers. A fifth ball broke the pallet on which the Marshal lay, and at the same time brought down the side of an oven in which five or six little children belonging to the peasant who had owned the hut, were discovered huddled together. The poor little things rushed out into the smoke and confusion in a great state of terror, much to the surprise of our men. There was something very touching in the way the little creatures clung to each other in the midst of the struggle. Fortunately our party came up just when things were going hardly with the besieged, for we had quickened our pace when we heard the firing, and the Cossacks, who had lost some fifty men killed and wounded, took to flight at our approach. We escorted the Marshal to Osmiana, where we halted for the night.

Here we found a division, consisting of some 12,000 fine

young recruits, who had just arrived from France as reserves, under General Loison. Alas! twenty-four hours of our temperature was enough to kill off half of them, for they were in summer clothing, and not yet acclimatised; and three days later, when we reached Wilna, not one survived of the poor fellows whose weeping mothers had watched them start so short a time ago. I have been told since by several Russians that if the wind had blown from the north with the temperature at from 25 to 30 degrees, not one of us would have escaped alive. When the murderous north wind is blowing, the Russians generally remain indoors all day and night in rooms heated by stoves, and if they ever do venture forth it is only after a good meal, cased in woollen garments and thick furs, with which in our inexperience few of us had provided ourselves. The French died off, but the Cossacks fared splendidly.

Some half a million French soldiers perished in the "terrible retreat" from Russia.

William H. Prescott

The Death of Montezuma

Montezuma, the last Aztec ruler of Mexico, perished during the invasion of the Spanish conquistadors under Hernando Cortes in 1520.

THE palace of Axayacatl, in which the Spaniards were quartered, was a vast, irregular pile of stone buildings, having but one floor, except in the centre, where another story was added, consisting of a suite of apartments which rose like turrets on the main building of the edifice. A vast area stretched around, encompassed by a stone wall of no great height. This was supported by towers or bulwarks at certain intervals, which gave it some degree of strength, not, indeed, as compared with European fortifications, but sufficient to resist the rude battering enginery of the Indians. The parapet had been pierced here and there with embrasures for the artillery, which consisted of thirteen guns; and smaller apertures were made in other parts for the convenience of the arquebusiers. The Spanish forces found accommodations within the great building; but the numerous body of Tlascalan auxiliaries could have had no other shelter than what was afforded by barracks or sheds hastily constructed for the purpose in the spacious court-yard. Most of them, probably, bivouacked under the open sky, in a climate milder than that to which they were accustomed among the rude hills of their native land. Thus crowded into a small and compact compass, the whole army could be assembled at a moment's notice; and, as the Spanish commander was careful to enforce the strictest discipline and vigilance, it was scarcely possible that he could be taken by surprise. No sooner, therefore, did the trumpet call to arms, as the approach of the enemy was announced, than every soldier was at his post, the cavalry mounted, the artillery-men at their guns, and the archers and arquebusiers stationed so as to give the assailants a warm reception.

On they came, with the companies, or irregular masses, into which the multitude was divided, rushing forward each in its own dense column, with many a gay banner displayed, and many a bright gleam of light reflected from helmet, arrow, and spear-head, as they were tossed about in their disorderly array. As they drew near the inclosure, the Aztecs set up a hideous yell, or rather that shrill whistle used in fight by the nations of Anahuac, which rose far above the sound of shell and atabal, and their other rude instruments of warlike melody. They followed this by a tempest of missiles – stones, darts, and arrows – which fell thick as rain on the besieged, while volleys of the same kind descended from the crowded terraces in the neighborhood.

The Spaniards waited until the foremost column had arrived within the best distance for giving effect to their fire, when a general discharge of artillery and arquebuses swept the ranks of the assailants, and mowed them down by hundreds. The Mexicans were familiar with the report of these formidable engines, as they had been harmlessly discharged on some holiday festival; but never till now had they witnessed their murderous power. They stood aghast for a moment, as with bewildered looks they staggered under the fury of the fire; but, soon rallying, the bold barbarians uttered a piercing cry, and rushed forward over the prostrate bodies of their comrades. A second and third volley checked their career, and threw them into disorder, but still they pressed on, letting off clouds of arrows; while their comrades on the roofs of the houses took more deliberate aim at the combatants in the court-yard. The Mexicans were particularly expert in the use of the sling; and the stones which they hurled from their elevated positions on the heads of their enemies did even greater execution than the arrows. They glanced, indeed, from the mail-covered bodies of the cavaliers, and from those who were sheltered under the cotton panoply, or *escaupil*. But some of the soldiers, especially the veterans of Cortés, and many of their Indian allies, had but slight defences, and suffered greatly under this stony tempest.

The Aztecs, meanwhile, had advanced close under the walls of the intrenchment; their ranks broken and disordered, and their limbs mangled by the unintermitting fire of the Christians. But they still pressed on, under the very muzzle of the guns. They endeavored to scale the parapet, which, from its moderate height, was in itself a work of no great difficulty. But the moment they showed their heads above the rampart, they were shot down by the unerring marksmen within, or stretched on the ground by a blow of a Tlascalan *maquahuitl*. Nothing daunted, others soon appeared to take the place of the fallen, and strove, by raising themselves on the writhing bodies of their dying comrades, or by fixing their spears in the crevices of the wall, to surmount the barrier. But the attempt proved equally vain.

Defeated here, they tried to effect a breach in the parapet by battering it with heavy pieces of timber. The works were not constructed on those scientific principles by which one part is made to overlook and protect another. The besiegers, therefore, might operate at their pleasure, with but little molestation from the garrison within, whose guns could not be brought into a position to bear on them, and who could mount no part of their own works for their defence, without exposing their persons to the missiles of the whole besieging army. The parapet, however, proved too strong for the efforts of the assailants. In their despair they endeavored to set the Christian quarters on fire, shooting burning arrows into them, and climbing up so as to dart their firebrands through the embrasures. The principal edifice was of stone. But the temporary defences of the Indian allies, and other parts of the exterior works, were of wood. Several of these took fire, and the flames spread rapidly among the light, combustible materials. This was a disaster for which the besieged were wholly unprepared. They had little water, scarcely enough for their own consumption. They endeavored to extinguish the flames by heaping on earth. But in vain. Fortunately the great building was of materials which defied the destroying element. But the fire raged in some of the outworks, connected with the parapet, with a fury which could only be checked by throwing down a part of the wall itself, thus laying open a formidable breach. This, by the general's order, was speedily protected by a battery of heavy guns, and a file of arquebusiers, who kept up an incessant volley through the opening on the assailants.

The fight now raged with fury on both sides. The walls around the palace belched forth an unintermitting sheet of flames and smoke. The groans of the wounded and dying were lost in the fiercer battle-cries of the combatants, the roar of the artillery, the sharper rattle of the musketry, and the hissing sound of Indian missiles. It was the conflict of the European with the American; of civilized man with the barbarian; of the science of the one with the rude weapons and warfare of the other: And as the ancient walls of Tenochtitlan shook under the thunders of the artillery – it announced that the white man, the destroyer, had set his foot within her precincts.

Night at length came, and drew her friendly mantle over the contest. The Aztec seldom fought by night. It brought little repose, however, to the Spaniards, in hourly expectation of an assault; and they found abundant occupation in restoring the breaches in their defences, and in repairing their battered armor. The beleaguering host lay on their arms through the night, giving token of their presence, now and then, by sending a stone or shaft over the battlements, or by a solitary cry of defiance from some warrior more determined than the rest, till all other sounds were lost in the vague, indistinct murmurs which float upon the air in the neighborhood of a vast assembly.

The ferocity shown by the Mexicans seems to have been a thing for which Cortés was wholly unprepared. His past experience, his uninterrupted career of victory with a much feebler force at his command, had led him to underrate the military efficiency, if not the valor, of the Indians. The apparent facility, with which the Mexicans had acquiesced in the outrages on their sovereign and themselves, had led him to hold their courage, in particular, too lightly. He could not believe the present assault to be anything more than a temporary ebullition of the populace, which would soon waste itself by its own fury. And he proposed, on the following day, to sally out and inflict such chastisement on his foes as should bring them to their senses, and show who was master in the capital.

With early dawn, the Spaniards were up and under arms; but not before their enemies had given evidence of their hostility by the random missiles, which, from time to time, were sent into the inclosure. As the grey light of morning advanced, it showed the besieging army far from being diminished in numbers, filling up the great square and neighboring avenues in more dense array than on the preceding evening. Instead of a confused, disorderly rabble, it had the appearance of something like a regular force, with its battalions distributed under their respective banners, the devices of which showed a contribution from the principal cities and districts in the Valley. High above the rest was conspicuous the ancient standard of Mexico, with its well known cognizance, an eagle pouncing on an ocelot, emblazoned on a rich mantle of feather-work. Here and there priests might be seen mingling in the ranks of the besiegers, and, with frantic gestures, animating them to avenge their insulted deities.

The greater part of the enemy had little clothing save the *maxtlatl*, or sash round the loins. They were variously armed, with long spears tipped with copper, or flint, or sometimes merely pointed and hardened in the fire. Some were provided with slings, and others with darts having two or three points, with long strings attached to them, by which, when discharged, they could be torn away again from the body of the wounded. This was a formidable weapon, much dreaded by the Spaniards. Those of a higher order wielded the terrible *maquahuitl*, with its sharp and brittle blades of obsidian. Amidst the motley bands of warriors, were seen many whose showy dress and air of authority intimated persons of high military consequence. Their breasts were protected by plates of metal, over which was thrown the gay surcoat of feather-work. They wore casques resembling, in their form, the head of some wild and ferocious animal, crested with bristly hair, or overshadowed by tall and graceful plumes of many a brilliant color. Some few were decorated with the red fillet bound round the hair, having tufts of cotton attached to it, which denoted by their number that of the victories they had won, and their own preeminent rank among the warriors of the nation. The motley assembly plainly showed that priest, warrior, and citizen had all united to swell the tumult.

Before the sun had shot his beams into the Castilian quarters, the enemy were in motion, evidently preparing to renew the assault of the preceding day. The Spanish commander determined to anticipate them by a vigorous sortie, for which he had already made the necessary dispositions. A general discharge of ordnance and musketry sent death far and wide into the enemy's ranks, and, before they had time to recover from their confusion, the gates were thrown open, and Cortés, sallying out at the head of his cavalry, supported by a large body of infantry and several thousand Tlascalans, rode at full gallop against them. Taken thus by surprise, it was scarcely possible to offer much resistance. Those who did were trampled down under the horses' feet, cut to pieces with the broadswords, or pierced with the lances of the riders. The infantry followed up the blow, and the rout for the moment was general.

But the Aztecs fled only to take refuge behind a barricade, or strong work of timber and earth, which had been thrown across the great street through which they were pursued. Rallying on the other side, they made a gallant stand, and poured in turn a volley of their light weapons on the Spaniards, who, saluted with a storm of missiles at the same time, from the terraces of the houses, were checked in their career, and thrown into some disorder.

Cortés, thus impeded, ordered up a few pieces of heavy ordnance, which soon swept away the barricades, and cleared a passage for the army. But it had lost the momentum acquired in its rapid advance. The enemy had time to rally and to meet the Spaniards on more equal terms. They were attacked in flank, too, as they advanced, by fresh battalions, who swarmed in from the adjoining streets and lanes. The canals were alive with boats filled with warriors, who, with their formidable darts searched every crevice or weak place in the armor of proof, and made havoc on the unprotected bodies of the Tlascalans. By repeated and vigorous charges, the Spaniards succeeded in driving the Indians before them; though many, with a desperation which showed they loved vengeance better than life, sought to embarrass the movements of their horses by clinging to their legs, or, more successfully strove to pull the riders from their saddles. And woe to the unfortunate cavalier who was thus dismounted – to be despatched by the brutal *maquahuitl*, or to be dragged on board a canoe to the bloody altar of sacrifice!

But the greatest annoyance which the Spaniards endured was from the missiles from the *azoteas*, consisting often of large stones, hurled with a force that would tumble the stoutest rider from his saddle. Galled in the extreme by these discharges, against which even their shields afforded no adequate protection, Cortés ordered fire to be set to the buildings. This was no very difficult matter, since, although chiefly of stone, they were filled with mats, cane-work, and other combustible materials, which were soon in a blaze. But the buildings stood separated from one another by canals and drawbridges, so that the flames did not easily communicate to the neighboring edifices. Hence, the labor of the Spaniards was incalculably increased and their progress in the work of destruction – fortunately for the city – was comparatively slow. They did not relax their efforts, however, till several hundred houses had been consumed, and the miseries of a conflagration, in which the wretched inmates perished equally with the defenders, were added to the other horrors of the scene.

The day was now far spent. The Spaniards had been everywhere victorious. But the enemy, though driven back on every point, still kept the field. When broken by the furious charges of the cavalry, he soon rallied behind the temporary defences, which, at different intervals, had been thrown across the streets, and, facing about, renewed the fight with undiminished courage, till the sweeping away of the barriers by the cannon of the assailants left a free passage for the movements of their horse. Thus the action was a succession of rallying and retreating, in which both parties suffered much, although the loss inflicted on the Indians was probably tenfold greater than that of the Spaniards. But the Aztecs could better afford the loss of a hundred lives than their antagonists that of one. And, while the Spaniards showed an array broken, and obviously thinned in numbers, the Mexican army, swelled by the tributary levies which flowed in upon it from the neighboring streets, exhibited, with all its losses, no sign of diminution. At length, sated with carnage, and exhausted by toil and hunger, the Spanish commander drew off his men, and sounded a retreat.

On his way back to his quarters, he beheld his friend, the secretary Duero, in a street adjoining, unhorsed, and hotly engaged with a body of Mexicans, against whom he was desperately defending himself with his poniard. Cortés, roused at the sight, shouted his war-cry, and, dashing into the midst of the enemy, scattered them like chaff by the fury of his onset; then, recovering his friend's horse, he enabled him to remount, and the two cavaliers, striking their spurs into their steeds, burst through their opponents and joined the main body of the army. Such displays of generous gallantry were not uncommon in these engagements, which called forth more feats of personal adventure than battles with antagonists better skilled in the science of war. The chivalrous bearing of the general was emulated in full measure by Sandoval, De Leon, Olid, Alvarado, Ordaz, and his other brave companions, who won such glory under the eye of their leader, as prepared the way for the independent commands which afterwards placed provinces and kingdoms at their disposal.

The undaunted Aztecs hung on the rear of their retreating foes, annoying them at every step by fresh flights of stones and arrows; and, when the Spaniards had re-entered their fortress, the Indian host encamped around it, showing the same dogged resolution as on the preceding evening. Though true to their ancient habits of inaction during the night, they broke the stillness of the hour by insulting cries and menaces, which reached the ears of the besieged. "The gods have delivered you, at last, into our hands," they said; "Huitzilopotchli has long cried for his victims. The stone sacrifice is ready. The knives are sharpened. The wild beasts in the palace are roaring for their offal. And the cages," they added, taunting the Tlascalans with their leanness, "are waiting for the false sons of Anahuac, who are to be fattened for the festival!" These dismal menaces, which sounded fearfully in the ears of the besieged, who understood too well their import, were mingled with piteous lamentations for their sovereign, whom they called on the Spaniards to deliver up to them.

Cortés suffered much from a severe wound which he had received in the hand in the late action. But the anguish of his mind must have been still greater, as he brooded over the dark prospect before him. He had mistaken the character of the Mexicans. Their long and patient endurance had been a violence to their natural temper, which, as their whole history proves, was arrogant and ferocious beyond that of most of the races of Anahuac. The restraint, which, in deference to their monarch, more than to their own fears, they had so long put on their natures, being once removed, their passions burst forth with accumulated violence. The Spaniards had encountered in the Tlascalan an open enemy, who had no grievance to complain of, no wrong to redress. He fought under the vague apprehension only of some coming evil to his country. But the Aztec, hitherto the proud lord of the land, was goaded by insult and injury, till he reached that pitch of self-devotion, which made life cheap, in comparison with revenge. Armed thus with the energy of despair, the savage is almost a match for the civilized man; and a whole nation, moved to its depths by a common feeling, which swallows up all selfish considerations of personal interest and safety, becomes, whatever be its resources, like the earthquake and the tornado, the most formidable among the agencies of nature.

Considerations of this kind may have passed through the mind of Cortés, as he reflected on his own impotence to restrain the fury of the Mexicans, and resolved, in despite of his late supercilious treatment of Montezuma, to employ his authority to allay the tumult – an authority so successfully exerted in behalf of Alvarado, at an earlier stage of the insurrection. He was the more confirmed in his purpose, on the following morning, when the assailants, redoubling their efforts, succeeded in scaling the works in one quarter, and effecting an entrance into the inclosure. It is true, they were met with so resolute a spirit, that not a man, of those who entered, was left alive. But, in the impetuosity of the assault, it seemed, for a few moments, as if the place was to be carried by storm.

Cortés now sent to the Aztec emperor to request his interposition with his subjects in behalf of the Spaniards. But Montezuma was not in the humor to comply. He had remained moodily in his quarters ever since the general's return. Disgusted with the treatment he had received, he had still further cause for mortification in finding himself the ally of those who were the open enemies of his nation. From his apartment he had beheld the tragical scenes in his capital, and seen another, the presumptive heir to his throne, taking the place which he should have occupied at the head of his warriors, and fighting the battles of his country. Distressed by his position, indignant at those who had placed him in it, he coldly answered, "What have I to do with Malinche? I do not wish to hear from him. I desire only to die. To what a state has my willingness to serve him reduced me!" When urged still further to comply by Olid and father Olmedo, he added, "It is of no use. They will neither believe me, nor the false words and promises of Malinche. You will never leave these walls alive." On being assured, however, that the Spaniards would willingly depart, if a way were opened to them by their enemies, he at length – moved, probably more by a desire to spare the blood of his subjects, than of the Christians – consented to expostulate with his people.

In order to give the greater effect to his presence, he put on his imperial robes. The *tilmatli*, his mantle of white and blue, flowed over his shoulders, held together by its rich clasp of the green *chalchivitl*. The same precious gem, with emeralds of uncommon size, set in gold, profusely ornamented other parts of his dress. His feet were shod with the golden sandals, and his brows covered by the *copilli*, or Mexican diadem, resembling in form the pontifical tiara. Thus attired, and surrounded by a guard of Spaniards and several Aztec nobles, and preceded by the golden wand, the symbol of sovereignty, the Indian monarch ascended the central turret of the palace. His presence was instantly recognised by the people, and, as the royal retinue advanced along the battlements, a change, as if by magic, came over the scene. The clang of instruments, the fierce cries of the assailants, were hushed, and a deathlike stillness pervaded the whole assembly, so fiercely agitated, but a few moments before, by the wild tumult of war! Many prostrated themselves on the ground; others bent the knee; and all turned with eager expectation towards the monarch, whom they had been taught to reverence with slavish awe, and from whose countenance they had been wont to turn away as from the intolerable splendors of divinity! Montezuma saw his advantage; and, while he stood thus confronted with his awe-struck people, he seemed to recover all his former authority and confidence, as he felt himself to be still a king. With a calm voice, easily heard over the silent assembly, he is said by the Castilian writers to have thus addressed them.

"Why do I see my people here in arms against the palace of my fathers? Is it that you think your sovereign a prisoner, and wish to release him? If so, you have acted rightly. But you are mistaken. I am no prisoner. The strangers are my guests. I remain with them only from choice, and can leave them when I list. Have you come to drive them from the city? That is unnecessary. They will depart of their own accord, if you will open a way for them. Return to your homes, then. Lay down your arms. Show your obedience to me who have a right to it. The white men shall go back to their own land; and all shall be well again within the walls of Tenochtitlan."

As Montezuma announced himself the friend of the detested strangers, a murmur ran through the multitude; a murmur of contempt for the pusillanimous prince who could show himself so insensible to the insults and injuries for which the nation was in arms! The swollen tide of their passions swept away all the barriers of ancient reverence, and, taking a new direction, descended on the head of the unfortunate monarch, so far degenerated from his warlike ancestors. "Base Aztec," they exclaimed, "woman, coward, the white men have made you a woman – fit only to weave and spin!" These bitter taunts were soon followed by still more hostile demonstrations. A chief, it is said, of high rank, bent a bow or brandished a javelin with an air of defiance against the emperor, when, in an instant, a cloud of stones and arrows descended on the spot where the royal train was gathered. The Spaniards appointed to protect his person had been thrown off their guard by the respectful deportment of the people during their lord's address. They now hastily interposed their bucklers. But it was too late. Montezuma was wounded by three of the missiles, one of which, a stone, fell with such violence on his head, near the temple, as brought him senseless to the ground. The Mexicans, shocked at their own sacrilegious act, experienced a sudden revulsion of feeling, and, setting up a dismal cry, dispersed panic-struck, in different directions. Not one of the multitudinous array remained in the great square before the palace!

The unhappy prince, meanwhile was borne by his attendants to his apartments below. On recovering from the insensibility caused by the blow, the wretchedness of his condition broke upon him. He had tasted the last bitterness of degradation. He had been reviled, rejected, by his people. The meanest of the rabble had raised their hands against him. He had nothing more to live for. It was in vain that Cortés and his officers endeavored to soothe the anguish of his spirit and fill him with better thoughts. He spoke not a word in answer. His wound, though dangerous, might still, with skilful treatment, not prove mortal. But Montezuma refused all the remedies prescribed for it. He tore off the bandages as often as they were applied, maintaining, all the while, the most determined silence. He sat with eyes dejected, brooding over his fallen fortunes, over the image of ancient majesty, and present humiliation. He had survived his honor. But a spark of his ancient spirit seemed to kindle in his bosom, as it was clear he did not mean to survive his disgrace. From this painful scene the Spanish general and his followers were soon called away by the new dangers which menaced the garrison.

Opposite to the Spanish quarters, at only a few rods' distance, stood the great *teocalli* of Huitzilopotchli. This pyramidal mound, with the sanctuaries that crowned it, rising altogether to the height of near a hundred and fifty feet, afforded an elevated position that completely commanded the palace of Axayacatl, occupied by the Christians. A body of five or six hundred Mexicans, many of them nobles and warriors of the highest rank, had got possession of the *teocalli*, whence they discharged such a tempest of arrows on the garrison, that no one could leave his defences for a moment without imminent danger; while the Mexicans, under shelter of the sanctuaries, were entirely covered from the fire of the besieged. It was obviously necessary to dislodge the enemy, if the Spaniards would remain longer in their quarters.

Cortés assigned this service to his chamberlain, Escobar, giving him a hundred men for the purpose, with orders to storm the *teocalli*, and set fire to the sanctuaries. But that officer was thrice repulsed in the attempt, and, after the most desperate efforts, was obliged to return with considerable loss, and without accomplishing his object.

Cortés, who saw the immediate necessity of carrying the place, determined to lead the storming party himself. He was then suffering much from the wound in his left hand, which had disabled it for the present. He made the arm serviceable, however, by fastening his buckler to it, and, thus crippled, sallied out at the head of three hundred chosen cavaliers, and several thousand of his auxiliaries.

In the court-yard of the temple he found a numerous body of Indians prepared to dispute his passage. He briskly charged them, but the flat, smooth stones of the pavement were so slippery, that the horses lost their footing, and many of them fell. Hastily dismounting, they sent back the animals to their quarters, and, renewing the assault, the Spaniards succeeded without much difficulty in dispersing the Indian warriors, and opening a free passage for themselves to the *teocalli*. This building, as the reader may remember, was a huge pyramidal structure, about three hundred feet square at the base. A flight of stone steps on the outside, at one of the angles of the mound, led to a platform, or terraced walk, which passed round the building until it reached a similar flight of stairs directly over the preceding, that conducted to another landing as before. As there were five bodies or divisions of the *teocalli*, it became necessary to pass round its whole extent four times, or nearly a mile, in order to reach the summit, which, it may be recollected, was an open area, crowned only by the two sanctuaries dedicated to the Aztec deities.

Cortés, having cleared a way for the assault, sprang up the lower stairway, followed by Alvarado, Sandoval, Ordaz, and the other gallant cavaliers of his little band, leaving a file of arquebusiers and a strong corps of Indian allies to hold the enemy in check at the foot of the monument. On the first landing, as well as on the several galleries above, and on the summit, the Aztec warriors were drawn up to dispute his passage. From their elevated position they showered down volleys of lighter missiles, together with heavy stones, beams, and burning rafters, which, thundering along the stairway, overturned the ascending Spaniards, and carried desolation through their ranks. The more fortunate, eluding or springing over these obstacles, succeeded in gaining the first terrace; where, throwing themselves on their enemies, they compelled them, after a short resistance, to fall back. The assailants pressed on, effectually supported by a brisk fire of the musketeers from below, which so much galled the Mexicans in their exposed situation, that they were glad to take shelter on the broad summit of the *teocalli*.

Cortés and his comrades were close upon their rear, and the two parties soon found themselves face to face on this aërial battle-field, engaged in mortal combat in presence of the whole city, as well as of the troops in the court-yard, who paused, as if by mutual consent, from their own hostilities, gazing in silent expectation on the issue of those above. The area, though somewhat smaller than the base of the *teocalli*, was large enough to afford a fair field of fight for a thousand combatants. It was paved with broad, flat stones. No impediment occurred over its surface, except the huge sacrificial block, and the temples of stone which rose to the height of forty feet, at the further extremity of the arena. One of these had been consecrated to the Cross. The other was still occupied by the Mexican war-god. The Christian and the Aztec contended for their religions under the very shadow of their respective shrines; while the Indian priests, running to and fro, with their hair wildly streaming over their sable mantles, seemed hovering in mid air, like so many demons of darkness urging on the work of slaughter!

The parties closed with the desperate fury of men who had no hope but in victory. Quarter was neither asked nor given; and to fly was impossible. The edge of the area was unprotected by parapet or battlement. The least slip would be fatal; and the combatants, as they struggled in mortal agony, were sometimes seen to roll over the sheer sides of the precipice together. Cortés himself is said to have had a narrow escape from this dreadful fate. Two warriors, of strong, muscular frames, seized on him, and were dragging him violently towards the brink of the pyramid. Aware of their intention, he struggled with all his force, and, before they could accomplish their purpose, succeeded in tearing himself from their grasp, and hurling one of them over the walls with his own arm! The story is not improbable in itself, for Cortés was a man of uncommon agility and strength. It has been often repeated; but not by contemporary history.

The battle lasted with unintermitting fury for three hours. The number of the enemy was double that of the Christians; and it seemed as if it were a contest which must be determined by numbers and brute force, rather than by superior science. But it was not so. The invulnerable armor of the Spaniard, his sword of matchless temper, and his skill in the use of it, gave him advantages which far outweighed the odds of physical strength and numbers. After doing all that the courage of despair could enable men to do, resistance grew fainter and fainter on the side of the Aztecs. One after another they had fallen. Two or three priests only survived to be led away in triumph by the victors. Every other combatant was stretched a corpse on the bloody arena, or had been hurled from the giddy heights. Yet the loss of the Spaniards was not inconsiderable. It amounted to forty-five of their best men, and nearly all the remainder were more or less injured in the desperate conflict.

The victorious cavaliers now rushed towards the sanctuaries. The lower story was of stone; the two upper were of wood. Penetrating into their recesses, they had the mortification to find the image of the Virgin and the Cross removed. But in the other edifice they still beheld the grim figure of Huitzilopotchli, with his censer of smoking hearts, and the walls of his oratory reeking with gore – not improbably of their own country-men! With shouts of triumph the Christians tore the uncouth monster from his niche, and tumbled him, in the presence of the horror-struck Aztecs, down the steps of the *teocalli*. They then set fire to the accursed building. The flames speedily ran up the slender towers, sending forth an ominous light over city, lake, and valley, to the remotest hut among the mountains. It was the funeral pyre of Paganism, and proclaimed the fall of that sanguinary religion which had so long hung like a dark cloud over the fair regions of Anahuac!

Having accomplished this good work, the Spaniards descended the winding slopes of the *teocalli* with more free and buoyant step, as if conscious that the blessing of Heaven now rested on their arms. They passed through the dusky files of Indian warriors in the court-yard, too much dismayed by the appalling scenes they had witnessed to offer resistance; and reached their own quarters in safety. That very night they followed up the blow by a sortie on the sleeping town, and burned three hundred houses, the horrors of conflagration being made still more impressive by occurring at the hour when the Aztecs, from their own system of warfare, were least prepared for them.

Hoping to find the temper of the natives somewhat subdued by these reverses, Cortés now determined, with his usual policy, to make them a vantage-ground for proposing terms of accommodation. He accordingly invited the enemy to a parley, and, as the principal chiefs, attended by their followers, assembled in the great square, he mounted the turret before occupied by Montezuma, and made signs that he would address them. Marina, as usual, took her place by his side, as his interpreter. The multitude gazed with earnest curiosity on the Indian girl, whose influence with the Spaniards was well known, and whose connection with the general, in particular, had led the Aztecs to designate him by her Mexican name of Malinche. Cortés, speaking through the soft, musical tones of his mistress, told his audience they must now be convinced, that they had nothing further to hope from opposition to the Spaniards. They had seen their gods trampled in the dust, their altars broken, their dwellings burned their warriors falling on all sides. "All this," continued he, "you have brought on yourselves by your rebellion. Yet for the affection the sovereign, whom you have so unworthily treated, still bears you, I would willingly stay my hand, if you will lay down your arms, and return once more to your obedience. But, if you do not," he concluded, "I will make your city a heap of ruins, and leave not a soul alive to mourn over it!"

But the Spanish commander did not yet comprehend the character of the Aztecs, if he thought to intimidate them by menaces. Calm in their exterior and slow to move, they were the more difficult to pacify when roused; and now that they had been stirred to their inmost depths, it was no human voice that could still the tempest. It may be, however, that Cortés did not so much misconceive the character of the people. He may have felt that an authoritative tone was the only one he could assume with any chance of effect, in his present position, in which milder and more conciliatory language would, by intimating a consciousness of inferiority, have too certainly defeated its own object.

It was true, they answered, he had destroyed their temples, broken in pieces their gods, massacred their countrymen. Many more, doubtless, were yet to fall under their terrible swords. But they were content so long as for every thousand Mexicans they could shed the blood of a single white man! "Look out," they continued, "on our terraces and streets, see them still thronged with warriors as far as your eyes can reach. Our numbers are scarcely diminished by our losses. Yours, on the contrary, are lessening every hour. You are perishing from hunger and sickness. Your provisions and water are falling. You must soon fall into our hands. *The bridges are broken down, and you cannot escape!* There will be too few of you left to glut the vengeance of our Gods!" As they concluded, they sent a volley of arrows over the battlements, which compelled the Spaniards to descend and take refuge in their defences.

The fierce and indomitable spirit of the Aztecs filled the besieged with dismay. All, then, that they had done and suffered, their battles by day, their vigils by night, the perils they had braved, even the victories they had won, were of no avail. It was too evident that they had no longer the spring of ancient superstition to work upon, in the breasts of the natives, who, like some wild beast that has burst the bonds of his keeper, seemed now to swell and exult in the full consciousness of their strength. The annunciation respecting the bridges fell like a knell on the ears of the Christians. All that they had heard was too true – and they gazed on one another with looks of anxiety and dismay.

The same consequences followed, which sometimes take place among the crew of a shipwrecked vessel. Subordination was lost in the dreadful sense of danger. A spirit of mutiny broke out, especially among the recent levies drawn from the army of Narvaez. They had come into the country from no motive of ambition, but attracted simply by the glowing reports of its opulence, and they had fondly hoped to return in a few months with their pockets well lined with the gold of the Aztec monarch. But how different had been their lot! From the first hour of their landing, they had experienced only trouble and disaster, privations of every description, sufferings unexampled, and they now beheld in perspective a fate yet more appalling. Bitterly did they lament the hour when they left the sunny fields of Cuba for these cannibal regions! And heartily did they curse their own folly in listening to the call of Velasquez, and still more, in embarking under the banner of Cortés!

They now demanded with noisy vehemence to be led instantly from the city, and refused to serve longer in defence of a place where they were cooped up like sheep in the shambles, waiting only to be dragged to slaughter. In all this they were rebuked by the more orderly, soldier-like conduct of the veterans of Cortés. These latter had shared with their general the day of his prosperity, and they were not disposed to desert him in the tempest. It was, indeed, obvious, on a little reflection, that the only chance of safety, in the existing crisis, rested on subordination and union; and that even this chance must be greatly diminished under any other leader than their present one.

Thus pressed by enemies without and by factions within, that leader was found, as usual, true to himself. Circumstances so appalling, as would have paralyzed a common mind, only stimulated his to higher action, and drew forth all its resources. He combined what is most rare, singular coolness and constancy of purpose, with a spirit of enterprise that might well be called romantic. His presence of mind did not now desert him. He calmly surveyed his condition, and weighed the difficulties which surrounded him, before coming to a decision. Independently of the hazard of a retreat in the face of a watchful and desperate foe, it was a deep mortification to surrender up the city, where he had so long lorded it as a master; to abandon the rich treasures which he had secured to himself and his followers; to forego the very means by which he hoped to propitiate the favor of his sovereign, and secure an amnesty for his irregular proceedings. This, he well knew, must, after all, be dependent on success. To fly now was to acknowledge himself further removed from the conquest than ever. What a close was this to a career so auspiciously begun! What a contrast to his magnificent vaunts! What a triumph would it afford to his enemies! The governor of Cuba would be amply revenged.

But, if such humiliating reflections crowded on his mind, the alternative of remaining, in his present crippled condition, seemed yet more desperate. With his men daily diminishing in strength and numbers, their provisions reduced so low that a small daily ration of bread was all the sustenance afforded to the soldier under his extraordinary fatigues, with the breaches every day widening in his feeble fortifications, with his ammunition, in fine, nearly expended, it would be impossible to maintain the place much longer – and none but men of iron constitutions and tempers, like the Spaniards, could have held it out so long – against the enemy. The chief embarrassment was as to the time and manner in which it would be expedient to evacuate the city. The best route seemed to be that of Tlacopan (Tacuba). For the causeway, the most dangerous part of the road, was but two miles long in that direction, and would, therefore, place the fugitives, much sooner than either of the other great avenues, on terra firma. Before his final departure, however, he proposed to make another sally in that direction, in order to reconnoitre the ground, and, at the same time, divert the enemy's attention from his real purpose by a show of active operations.

For some days, his workmen had been employed in constructing a military machine of his own invention. It was called a *manta*, and was contrived somewhat on the principle of the mantelets used in the wars of the Middle Ages. It was, however, more complicated, consisting of a tower made of light beams and planks, having two chambers, one over the other. These were to be filled with musketeers, and the sides were provided with loop-holes, through which a fire could be kept up on the enemy. The great advantage proposed by this contrivance was to afford a defence to the troops against the missiles hurled from the terraces. These machines, three of which were made, rested on rollers, and were provided with strong ropes, by which they were to be dragged along the streets by the Tlascalan auxiliaries.

The Mexicans gazed with astonishment on this warlike machinery, and, as the rolling fortresses advanced, belching forth fire and smoke from their entrails, the enemy, incapable of making an impression on those within, fell back in dismay. By bringing the *mantas* under the walls of the houses, the Spaniards were enabled to fire with effect on the mischievous tenants of the *azoteas*, and when this did not silence them, by letting a ladder, or light drawbridge, fall on the roof from the top of the *manta*, they opened a passage to the terrace, and closed with the combatants hand to hand. They could not, however, thus approach the higher buildings, from which the Indian warriors threw down such heavy masses of stone and timber as dislodged the planks that covered the machines, or, thundering against their sides, shook the frail edifices to their foundation, threatening all within with indiscriminate ruin. Indeed, the success of the experiment was doubtful, when the intervention of a canal put a stop to their further progress.

The Spaniards now found the assertion of their enemies too well confirmed. The bridge which traversed the opening had been demolished; and, although the canals which intersected the city were, in general, of no great width or depth, the removal of the bridges not only impeded the movements of the general's clumsy machines, but effectually disconcerted those of his cavalry. Resolving to abandon the *mantas*, he gave orders to fill up the chasm with stone, timber, and other rubbish drawn from the ruined buildings, and to make a new passage-way for the army. While this labor was going on, the Aztec slingers and archers on the other side of the opening kept up a galling discharge on the Christians, the more defenceless from the nature of their occupation. When the work was completed, and a safe passage secured, the Spanish cavaliers rode briskly against the enemy, who, unable to resist the shock of the steel-clad column, fell back with precipitation to where another canal afforded a similar strong position for defence.

There were no less than seven of these canals, intersecting the great street of Tlacopan, and at every one the same scene was renewed, the Mexicans making a gallant stand, and inflicting some loss, at each, on their persevering antagonists. These operations consumed two days, when, after incredible toil, the Spanish general had the satisfaction to find the line of communication completely re-established through the whole length of the avenue, and the principal bridges placed under strong detachments of infantry. At this juncture, when he had driven the foe before him to the furthest extremity of the street, where it touches on the causeway, he was informed, that the Mexicans, disheartened by their reverses, desired to open a parley with him respecting the terms of an accommodation, and that their chiefs awaited his return for that purpose at the fortress. Overjoyed at the intelligence, he instantly rode back, attended by Alvarado, Sandoval, and about sixty of the cavaliers, to his quarters.

The Mexicans proposed that he should release the two priests captured in the temple, who might be the bearers of his terms, and serve as agents for conducting the negotiation. They were accordingly sent with the requisite instructions to their countrymen. But they did not return. The whole was an artifice of the enemy, anxious to procure the liberation of their religious leaders, one of whom was their *teoteuctli*, or high-priest, whose presence was indispensable in the probable event of a new coronation.

Cortés, meanwhile, relying on the prospects of a speedy arrangement, was hastily taking some refreshment with his officers, after the fatigues of the day, when he received the alarming tidings that the enemy were in arms again, with more fury than ever; that they had overpowered the detachments posted under Alvarado at three of the bridges, and were busily occupied in demolishing them. Stung with shame at the facility with which he had been duped by his wily foe, or rather by his own sanguine hopes, Cortés threw himself into the saddle, and, followed by his brave companions, galloped back at full speed to the scene of action. The Mexicans recoiled before the impetuous charge of the Spaniards. The bridges were again restored; and Cortés and his cavalry rode down the whole extent of the great street, driving the enemy, like frightened deer, at the points of their lances. But, before he could return on his steps, he had the mortification to find that the indefatigable foe, gathering from the adjoining lanes and streets, had again closed on his infantry, who, worn down by fatigue, were unable to maintain their position at one of the principal bridges. New swarms of warriors now poured in on all sides, overwhelming the little band of Christian cavaliers with a storm of stones, darts, and arrows, which rattled like hail on their armor and on that of their well-barbed horses. Most of the missiles, indeed, glanced harmless from the good panoplies of steel, or thick quilted cotton, but, now and then, one better aimed penetrated the joints of the harness, and stretched the rider on the ground.

The confusion became greater around the broken bridge. Some of the horsemen were thrown into the canal, and their steeds floundered wildly about without a rider. Cortés himself, at this crisis, did more than any other to cover the retreat of his followers. While the bridge was repairing, he plunged boldly into the midst of the barbarians, striking down an enemy at every vault of his charger, cheering on his own men, and spreading terror through the ranks of his opponents by the well-known sound of his battle-cry. Never did he display greater hardihood, or more freely expose his person, emulating, says an old chronicler, the feats of the Roman Cocles. In this way he stayed the tide of assailants, till the last man had crossed the bridge, when, some of the planks having given way, he was compelled to leap a chasm of full six feet in width, amidst a cloud of missiles, before he could place himself in safety. A report ran through the army that the general was slain. It soon spread through the city, to the great joy of the Mexicans, and reached the fortress, where the besieged were thrown into no less consternation. But, happily for them, it was false. He, indeed, received two severe contusions on the knee, but in other respects remained uninjured. At no time, however, had he been in such extreme danger; and his escape, and that of his companions, was esteemed little less than a miracle. More than one grave historian refers the preservation of the Spaniards to the watchful care of their patron Apostle, St. James, who, in these desperate conflicts, was beheld careering on his milk-white steed at the head of the Christian squadrons, with his sword flashing lightning, while a lady robed in white – supposed to be the Virgin – was distinctly seen by his side, throwing dust in the eyes of the infidel! The fact is attested both by Spaniards and Mexicans – by the latter after their conversion to Christianity. Surely, never was there a time when the interposition of their tutelar saint was more strongly demanded.

The coming of night dispersed the Indian battalions, which, vanishing like birds of ill omen from the field, left the well-contested pass in possession of the Spaniards. They returned, however, with none of the joyous feelings of conquerors to their citadel, but with slow step and dispirited, with weapons hacked, armor battered, and fainting under the loss of blood, fasting, and fatigue. In this condition they had yet to learn the tidings of a fresh misfortune in the death of Montezuma.

The Indian monarch had rapidly declined, since he had received his injury, sinking, however, quite as much under the anguish of a wounded spirit, as under disease. He continued in the same moody state of insensibility as that already described; holding little communication with those around him, deaf to consolation, obstinately rejecting all medical remedies as well as nourishment. Perceiving his end approach, some of the cavaliers present in the fortress, whom the kindness of his manners had personally attached to him, were anxious to save the soul of the dying prince from the sad doom of those who perish in the darkness of unbelief. They accordingly waited on him, with father Olmedo at their head, and in the most earnest manner implored him to open his eyes to the error of his creed, and consent to be baptized. But Montezuma – whatever may have been suggested to the contrary – seems never to have faltered in his hereditary faith, or to have contemplated becoming an apostate; for surely he merits that name in its most odious application, who, whether Christian or pagan, renounces his religion without conviction of its falsehood. Indeed, it was a too implicit reliance on its oracles, which had led him to give such easy confidence to the Spaniards. His intercourse with them had, doubtless, not sharpened his desire to embrace their communion; and the calamities of his country he might consider as sent by his gods to punish him for his hospitality to those who had desecrated and destroyed their shrine.

When father Olmedo, therefore, kneeling at his side, with the uplifted crucifix, affectionately besought him to embrace the sign of man's redemption, he coldly repulsed the priest, exclaiming, "I have but a few moments to live, and will not at this hour desert the faith of my fathers." One thing, however, seemed to press heavily on Montezuma's mind. This was the fate of his children, especially of three daughters, whom he had by his two wives; for there were certain rites of marriage, which distinguished the lawful wife from the concubine. Calling Cortés to his bedside, he earnestly commended these children to his care, as "the most precious jewels that he could leave him." He besought the general to interest his master, the emperor, in their behalf, and to see that they should not be left destitute, but be allowed some portion of their rightful inheritance. "Your lord will do this," he concluded, "if it were only for the friendly offices I have rendered the Spaniards, and for the love I have shown them – though it has brought me to this condition! But for this I bear them no ill-will." Such, according to Cortés himself, were the words of the dying monarch. Not long after, on the 30th of June, 1520, he expired in the arms of some of his own nobles, who still remained faithful in their attendance on his person. "Thus," exclaims a native historian, one of his enemies, a Tlascalan, "thus died the unfortunate Montezuma, who had swayed the sceptre with such consummate policy and wisdom; and who was held in greater reverence and awe than any other prince of his lineage, or any, indeed, that ever sat on a throne in this Western World. With him may be said to have terminated the royal line of the Aztecs, and the glory to have passed away from the empire, which under him had reached the zenith of its prosperity." "The tidings of his death," says the old Castilian chronicler, Diaz, "were received with real grief by every cavalier and soldier in the army who had had access to his person; for we all loved him as a father – and no wonder, seeing how good he was." This simple, but emphatic, testimony to his desert, at such a time, is in itself the best refutation of the suspicions occasionally entertained of his fidelity to the Christians.

Rudolf Hoss

I, The Commandant of Auschwitz

Hoss was an early member of the SS, and in 1940 became the commandant of the extermination camp at Auschwitz, Poland.

By the will of the Reichsführer SS, Auschwitz became the greatest human extermination centre of all time.

When in the summer of 1941 he himself gave me the order to prepare installations at Auschwitz where mass exterminations could take place, and personally to carry out these exterminations, I did not have the slightest idea of their scale or consequences. It was certainly an extraordinary and monstrous order. Nevertheless the reasons behind the extermination programme seemed to me right. I did not reflect on it at the time: I had been given an order, and I had to carry it out. Whether this mass extermination of the Jews was necessary or not was something on which I could not allow myself to form an opinion, for I lacked the necessary breadth of view.

If the Führer had himself given the order for the "final solution of the Jewish question," then, for a veteran National-Socialist and even more so for an SS officer, there could be no question of considering its merits. "The Führer commands, we follow" was never a mere phrase or slogan. It was meant in bitter earnest.

Since my arrest it has been said to me repeatedly that I could have disobeyed this order, and that I might even have assassinated Himmler. I do not believe that of all the thousands of SS officers there could have been found a single one capable of such a thought. It was completely impossible. Certainly many SS officers grumbled and complained about some of the harsh orders that came from the Reichsführer SS, but they nevertheless always carried them out.

Many orders of the Reichsführer SS deeply offended a great number of his SS officers, but I am perfectly certain that not a single one of them would have dared to raise a hand against him, or would have even contemplated doing so in his most secret thoughts. As Reichsführer SS, his person was inviolable. His basic orders, issued in the name of the Führer, were sacred. They brooked no consideration, no argument, no interpretation. They were carried out ruthlessly and regardless of consequences, even though these might well mean the death of the officer concerned, as happened to not a few SS officers during the war.

It was not for nothing that during training the self-sacrifice of the Japanese for their country and their emperor, who was also their god, was held up as a shining example to the SS.

SS training was not comparable to a university course which can have as little lasting effect on the students as water on a duck's back. It was on the contrary something that was deeply engrained, and the Reichsführer SS knew very well what he could demand of his men.

But outsiders simply cannot understand that there was not a single SS officer who would disobey an order from the Reichsführer SS, far less consider getting rid of him because of the gruesomely hard nature of one such order.

What the Führer, or in our case his second-in-command, the Reichsführer SS, ordered was always right.

Democratic England also has a basic national concept: "My country, right or wrong!" and this is adhered to by every nationally-conscious Englishman.

Before the mass extermination of the Jews began, the Russian *politruks* and political commissars were liquidated in almost all the concentration camps during 1941 and 1942.

In accordance with a secret order issued by Hitler, these Russian *politruks* and political commissars were combed out of all the prisoner-of-war camps by special detachments from the Gestapo. When identified, they were transferred to the nearest concentration camp for liquidation. It was made known that these measures were taken because the Russians had been killing all German soldiers who were party members or belonged to special sections of the NSDAP, especially members of the SS, and also because the political officials of the Red Army had been ordered, if taken prisoner, to create every kind of disturbance in the prisoner-of-war camps and their places of employment and to carry out sabotage wherever possible.

The political officials of the Red Army thus identified were brought to Auschwitz for liquidation. The first, smaller transports of them were executed by firing squads.

While I was away on duty, my deputy, Fritzsch, the commander of the protective custody camp, first tried gas for these killings. It was a preparation of prussic acid, called Cyclon B, which was used in the camp as an insecticide and of which there was always a stock on hand. On my return, Fritzsch reported this to me, and the gas was used again for the next transport.

The gassing was carried out in the detention cells of Block II. Protected by a gas-mask, I watched the killing myself. In the crowded cells death came instantaneously the moment the Cyclon B was thrown in. A short, almost smothered cry, and it was all over. During this first experience of gassing people, I did not fully realise what was happening, perhaps because I was too impressed by the whole procedure. I have a clearer recollection of the gassing of nine hundred Russians which took place shortly afterwards in the old crematorium, since the use of Block II for this purpose caused too much trouble. While the transport was detraining, holes were pierced in the earth and concrete ceiling of the mortuary. The Russians were ordered to undress in an anteroom; they then quietly entered the mortuary, for they had been told they were to be deloused. The whole transport exactly filled the mortuary to capacity. The doors were then sealed and the gas shaken down through the holes in the roof. I do not know how long this killing took. For a little while a humming sound could be heard. When the powder was thrown in, there were cries of "Gas!," then a great bellowing, and the trapped prisoners hurled themselves against both the doors. But the doors held. They were opened several hours later, so that the place might be aired. It was then that I saw, for the first time, gassed bodies in the mass.

It made me feel uncomfortable and I shuddered, although I had imagined that death by gassing would be worse than it was. I had always thought that the victims would experience a terrible choking sensation. But the bodies, without exception, showed no signs of convulsion. The doctors explained to me that the prussic acid had a paralysing effect on the lungs, but its action was so quick and strong that death came before the convulsions could set in, and in this its effects differed from those produced by carbon monoxide or by a general oxygen deficiency.

The killing of these Russian prisoners-of-war did not cause me much concern at the time. The order had been given, and I had to carry it out. I must even admit that this gassing set my mind at rest, for the mass extermination of the Jews was to start soon and at that time neither Eichmann nor I was certain how these mass killings were to be carried out. It would be by gas, but we did not know which gas or how it was to be used. Now we had the gas, and we had established a procedure. I always shuddered at the prospect of carrying out exterminations by shooting, when I thought of the vast numbers concerned, and of the women and children. The shooting of hostages, and the group executions ordered by the Reichsführer SS or by the Reich Security Head Office had been enough for me. I was therefore relieved to think that we were to be spared all these blood-baths, and that the victims too would be spared suffering until their last moment came. It was precisely this which had caused me the greatest concern when I had heard Eichmann's description of Jews being mown down by the Special Squads armed with machine-guns and machine-pistols. Many gruesome scenes are said to have taken place, people running away after being shot, the finishing off of the wounded and particularly of the women and children. Many members of the *Einsatzkommandos*, unable to endure wading through blood any longer, had committed suicide. Some had even gone mad. Most of the members of these *Kommandos* had to rely on alcohol when carrying out their horrible work. According to Höfle's description, the men employed at Globocnik's extermination centres consumed amazing quantities of alcohol.

In the spring of 1942 the first transports of Jews, all earmarked for extermination, arrived from Upper Silesia.

They were taken from the detraining platform to the "Cottage" – to Bunker I – across the meadows where later Building Site II was located. The transport was conducted by Aumeier and Palitzsch and some of the block leaders. They talked with the Jews about general topics, enquiring concerning their qualifications and trades, with a view to misleading them. On arrival at the "Cottage," they were told to undress. At first they went calmly into the rooms where they were supposed to be disinfected. But some of them showed signs of alarm, and spoke of death by suffocation and of annihilation. A sort of panic set in at once. Immediately all the Jews still outside were pushed into the chambers, and the doors were screwed shut. With subsequent transports the difficult individuals were picked out early on and most carefully supervised. At the first signs of unrest, those responsible were unobtrusively led behind the building and killed with a small-calibre gun, that was inaudible to the others. The presence and calm behavior of the Special Detachment served to reassure those who were worried or who suspected what was about to happen. A further calming effect was obtained by members of the Special Detachment accompanying them into the rooms and remaining with them until the last moment, while an SS-man also stood in the doorway until the end.

It was most important that the whole business of arriving and undressing should take place in an atmosphere of the greatest possible calm. People reluctant to take off their clothes had to be helped by those of their companions who had already undressed, or by men of the Special Detachment.

The refractory ones were calmed down and encouraged to undress. The prisoners of the Special Detachment also saw to it that the process of undressing was carried out quickly, so that the victims would have little time to wonder what was happening.

The eager help given by the Special Detachment in encouraging them to undress and in conducting them into the gas-chambers was most remarkable. I have never known, nor heard, of any of its members giving these people who were about to be gassed the slightest hint of what lay ahead of them. On the contrary, they did everything in their power to deceive them and particularly to pacify the suspicious ones. Though they might refuse to believe the SS-men, they had complete faith in these members of their own race, and to reassure them and keep them calm the Special Detachments therefore always consisted of Jews who themselves came from the same districts as did the people on whom a particular action was to be carried out.

They would talk about life in the camp, and most of them asked for news of friends or relations who had arrived in earlier transports. It was interesting to hear the lies that the Special Detachment told them with such conviction, and to see the emphatic gestures with which they underlined them.

Many of the women hid their babies among the piles of clothing. The men of the Special Detachment were particularly on the look-out for this, and would speak words of encouragement to the woman until they had persuaded her to take the child with her. The women believed that the disinfectant might be bad for their smaller children, hence their efforts to conceal them.

The smaller children usually cried because of the strangeness of being undressed in this fashion, but when their mothers or members of the Special Detachment comforted them, they became calm and entered the gas chambers, playing or joking with one another and carrying their toys.

I noticed that women who either guessed or knew what awaited them nevertheless found the courage to joke with the children to encourage them, despite the mortal terror visible in their own eyes.

One woman approached me as she walked past and, pointing to her four children who were manfully helping the smallest ones over the rough ground, whispered:

"How can you bring yourself to kill such beautiful, darling children? Have you no heart at all?"

One old man, as he passed by me, hissed:

"Germany will pay a heavy penance for this mass murder of the Jews."

His eyes glowed with hatred as he said this. Nevertheless he walked calmly into the gas-chamber, without worrying about the others.

One young woman caught my attention particularly as she ran busily hither and thither, helping the smallest children and the old women to undress. During the selection she had had two small children with her, and her agitated behaviour and appearance had brought her to my notice at once. She did not look in the least like a Jewess. Now her children were no longer with her. She waited until the end, helping the women who were not undressed and who had several children with them, encouraging them and calming the children. She went with the very last ones into the gas-chamber. Standing in the doorway, she said:

"I knew all the time that we were being brought to Auschwitz to be gassed. When the selection took place I avoided being put with the able-bodied ones, as I wished to look after the children. I wanted to go through it all, fully conscious of what was happening. I hope that it will be quick. Goodbye!"

From time to time women would suddenly give the most terrible shrieks while undressing, or tear their hair, or scream like maniacs. These were immediately led away behind the building and shot in the back of the neck with a small-calibre weapon.

It sometimes happened that, as the men of the Special Detachment left the gas-chamber, the women would suddenly realise what was happening, and would call down every imaginable curse upon our heads.

I remember, too, a woman who tried to throw her children out of the gas-chamber, just as the door was closing. Weeping she called out:

"At least let my precious children live."

There were many such shattering scenes, which affected all who witnessed them.

During the spring of 1942 hundreds of vigorous men and women walked all unsuspecting to their death in the gas-chambers, under the blossom-laden fruit trees of the "Cottage" orchard. This picture of death in the midst of life remains with me to this day.

The process of selection, which took place on the unloading platforms, was in itself rich in incident.

The breaking up of families, and the separation of the men from the women and children, caused much agitation and spread anxiety throughout the whole transport. This was increased by the further separation from the others of those capable of work. Families wished at all costs to remain together. Those who had been selected ran back to rejoin their relations. Mothers with children tried to join their husbands, or old people attempted to find those of their children who had been selected for work, and who had been led away.

Often the confusion was so great that the selections had to be begun all over again. The limited area of standing-room did not permit better sorting arrangements. All attempts to pacify these agitated mobs were useless. It was often necessary to use force to restore order.

As I have already frequently said, the Jews have strongly developed family feelings. They stick together like limpets. Nevertheless, according to my observations, they lack solidarity. One would have thought that in a situation such as this they would inevitably help and protect one another. But no, quite the contrary. I have often known and heard of Jews, particularly those from Western Europe, who revealed the addresses of those members of their race still in hiding.

One woman, already in the gas-chamber, shouted out to a

non-commissioned officer the address of a Jewish family. A man who, to judge by his clothes and deportment appeared to be of very good standing, gave me, while actually undressing, a piece of paper on which was a list of the addresses of Dutch families who were hiding Jews.

I do not know what induced the Jews to give such information. Was it for reasons of personal revenge, or were they jealous that those others should survive?

The attitude of the men of the Special Detachment was also strange. They were all well aware that once the actions were completed they, too, would meet exactly the same fate as that suffered by these thousands of their own race, to whose destruction they had contributed so greatly. Yet the eagerness with which they carried out their duties never ceased to amaze me. Not only did they never divulge to the victims their impending fate, and were considerately helpful to them while they undressed, but they were also quite prepared to use violence on those who resisted. Then again, when it was a question of removing the trouble-makers and holding them while they were shot, they would lead them out in such a way that the victims never saw the non-commissioned officer standing there with his gun ready, and he was able to place its muzzle against the back of their necks without their noticing it. It was the same story when they dealt with the sick and the invalids, who could not be taken into the gas-chambers. And it was all done in such a matter-of-course manner that they might themselves have been the exterminators.

Then the bodies had to be taken from the gas-chambers, and after the gold teeth had been extracted, and the hair cut off, they had to be dragged to the pits or to the crematoria. Then the fires in the pits had to be stoked, the surplus fat drained off, and the mountain of burning corpses constantly turned over so that the draught might fan the flames.

They carried out all these tasks with a callous indifference as though it were all part of an ordinary day's work. While they dragged the corpses about, they ate or they smoked. They did not stop eating even when engaged on the grisly job of burning corpses which had been lying for some time in mass graves.

It happened repeatedly that Jews of the Special Detachment would come upon the bodies of close relatives among the corpses, and even among the living as they entered the gas-chambers. They were obviously affected by this, but it never led to any incident.

I myself saw a case of this sort. Once when bodies were being carried from a gas-chamber to the fire-pit, a man of the Special Detachment suddenly stopped and stood for a moment as though rooted to the spot. Then he continued to drag out a body with his comrades. I asked the Capo what was up. He explained that the corpse was that of the Jew's wife. I watched him for a while, but noticed nothing peculiar in his behaviour. He continued to drag corpses along, just as he had done before. When I visited the Detachment a little later, he was sitting with the others and eating, as though nothing had happened. Was he really able to hide his emotions so completely, or had he become too brutalised to care even about this?

Where did the Jews of the Special Detachment derive the strength to carry on night and day with their grisly work? Did they hope that some whim of fortune might at the last moment snatch them from the jaws of death? Or had they become so dulled by the accumulation of horror that they were no longer capable even of ending their own lives and thus escaping from this "existence"?

I have certainly watched them closely enough, but I have never really been able to get to the bottom of their behaviour.

The Jew's way of living and of dying was a true riddle that I never managed to solve.

All these experiences and incidents which I have described could be multiplied many times over. They are excerpts only, taken from the whole vast business of the extermination, side-lights as it were.

This mass extermination, with all its attendant circumstances, did not, as I know, fail to affect those who took a part in it. With very few exceptions, nearly all of those detailed to do this monstrous "work," this "service," and who, like myself, have given sufficient thought to the matter, have been deeply marked by these events.

Many of the men involved approached me as I went my rounds through the extermination buildings, and poured out their anxieties and impressions to me, in the hope that I could allay them.

Again and again during these confidential conversations I was asked: is it necessary that we do all this? Is it necessary that hundreds of thousands of women and children be destroyed? And I, who in my innermost being had on countless occasions asked myself exactly this question, could only fob them off and attempt to console them by repeating that it was done on Hitler's order. I had to tell them that this extermination of Jewry had to be, so that Germany and our posterity might be freed for ever from their relentless adversaries.

There was no doubt in the mind of any of us that Hitler's order had to be obeyed regardless, and that it was the duty of the SS to carry it out. Nevertheless we were all tormented by secret doubts.

I myself dared not admit to such doubts. In order to make my subordinates carry on with their task, it was psychologically essential that I myself appear convinced of the necessity for this gruesomely harsh order.

Everyone watched me. They observed the impression produced upon me by the kind of scenes that I have described above, and my reactions. Every word I said on the subject was discussed. I had to exercise intense self-control in order to prevent my innermost doubts and feelings of oppression from becoming apparent.

I had to appear cold and indifferent to events that must have wrung the heart of anyone possessed of human feelings. I might not even look away when afraid lest my natural emotions got the upper hand. I had to watch coldly, while the mothers with laughing or crying children went into the gas-chambers.

On one occasion two small children were so absorbed in some game that they quite refused to let their mother tear them away from it. Even the Jews of the Special Detachment were reluctant to pick the children up. The imploring look in the eyes of the mother, who certainly knew what was happening, is something I shall never forget. The people were already in the gas-chamber and becoming restive, and I had to act. Everyone was looking at me. I nodded to the junior non-commissioned officer on duty and he picked up the screaming, struggling children in his arms and carried them into the gas-chamber, accompanied by their mother who was weeping in the most heart-rending fashion. My pity was so great that I longed to vanish from the scene: yet I might not show the slightest trace of emotion.

I had to see everything. I had to watch hour after hour, by day and by night, the removal and burning of the bodies, the extraction of the teeth, the cutting of the hair, the whole grisly, interminable business. I had to stand for hours on end in the ghastly stench, while the mass graves were being opened and the bodies dragged out and burned.

I had to look through the peep-hole of the gas-chambers and watch the process of death itself, because the doctors wanted me to see it.

I had to do all this because I was the one to whom everyone looked, because I had to show them all that I did not merely issue the orders and make the regulations but was also prepared myself to be present at whatever task I had assigned to my subordinates.

The Reichsführer SS sent various high-ranking Party leaders and SS officers to Auschwitz so that they might see for themselves the process of extermination of the Jews. They were all deeply impressed by what they saw. Some who had previously spoken most loudly about the necessity for this extermination fell silent once they had actually seen the "final solution of the Jewish problem." I was repeatedly asked how I and my men could go on watching these operations, and how we were able to stand it.

My invariable answer was that the iron determination with which we must carry out Hitler's orders could only be obtained by a stifling of all human emotions. Each of these gentlemen declared that he was glad the job had not been given to him.

Even Mildner and Eichmann, who were certainly tough enough, had no wish to change places with me. This was one job which nobody envied me.

I had many detailed discussions with Eichmann concerning all matters connected with the "final solution of the Jewish problem," but without ever disclosing my inner anxieties. I tried in every way to discover Eichmann's innermost and real convictions about this "solution."

Yes, every way. Yet even when we were quite alone together and the drink had been flowing freely so that he was in his most expansive mood, he showed that he was completely obsessed with the idea of destroying every single Jew that he could lay his hands on. Without pity and in cold blood we must complete this extermination as rapidly as possible. Any compromise, even the slightest, would have to be paid for bitterly at a later date.

In the face of such grim determination I was forced to bury all my human considerations as deeply as possible.

Indeed, I must freely confess that after these conversations with Eichmann I almost came to regard such emotions as a betrayal of the Führer.

There was no escape for me from this dilemma.

I had to go on with this process of extermination. I had to continue this mass murder and coldly to watch it, without regard for the doubts that were seething deep inside me.

I had to observe every happening with a cold indifference. Even those petty incidents that others might not notice I found hard to forget. In Auschwitz I truly had no reason to complain that I was bored.

If I was deeply affected by some incident, I found it impossible to go back to my home and my family. I would mount my horse and ride, until I had chased the terrible picture away. Often, at night, I would walk through the stables and seek relief among my beloved animals.

It would often happen, when at home, that my thoughts suddenly turned to incidents that had occurred during the extermination. I then had to go out. I could no longer bear to be in my homely family circle. When I saw my children happily playing, or observed my wife's delight over our youngest, the thought would often come to me: how long will our happiness last? My wife could never understand these gloomy moods of mine, and ascribed them to some annoyance connected with my work.

When at night I stood out there beside the transports, or by the gas-chambers or the fires, I was often compelled to think of my wife and children, without, however, allowing myself to connect them closely with all that was happening.

It was the same with the married men who worked in the crematoria or at the fire-pits.

When they saw the women and children going into the gas-chambers, their thoughts instinctively turned to their own families.

I was no longer happy in Auschwitz once the mass exterminations had begun.

I had become dissatisfied with myself. To this must be added that I was worried because of anxiety about my principal task, the never-ending work, and the untrustworthiness of my colleagues.

Then the refusal to understand, or even to listen to me, on the part of my superiors. It was in truth not a happy or desirable state of affairs. Yet everyone in Auschwitz believed that the commandant lived a wonderful life.

My family, to be sure, were well provided for in Auschwitz. Every wish that my wife or children expressed was granted them. The children could live a free and untrammelled life. My wife's garden was a paradise of flowers. The prisoners never missed an opportunity for doing some little act of kindness to my wife or children, and thus attracting their attention.

No former prisoner can ever say that he was in any way or at any time badly treated in our house. My wife's greatest pleasure would have been to give a present to every prisoner who was in any way connected with our household.

The children were perpetually begging me for cigarettes for the prisoners. They were particularly fond of the ones who worked in the garden.

My whole family displayed an intense love of agriculture and particularly for animals of all sorts. Every Sunday I had to walk them all across the fields, and visit the stables, and we might never miss out the kennels where the dogs were kept. Our two horses and the foal were especially beloved.

The children always kept animals in the garden, creatures the prisoners were forever bringing them. Tortoises, martens, cats, lizards: there was always something new and interesting to be seen there. In summer they splashed in the paddling pool in the garden, or in the Sola. But their greatest joy was when Daddy bathed with them. He had, however, so little time for all these childish pleasures. Today I deeply regret that I did not devote more time to my family. I always felt that I had to be on duty the whole time. This exaggerated sense of duty has always made life more difficult for me than it actually need have been. Again and again my wife reproached me and said: "You must think not only of the service always, but of your family too."

Yet what did my wife know about all that lay so heavily on my mind? She has never been told.

Lord Russell

The Bataan Death March

An account of one of the many war crimes committed by Nipponese forces, 1941–45.

At 2 a.m. on 9th April 1942 Major-General King, who was in command of the American-Filipino forces in Bataan, sent two of his staff officers forward under a flag of truce to make an appointment for him to meet the commander of the Japanese force with a view to surrendering. Shortly after dawn the American officers returned, having made contact with the enemy, and Major-General King went forward with his two aides in one car followed by the two staff officers in another. During their journey, although both cars carried large white flags, the American parlementaires were repeatedly attacked with light bombs and machine-gun fire from low-flying Japanese aircraft, and had to take cover.

By 10 a.m. they reached Lamao, the headquarters of a Japanese infantry division. Their commander interviewed Major-General King and explained that he had no authority to treat with the American commander, but that he had informed General Homma who would send an officer with full authority to negotiate the surrender terms.

An hour or so later General Homma's chief-of-staff arrived to discuss the surrender on behalf of his commander-in-chief. The rest of the story is told in Major-General King's own words:

> I was concerned only with the treatment that my men would receive, and whether they would be treated as prisoners of war. The Japanese officer demanded my unconditional surrender. I attempted to secure from him an assurance that my men would be treated as prisoners of war. He accused me of declining to surrender unconditionally and

> of trying to make a condition. We talked back and forth in this vein for some time, I should guess about half an hour. Finally he said to me, through the interpreter, "The Imperial Japanese Army are not barbarians". With that assurance I had to content myself and surrender.
>
> In destroying arms and equipment in preparation for surrender I had reserved enough motor transportation and gasoline to transport all my troops out of Bataan. I endeavoured, prior to surrender, to secure an assurance that this might be done. I pleaded, after my surrender, that this should be done, offering to furnish personnel as might be required by the Japanese for this purpose or to assist in any way they might require. The Japanese told me that they would handle the movement of the prisoners as they desired, that I would have nothing to do with it, and that my wishes in that connection could not be considered.

The prisoners were marched in intense heat along the road to San Fernando, Pampanga, a distance of about seventy-five miles. They had all been on short rations for a considerable period before their capture, and there was a high percentage of sick and wounded prisoners among them; nevertheless, the sick and wounded were also forced to march with the others.

Those who fell by the roadside and were unable to continue, and they were many in number, were shot or bayoneted. Others were taken from the ranks, beaten, tortured and killed. The march lasted nine days, the Japanese guard being relieved at five-kilometre intervals by fresh guards who had been transported in the American trucks.

For the first five days the prisoners received no food, and never any water except what they were able to drink out of caribou wallows and ditches along the highway. Some food was thrown them by Filipinos, and occasionally they broke ranks and grouped themselves round a well, a wallow or a ditch to slake their thirst. Whenever this occurred the Japanese guards opened fire on them.

Throughout the march their escort maltreated them. They were beaten, bayoneted and kicked with hobnailed boots. Dead bodies littered the side of the road.

Staff-Sergeant Samuel Moody of the United States Regular Army has described what happened to one of his friends, a Sergeant Jones:

> My friend Sergeant Jones had a severe case of dysentery caused from drinking the muddy caribou wallow water. When he fell to the rear due to his condition he was beaten and stuck with a bayonet. Later he died from his wounds.

A graphic description of this march was given in evidence at the Tokyo trial by a soldier of the United States Army, D. F. Ingle, who at the time of the American surrender was a patient in a field hospital. He had been admitted to hospital suffering from pneumonia, but shortly after admission he was slightly wounded during an attack made on the hospital by a Japanese aircraft, although it was clearly marked by a Red Cross.

When the Japanese arrived, Ingle was lying on a stretcher with a temperature of 105 degrees. A Japanese soldier prodded him in the back with a bayonet and ordered him to sit up, which Ingle did as quickly as possible. The Japanese then proceeded to relieve him of his watch, his ring, his wallet and all his personal belongings, with the exception of two photographs which Ingle managed to convince the soldier were of Ingle's mother.

Despite the fact that the American was obviously very ill he was forced to join the death march. During the whole nine days he received neither food nor water except what he obtained, liked the other prisoners, from the caribou wallows, and other water holes. Ingle's testimony continued:

> The water in the ponds and in the ditches was so polluted that it was highly dangerous to drink, and that which came from the artesian wells was of such a small amount that when great numbers of men tried to get it, the Japanese troops would simply raise their rifles and fire into the group, and when the smoke and dust had cleared away it showed that pure water could cause your death as well as that which was polluted.
>
> The Filipino civilians tried on many occasions to give us food, but they did so at the risk of their lives and, indeed, many lost their lives so doing. Apart from that, only an occasional sugar cane patch offered the chance of food, but to try and get some was courting death.
>
> I remember particularly an Episcopalian chaplain named Day. He had contracted dysentery by drinking foul water from a stream or pond beside the highway, and it had become necessary for him to answer the call of

> nature every few minutes. His usual procedure was to step smartly out of the ranks, relieve himself, and slip quickly back into the column. He had perfected the drill and it had become only a matter of seconds. On one occasion, however, he was spotted by one of the guards and bayoneted. From then onwards he had to be helped, and I was one of those who did so. Taking it in turns, for the remainder of the march, two men at a time had to assist the chaplain to keep up with the others. He was given no medical attention, and had it not been for the help given him by his comrades he would have been left at the roadside to die, or shot and his body thrown on the side of the road.

Ingle was unable to say how many he saw shot or bayoneted because it had become such a commonplace occurrence, and after the first few hundred he stopped counting.

On the sixth day of the march the prisoners were informed, through an interpreter, that if they would hand over their watches, rings and other valuables they would be given food. By then, however, few of them had anything left, for most of them had been "frisked" by the guards before the march began. Those lucky enough to have any valuables left willingly parted with them, with the result that on the evening of the sixth day the prisoners each received one teacupful of rice.

On the ninth day the prisoners received the welcome news that they would have to march no further. They were going to ride the rest of the way to Camp O'Donnell. Their relief on hearing the glad tidings was short lived, however, for they were then crowded into very small Filipino railway coaches, a hundred men to each coach. So overcrowded were they, that there were many who, during the whole trip, never touched the floor. Hundreds fainted from lack of air, and many died of suffocation.

It is not known exactly how many died on the move from Bataan to Camp O'Donnell, but the evidence indicates that not less than eight thousand American and Filipino prisoners lost their lives during the journey.

Murata, who had been sent to the Philippines in February 1942 by War Minister Tojo as adviser on civil affairs, drove along the Bataan–San Fernando road, and saw the dead bodies on the side of the highway in such great numbers that he even spoke to General Homma about it. After that Homma, at least, could not plead ignorance.

Tojo admitted that he, also, heard about the march in 1942 from many different sources. He was told, he said, that the prisoners had been forced to march long distances in the heat, and that many deaths had occurred. He also admitted that the United States Government's protest against this unlawful treatment of prisoners of war had been discussed at the bi-weekly meetings of the Bureaux Chiefs in the War Ministry soon after the march took place, but no decision was arrived at, and he left the matter to the discretion of the heads of departments concerned.

The Japanese forces in the Philippines were never called upon to make a report on the incident, and Tojo did not even discuss the march with General Homma when the General paid a visit to Japan early in 1943.

The first time Tojo ever made any inquiries was on his visit to the Philippines in May 1943, when he discussed it with Homma's chief-of-staff, who gave him all the details. Tojo took no action, however, and at his trial explained his failure to do so in these words, "it is the Japanese custom for the commander of an expeditionary army in the field to be given a mission, in the performance of which he is not subject to specific orders from Tokyo, but has considerable autonomy."

This can only mean that according to the Japanese method of waging war such atrocities were expected to occur, or were at least permitted, and that the Japanese Government was not concerned to prevent them.

Anne Frank

Diary of a Jewish Girl in Hiding

Following the Nazi occupation of Holland, Frank hid with her family and four others in a sealed-off office room in Amsterdam, until their betrayal in August 1944. She died in Belsen concentration camp.

Saturday, 20 June 1942
My father was thirty-six when he married my mother, who was then twenty-five. My sister Margot was born in 1926 in Frankfort-on-Main. I followed on 12 June 1929, and, as we are Jewish, we emigrated to Holland in 1933, where my father was appointed Managing Director of Travies N.V. This firm is in close relationship with the firm of "Kolen & Co." in the same building, of which my father is a partner.

The rest of our family, however, felt the full impact of Hitler's anti-Jewish laws, so life was filled with anxiety. In 1938 after the pogroms, my two uncles (my mother's brothers) escaped to the U.S.A. My old grandmother came to us; she was then seventy-three. After May, 1940, good times rapidly fled: first the war, then the capitulation, followed by the arrival of the Germans. That is when the sufferings of us Jews really began. Anti-Jewish decrees followed each other in quick succession. Jews must wear a yellow star, Jews must hand in their bicycles, Jews are banned from trams and are forbidden to drive. Jews are only allowed to do their shopping between three and five o'clock and then only in shops which bear the placard "Jewish shop". Jews must be indoors by eight o'clock and cannot even sit in their own gardens after that hour. Jews are forbidden to visit theatres, cinemas, and other places of entertainment. Jews may not take part in public sports. Swimming baths, tennis courts, hockey fields, and other sports grounds are all prohibited to them. Jews may not visit Christians. Jews must go to Jewish schools, and many more restrictions of a similar kind.

Thursday, 19 November 1942
Apart from that, all goes well. Dussel has told us a lot about the outside world, which we have missed for so long now. He had very sad news. Countless friends and acquaintances have gone to a terrible fate. Evening after evening the green and grey army lorries trundle past. The Germans ring at every front door to enquire if there are any Jews living in the house. If there are, then the whole family has to go at once. If they don't find any, they go on to the next house. No one has a chance of evading them unless one goes into hiding. Often they go round with lists, and only ring when they know they can get a good haul. Sometimes they let them off for cash – so much per head. It seems like the slave hunts of olden times. But it's certainly no joke; it's much too tragic for that. In the evenings, when it's dark, I often see rows of good, innocent people accompanied by crying children, walking on and on, in charge of a couple of these chaps, bullied and knocked about until they almost drop. No one is spared – old people, babies, expectant mothers, the sick – each and all join in the march of death.

How fortunate we are here, so well cared for and undisturbed. We wouldn't have to worry about all this misery were it not that we are so anxious about all those dear to us whom we can no longer help.

I feel wicked sleeping in a warm bed, while my dearest friends have been knocked down or have fallen into a gutter somewhere out in the cold night. I get frightened when I think of close friends who have now been delivered into the hands of the cruellest brutes that walk the earth. And all because they are Jews!

Wednesday, 13 January 1943
Everything has upset me again this morning, so I wasn't able to finish a single thing properly.

It is terrible outside. Day and night more of those poor miserable people are being dragged off, with nothing but a rucksack and a little money. On the way they are deprived even of these possessions. Families are torn apart, the men, women and children all being separated. Children coming home from school find that their parents have disappeared. Women return from shopping to find their homes shut up and their families gone.

Wednesday, 29 March 1944

People have to queue for vegetables and all kinds of other things; doctors are unable to visit the sick, because if they turn their backs on their cars for a moment they are stolen; burglaries and thefts abound, so much so that you wonder what has taken hold of the Dutch for them suddenly to have become such thieves. Little children of eight and eleven years break the windows of people's homes and steal whatever they can lay their hands on. No one dares to leave his house unoccupied for five minutes, because if you go, your things go too. Every day there are announcements in the newspapers offering rewards for the return of lost property, typewriters, Persian rugs, electric clocks, cloth, etc. Electric clocks in the streets are dismantled, public telephones are pulled to pieces – down to the last thread. Morale amongst the population can't be good, the weekly rations are not enough to last for two days except the coffee substitute. The invasion is a long time coming, and the men have to go to Germany. The children are ill or under-nourished, everyone is wearing old clothes and old shoes. A new sole costs 7.50 florins in the black market; moreover, hardly any of the shoemakers will accept shoe repairs, or if they do, you have to wait four months, during which time the shoes often disappear.

There's one good thing in the midst of it all, which is that as the food gets worse and the measures against the people more severe, so sabotage against the authorities steadily increases. The people in the food offices, the police, officials, they all either work with their fellow-citizens and help them or they tell tales on them and have them sent to prison. Fortunately, only a small percentage of Dutch people are on the wrong side.

Tuesday, 11 April 1944

We have been pointedly reminded that we are in hiding, that we are Jews in chains, chained to one spot, without any rights, but with a thousand duties. We Jews mustn't show our feelings, must be brave and strong, must accept all inconveniences and not grumble, must do what is within our power and trust in God. Some time this terrible war will be over. Surely the time will come when we are people again, and not just Jews.

Who has inflicted this upon us? Who has made us Jews

different to all other people? Who has allowed us to suffer so terribly up till now? It is God that has made us as we are, but it will be God, too, who will raise us up again. If we bear all this suffering and if there are still Jews left, when it is over, then Jews, instead of being doomed, will be held up as an example. Who knows, it might even be our religion, from which the world and all peoples learn good, and for that reason and that reason only, do we have to suffer now. We can never become just Netherlanders, or just English, or representatives of any country for that matter, we will always remain Jews, but we want to, too.

Be brave! Let us remain aware of our task and not grumble, a solution will come; God has never deserted our people. Right through the ages there have been Jews, through all the ages they have had to suffer, but it has made them strong too; the weak fall, but the strong will remain and never go under!

Black Elk

Wounded Knee

Wounded Knee in South Dakota was the site of the last major conflict in the Indian Wars. On 29 December 1890 the 7th Cavalry killed 53 Sioux for the loss of 25 troopers. Black Elk, a cousin of Crazy Horse, was an eyewitness.

That evening before it happened, I went in to Pine Ridge and heard these things, and while I was there, soldiers started for where the Big Foots were. These made about five hundred soldiers that were there next morning. When I saw them starting I felt that something terrible was going to happen. That night I could hardly sleep at all. I walked around most of the night.

In the morning I went out after my horses, and while I was out I heard shooting off toward the east, and I knew from the sound that it must be wagon-guns (cannon) going off. The sounds went right through my body, and I felt that something terrible would happen.

When I reached camp with the horses, a man rode up to me and said: "Hey-hey-hey! The people that are coming are fired on! I know it!"

I saddled up my buckskin and put on my sacred shirt. It was one I had made to be worn by no one but myself. It had a spotted eagle outstretched on the back of it, and the daybreak star was on the left shoulder, because when facing south that shoulder is toward the east. Across the breast, from the left shoulder to the right hip, was the flaming rainbow, and there was another rainbow around the neck, like a necklace, with a star at the bottom. At each shoulder, elbow, and wrist was an eagle feather; and over the whole shirt were red streaks of lightning. You will see that this was from my great vision, and you will know how it protected me that day.

I painted my face all red, and in my hair I put one eagle feather for the One Above.

It did not take me long to get ready, for I could still hear the shooting over there.

I started out alone on the old road that ran across the hills to Wounded Knee. I had no gun. I carried only the sacred bow of the west that I had seen in my great vision. I had gone only a little way when a band of young men came galloping after me. The first two who came up were Loves War and Iron Wasichu. I asked what they were going to do, and they said they were just going to see where the shooting was. Then others were coming up, and some older men.

We rode fast, and there were about twenty of us now. The shooting was getting louder. A horseback from over there came galloping very fast toward us, and he said: "Hey-hey-hey! They have murdered them!" Then he whipped his horse and rode away faster toward Pine Ridge.

In a little while we had come to the top of the ridge where, looking to the east, you can see for the first time the monument and the burying ground on the little hill where the church is. That is where the terrible thing started. Just south of the burying ground on the little hill a deep dry gulch runs about east and west, very crooked, and it rises westward to nearly the top of the ridge where we were. It had no name, but the Wasichus [white men], sometimes call it Battle Creek now. We stopped on the ridge not far from the head of the dry gulch. Wagon-guns were still going off over there on the little hill, and they were going off again where they hit along the gulch. There was much shooting down yonder, and there were many cries, and we could see cavalrymen scattered over the hills ahead of us. Cavalrymen were riding along the gulch and shooting into it, where the women and children were running away and trying to hide in the gullies and the stunted pines.

A little way ahead of us, just below the head of the dry gulch, there were some women and children who were huddled under a clay bank, and some cavalrymen were there pointing guns at them.

We stopped back behind the ridge, and I said to the others: "Take courage. These are our relatives. We will try to get them back." Then we all sang a sang which went like this:

A thunder being nation I am, I have said.
A thunder being nation I am, I have said.
You shall live.
You shall live.
You shall live.
You shall live.

Then I rode over the ridge and the others after me, and we were crying: "Take courage! It is time to fight!" The soldiers who were guarding our relatives shot at us and then ran away fast, and some more cavalrymen on the other side of the gulch did too. We got our relatives and sent them across the ridge to the northwest where they would be safe.

I had no gun, and when we were charging, I just held the sacred bow out in front of me with my right hand. The bullets did not hit us at all.

We found a little baby lying all alone near the head of the gulch. I could not pick her up just then, but I got her later and some of my people adopted her. I just wrapped her up tighter in a shawl that was around her and left her there. It was a safe place, and I had other work to do.

The soldiers had run eastward over the hills where there were some more soldiers, and they were off their horses and lying down. I told the others to stay back, and I charged upon them holding the sacred bow out toward them with my right hand. They all shot at me, and I could hear bullets all around me, but I ran my horse right close to them, and then swung around. Some soldiers across the gulch began shooting at me too, but I got back to the others and was not hurt at all.

By now many other Lakotas, who had heard the shooting, were coming up from Pine Ridge, and we all charged on the soldiers. They ran eastward toward where the trouble began. We followed down along the dry gulch, and what we saw was terrible. Dead and wounded women and children and little babies were scattered all along there where they had been trying to run away. The soldiers had followed along the gulch, as they ran, and murdered them in there. Sometimes they were in heaps because they had huddled together, and some were scattered all along. Sometimes bunches of them had been killed and torn to pieces where the wagon-guns hit them. I saw a little baby trying to suck its mother, but she was bloody and dead.

There were two little boys at one place in this gulch. They

had guns and they had been killing soldiers all by themselves. We could see the soldiers they had killed. The boys were all alone there, and they were not hurt. These were very brave little boys.

When we drove the soldiers back, they dug themselves in, and we were not enough people to drive them out from there. In the evening they marched off up Wounded Knee Creek, and then we saw all that they had done there.

Men and women and children were heaped and scattered all over the flat at the bottom of the little hill where the soldiers had their wagon-guns, and westward up the dry gulch all the way to the high ridge, the dead women and children and babies were scattered.

When I saw this I wished that I had died too, but I was not sorry for the women and children. It was better for them to be happy in the other world, and I wanted to be there too. But before I went there I wanted to have revenge. I thought there might be a day, and we should have revenge.

After the soldiers marched away, I heard from my friend, Dog Chief, how the trouble started, and he was right there by Yellow Bird when it happened. This is the way it was:

In the morning the soldiers began to take all the guns away from the Big Foots, who were camped in the flat below the little hill where the monument and burying ground are now. The people had stacked most of their guns, and even their knives, by the tepee where Big Foot was lying sick. Soldiers were on the little hill and all around, and there were soldiers across the dry gulch to the south and over east along Wounded Knee Creek too. The people were nearly surrounded, and the wagon-guns were pointing at them.

Some had not yet given up their guns, and so the soldiers were searching all the tepees, throwing things around and poking into everything. There was a man called Yellow Bird, and he and another man were standing in front of the tepee where Big Foot was lying sick. They had white sheets around and over them, with eyeholes to look through, and they had guns under these. An officer came to search them. He took the other man's gun, and then started to take Yellow Bird's. But Yellow Bird would not let go. He wrestled with the officer, and while they were wrestling, the gun went off and killed the officer. Wasichus and some others have said he meant to do this, but Dog Chief was standing right there, and he told me it was not so. As soon as the gun went off, Dog Chief told me, an officer shot and killed Big Foot who was lying sick inside the tepee.

Then suddenly nobody knew what was happening, except that the soldiers were all shooting and the wagon-guns began going off right in among the people.

Many were shot down right there. The women and children ran into the gulch and up west, dropping all the time, for the soldiers shot them as they ran. There were only about a hundred warriors and there were nearly five hundred soldiers. The warriors rushed to where they had piled their guns and knives. They fought soldiers with only their hands until they got their guns.

Dog Chief saw Yellow Bird run into a tepee with his gun, and from there he killed soldiers until the tepee caught fire. Then he died full of bullets.

It was a good winter day when all this happened. The sun was shining. But after the soldiers marched away from their dirty work. a heavy snow began to fall. The wind came up in the night. There was a big blizzard, and it grew very cold. The snow drifted deep in the crooked gulch, and it was one grave of butchered women and children and babies, who had never done any harm and were only trying to run away.

Nicholas Tomalin

Killing Cong

Tomalin was a war correspondent for the London Sunday Times. *His celebrated account of US General James F. Hollingsworth's "zapping" of Viet Cong is from 1966. Tomalin was killed seven years later covering the Yom Kippur War.*

After a light lunch last Wednesday [1 June 1966], General James F. Hollingsworth, of Big Red One, took off in his personal helicopter and killed more Vietnamese than all the troops he commanded.

The story of the General's feat begins in the divisional office, at Ki-Na, twenty miles north of Saigon, where a Medical Corps colonel is telling me that when they collect enemy casualties they find themselves with more than four injured civilians for every wounded Viet Cong – unavoidable in this kind of war.

The General strides in, pins two medals for outstanding gallantry to the chest of one of the colonel's combat doctors. Then he strides off again to his helicopter, and spreads out a polythene-covered map to explain our afternoon's trip.

The General has a big, real American face, reminiscent of every movie general you have seen. He comes from Texas, and is forty-eight. His present rank is Brigadier General, Assistant Division Commander, 1st Infantry Division, United States Army which is what the big red figure one on his shoulder flash means.

"Our mission today", says the General, "is to push those goddam VCs right off Routes 13 and 16. Now you see Routes 13 and 16 running north from Saigon toward the town of Phuoc Vinh, where we keep our artillery. When we got here first we prettied up those roads, and cleared Charlie Cong right out so we could run supplies up.

"I guess we've been hither and thither with all our operations since, an' the ol' VC he's reckoned he could creep back. He's been puttin' out propaganda he's goin' to interdict our right of passage along those routes. So this day we aim to zapp him, and zapp him, and zapp him again till we've zapped him right back where he came from. Yes, sir. Let's go."

The General's UH18 helicopter carries two pilots, two 50-calibre machine-gunners, and his aide, Dennis Gillman, an apple-cheeked subaltern from California. It also carries the General's own M16 carbine (hanging on a strut), two dozen smoke bombs, and a couple of CS anti-personnel gas-bombs, each as big as a small dustbin. Just beside the General is a radio console where he can tune in on orders issued by battalion commanders flying helicopters just beneath him, and company commanders in helicopters just below them. Under this interlacing of helicopters lies the apparently peaceful landscape beside Routes 13 and 16, filled with farmhouses and peasants hoeing rice and paddy fields.

So far today, things haven't gone too well. Companies Alpha, Bravo and Charlie have assaulted a suspected Viet Cong HQ, found a few tunnels but no enemy. The General sits at the helicopter's open door, knees apart, his shiny black toecaps jutting out into space, rolls a filtertip cigarette to-and-fro in his teeth, and thinks.

"Put me down at Battalion HQ," he calls to the pilot.

"There's sniper fire reported on choppers in that area, General."

"Goddam the snipers, just put me down."

Battalion HQ at the moment is a defoliated area of four acres packed with tents, personnel carriers, helicopters and milling GIs. We settle into the smell of crushed grass. The General leaps out and strides through his troops.

"Why General, excuse us, we didn't expect you here," says a sweating major.

"You killed any 'Cong yet?"

"Well no General, I guess he's just too scared of us today. Down the road a piece we've hit trouble, a bulldozer's fallen through a bridge, and trucks coming through a village knocked the canopy off a Buddhist pagoda. Saigon radioed us to repair that temple before proceeding – in the way of civic action, General. That put us back an hour. . . ."

"Yeah. Well Major, you spread out your perimeter here a bit, then get to killin' VCs will you?"

Back through the crushed grass to the helicopter.

"I don't know how you think about war. The way I see it, I'm just like any other company boss, gingering up the boys all that time, except I don't make money. I just kill people, and save lives."

In the air the General chews two more filtertips and looks increasingly forlorn. No action on Route 16, and another Big Red One general has got his helicopter in to inspect the collapsed bridge before ours.

"Swing us back along again," says the General.

"Reports of fire on choppers ahead, sir. Smoke flare near spot. Strike coming in."

"Go find that smoke."

A plume of white rises in the midst of dense tropical forest, with a Bird Dog spotter plane in attendance. Route 16 is to the right; beyond it a large settlement of red-tiled houses.

Two F105 jets appear over the horizon in formation, split, then one passes over the smoke, dropping the trail of silver, fish-shaped canisters. After four seconds' silence, light orange fire explodes in patches along an area fifty yards wide by three-quarters of a mile long. Napalm.

The trees and bushes burn, pouring dark oily smoke into the sky. The second plane dives and fire covers the entire strip of dense forest.

"Aaaah," cries the General. "Nice. Nice. Very neat. Come in low, let's see who's left down there."

"How do you know for sure the Viet Cong snipers were in that strip you burned?"

"We don't. The smoke position was a guess. That's why we zapp the whole forest."

"But what if there was someone, a civilian, walking through there?"

"Aw come son, you think there's folks just sniffing flowers in tropical vegetation like that? With a big operation on hereabouts? Anyone left down there, he's Charlie Cong all right."

I point at a paddy field full of peasants less than half a mile away.

"That's different son. We know they're genuine."

The pilot shouts: "General, half-right, two running for that bush."

"I see them. Down, down, goddam you."

In one movement he yanks his M16 off the hanger, slams in a

clip of cartridges and leans right out of the door, hanging on his seatbelt to fire one long burst in the general direction of the bush.

"General, there's a hole, maybe a bunker, down there."

"Smokebomb, circle, shift it."

"But General, how do you know those aren't just frightened peasants?"

"Running? Like that? Don't give me a pain. The clips, the clips, where in hell are the cartridges in this ship?"

The aide drops a smoke canister, the General finds his ammunition and the starboard machine-gunner fires rapid bursts into the bush, his tracer bouncing up off the ground round it.

We turn clockwise in ever tighter, lower circles, everyone firing. A shower of spent cartridge cases leaps from the General's carbine to drop, lukewarm, on my arm.

"I . . . WANT . . . YOU . . . TO . . . SHOOT . . . RIGHT . . . UP . . . THE . . . ASS . . . OF . . . THAT . . . HOLE . . . GUNNER."

Fourth time round the tracers flow right inside the tiny sand-bagged opening, tearing the bags, filling it with sand and smoke.

The General falls back off his seatbelt into his chair, suddenly relaxed, and lets out an oddly feminine, gentle laugh. "That's it," he says, and turns to me, squeezing his thumb and finger into the sign of a French chef's ecstasy.

We circle now above a single-storey building made of dried reeds. The first burst of fire tears the roof open, shatters one wall into fragments of scattered straw, and blasts the farmyard full of chickens into dismembered feathers.

"Zapp, zapp, zapp", cries the General. He is now using semi-automatic fire, the carbine bucking in his hands.

Pow, pow, pow, sounds the gun. All the noises of this war have an unaccountably Texan ring. . . .

"There's nothing alive in there", says the General. "Or they'd be skedaddling. Yes there is, by golly."

For the first time I see the running figure, bobbing and sprinting across the farmyard towards a clump of trees dressed in black pyjamas. No hat. No shoes.

"Now hit the tree."

We circle five times. Branches drop off the tree, leaves fly, its trunk is enveloped with dust and tracer flares. Gillman and the

General are now firing carbines side by side in the doorway. Gillman offers me his gun: No thanks.

Then a man runs from the tree, in each hand a bright red flag which he waves desperately above his head.

"Stop, stop, he's quit", shouts the General, knocking the machine-gun so tracers erupt into the sky.

"I'm going down to take him. Now watch it everyone, keep firing round-about, this may be an ambush."

We sink swiftly into the field beside the tree, each gunner firing cautionary bursts into the bushes. The figure walks towards us.

"That's a Cong for sure," cries the General in triumph and with one deft movement grabs the man's short black hair and yanks him off his feet, inboard. The prisoner falls across Lieutenant Gillman and into the seat beside me.

The red flags I spotted from the air are his hands, bathed solidly in blood. Further blood is pouring from under his shirt, over his trousers.

Now we are safely in the air again. Our captive cannot be more than sixteen years old, his head comes just about up to the white name patch – Hollingsworth – on the General's chest. He is dazed, in shock. His eyes calmly look first at the General, then at the Lieutenant, then at me. He resembles a tiny, fine-boned wild animal. I have to keep my hand firmly pressed against his shoulder to hold him upright. He is quivering. Sometimes his left foot, from some nervous impulse, bangs hard against the helicopter wall. The Lieutenant applies a tourniquet to his right arm.

"Radio base for an ambulance. Get the information officer with a camera. I want this Commie bastard alive till we get back . . . just stay with us till we talk to you, baby."

The General pokes with his carbine first at the prisoner's cheek to keep his head upright, then at the base of his shirt.

"Look at that now," he says, turning to me. "You still thinking about innocent peasants? Look at the weaponry."

Around the prisoner's waist is a webbing belt, with four clips of ammunition, a water bottle (without stopper); a tiny roll of bandages, and a propaganda leaflet which later turns out to be a set of Viet Cong songs, with a twenty piastre note (about *1s 6d*) folded in it.

Lieutenant Gillman looks concerned. "It's OK, you're OK", he mouths at the prisoner, who at that moment turns to me and with a surprisingly vigorous gesture waves his arm at my seat. He wants to lie down.

By the time I have fastened myself into yet another seat we are back at the landing pad. Ambulance orderlies come aboard, administer morphine, and rip open his shirt. Obviously a burst of fire has shattered his right arm up at the shoulder. The cut shirt now allows a large bulge of blue-red tissue to fall forward, its surface streaked with white nerve fibres and chips of bone (how did he ever manage to wave that arm in surrender?).

When the ambulance has driven off the General gets us all posed round the nose of the chopper for a group photographer like a gang of successful fishermen, then clambers up into the cabin again, at my request, for a picture to show just how he zapped those VCs. He is euphoric.

"Jeez I'm so glad you was along, that worked out just dandy. I've been written up time and time again back in the States for shootin' up VCs, but no one's been along with me like you before.

"I'll say perhaps your English generals wouldn't think my way is all that conventional, would they? Well, this is a new kind of war, flexible, quickmoving. Us generals must be on the spot to direct our troops. The helicopter adds a new dimension to battle.

"There's no better way to fight than goin' out to shoot VCs. An' there's nothing I love better than killin' Cong. No sir."

Claus Fuhrman

Götterdämmerung: The Fall of Berlin

A citizen's eye view of the fall of the capital of the Reich to Russian forces in April–May 1945.

Panic had reached its peak in the city. Hordes of soldiers stationed in Berlin deserted and were shot on the spot or hanged on the nearest tree. A few clad only in underclothes were dangling on a tree quite near our house. On their chests they had placards reading: "We betrayed the Führer." The Werewolf pasted leaflets on the houses: "Dirty cowards and defeatists/We've got them all on our lists!"

The SS went into underground stations, picked out among the sheltering crowds a few men whose faces they did not like, and shot them then and there.

The scourge of our district was a small one-legged Hauptscharführer of the SS, who stumped through the street on crutches, a machine pistol at the ready, followed by his men. Anyone he didn't like the look of he instantly shot. The gang went down cellars at random and dragged all the men outside, giving them rifles and ordering them straight to the front. Anyone who hesitated was shot.

The front was a few streets away. At the street corner diagonally opposite our house Walloon Waffen SS had taken up position; wild, desperate men who had nothing to lose and who fought to their last round of ammunition. Armed Hitler Youth were lying next to men of the Vlassov White Russian Army.

The continual air attacks of the last months had worn down our morale; but now, as the first shells whistled over our heads, the terrible pressure began to give way. It could not take much longer now, whatever the Walloon and French Waffen SS or the fanatic Hitler Youth with their 2-cm. anti-aircraft guns could do. The end was coming and all we had to do was to try to survive this final stage.

But that was by no means simple. Everything had run out. The only water was in the cellar of a house several streets away. To get bread one had to join a queue of hundreds, grotesquely adorned with steel helmets, outside the baker's shop at 3 a.m. At 5 a.m. the Russians started and continued uninterruptedly until 9 or 10. The crowded mass outside the baker's shop pressed closely against the walls, but no one moved from his place. Often the hours of queuing had been spent in vain; bread was sold out before one reached the shop. Later one could buy bread only if one brought half a bucket of water.

Russian low-flying wooden biplanes machine-gunned people as they stood apathetically in their queues and took a terrible toll of the waiting crowds. In every street dead bodies were left lying where they had fallen.

At the last moment the shopkeepers, who had been jealously hoarding their stocks, not knowing how much longer they would be allowed to, now began to sell them. Too late! For a small packet of coffee, half a pound of sausages, thousands laid down their lives. A salvo of the heavy calibre shells tore to pieces hundreds of women who were waiting in the market hall. Dead and wounded alike were flung on wheelbarrows and carted away; the surviving women continued to wait, patient, resigned, sullen, until they had finished their miserable shopping.

The pincers began to narrow on the capital. Air raids ceased; the front lines were too loose now for aircraft to distinguish between friend and foe. Slowly but surely the T.52 tanks moved forward through Prenzlauer Allee, through Schonhauser Allee, through Kaiserstrasse. The artillery bombardment poured on the city from three sides in unbroken intensity. Above it, one could hear sharply close and distinct, the rattling of machine-guns and the whine of bullets.

Now it was impossible to leave the cellar. And now the bickering and quarrelling stopped and we were suddenly all of one accord. Almost all the men had revolvers; we squatted in the farthest corner of the cellar in order to avoid being seen by patrolling SS, and were firmly determined to make short shrift of any Volkssturm men who might try to defend our house.

Under the direction of a master mason who had been a soldier in Russia for two years we "organized" our supplies. We made a roster for parties of two or three to go out and get water and bread. We procured steel helmets; under artillery fire we heaped up mountains of rubble outside the cellar walls in order to safeguard against shells from tanks.

The Nazis became very quiet. No one took the Wehrmacht communiqué seriously now, although Radio Berlin went on broadcasting it until 24 April. A tiny sheet of paper, the last newspaper of the Goebbels press, *Der Panzerbär* [the tank bear] announced Goering's deposition and the removal of the "government" seat to Flensburg.

We left the cellar at longer and longer intervals and often we could not tell whether it was night or day. The Russians drew nearer; they advanced through the underground railway tunnels, armed with flame-throwers; their advance snipers had taken up positions quite near us; and their shots ricocheted off the houses opposite. Exhausted German soldiers would stumble in and beg for water – they were practically children; I remember one with a pale, quivering face who said, "We shall do it all right; we'll make our way to the north west yet." But his eyes belied his words and he looked at me despairingly. What he wanted to say was, "Hide me, give me shelter. I've had enough of it." I should have liked to help him; but neither of us dared to speak. Each might have shot the other as a "defeatist".

An old man who had lived in our house had been hit by a shell splinter a few days ago and had bled to death. His corpse lay near the entrance and had already begun to smell. We threw him on a cart and took him to a burnt-out school building where there was a notice: "Collection point for Weinmeisterstrasse corpses." We left him there; one of us took the opportunity of helping himself to a dead policeman's boots.

The first women were fleeing from the northern parts of the city and some of them sought shelter in our cellar, sobbing that the Russians were looting all the houses, abducting the men and raping all the women and girls. I got angry, shouted I had had enough of Goebbels' silly propaganda, the time for that was past. If that was all they had to do, let them go elsewhere.

Whilst the city lay under savage artillery and rifle fire the citizens now took to looting the shops. The last soldiers withdrew farther and farther away. Somewhere in the ruins of the burning city SS-men and Hitler Youth were holding out fanatically. The crowds burst into cellars and storehouses. While bullets were whistling through the air they scrambled for a tin of fish or a pouch of tobacco.

On the morning of 1 May our flat was hit by a 21-cm. shell and almost entirely destroyed. On the same day water carriers reported that they had seen Russian soldiers. They could not be located exactly; they were engaged in house-to-house fighting which was moving very slowly. The artillery had been silent for some time when at noon on 2 May rifle fire too ceased in our district. We climbed out of our cellar.

From the street corner Russian infantry were slowly coming forward, wearing steel helmets with hand grenades in their belts and boots. The SS had vanished. The Hitler Youth had surrendered.

Bunny rushed and threw her arms round a short slit-eyed Siberian soldier who seemed more than a little surprised. I at once went off with two buckets to fetch water, but I did not get beyond the first street corner. All men were stopped there, formed into a column and marched off towards the east.

A short distance behind Alexanderplatz everything was in a state of utter turmoil and confusion. Russian nurses armed with machine-pistols were handing out loaves of bread to the German population. I took advantage of this turmoil to disappear and got back home safely. God knows where the others went.

After the first wave of combatant troops there followed reserves and supply troops who "liberated" us in the true Russian manner. At our street corner I saw two Russian soldiers assaulting a crying elderly woman and then raping her in full view of the stunned crowd. I ran home as fast as I could. Bunny was all right so far. We had barricaded the one remaining room of our flat with rubble and charred beams in such a manner that no one outside could suspect that anyone lived there.

Every shop in the district was looted. As I hurried to the market I was met by groups of people who were laden with sacks and boxes. Vast food reserves belonging to the armed forces had been stored there. The Russians had forced the doors open and let the Germans in.

The cellars, which were completely blacked out, now became the scene of an incredible spectacle. The starving people flung themselves like beasts over one another, shouting, pushing and struggling to lay their hands on whatever they could. I caught hold of two buckets of sugar, a few boxes of preserves, sixty packages of tobacco and a small sack of coffee which I quickly took back home before returning for more.

The second raid was also successful. I found noodles, tins of butter and a large tin of sardines. But now things were getting out of hand. In order not to be trampled down themselves the Russians fired at random into the crowds with machine-pistols, killing several.

I cannot remember how I extricated myself from this screaming, shouting chaos; all I remember is that even here in this utter confusion, Russian soldiers were raping women in one of the corners.

Bunny had meanwhile made me promise not to try to interfere if anything were to happen to her. There were stories in our district of men being shot trying to protect their wives. In the afternoon two Russians entered our flat, while Bunny was sitting on the bed with the child. They looked her over for some time; evidently they were not very impressed with her. We had not washed for a fortnight, and I had expressly warned Bunny not to make herself tidy, for I thought the dirtier and more neglected she looked the safer.

But the two gentlemen did not seem to have a very high standard as far as cleanliness was concerned. With the usual words, "Frau komm!" spoken in a menacing voice, one of them went towards her. I was about to interfere; but the other shouted "Stoi" and jammed his machine-pistol in my chest. In my despair I shouted "Run away, quick"; but that was, of course, impossible. I saw her quietly lay the baby aside, then she said, "Please don't look, darling." I turned to the wall.

When the first Russian had had enough they changed places. The second was chattering in Russian all the time. At last it was over. The man patted me on the shoulder: "Nix Angst! Russki Soldat gut!"

Terence Robertson

The Hunting of *U-99*

Kapitan-Leutnant Otto Kretschmer was the leading U-boat ace of the Second World War. By the time his infamous U-99 *was scuttled in March 1941 he had sunk 350,000 tons of Allied shipping in the cruel waters of the Atlantic.*

On the evening of March 15th, 1941, *U-99* received a signal from *U-110*, commanded by Lieutenant-Commander Lemp, who reported sighting a convoy between Iceland and the 61st parallel. Petersen, Kretschmer's Second Lieutenant, worked out an interception course and they raced for the new sighting position at full speed. Further to the west, Schepke in *U-100* also received the report and followed suit.

By early morning on the 16th Kretschmer was in the convoy's estimated position, but with nothing in sight. He dived and heard faint propeller noises on the hydrophones bearing to southwards. They surfaced, but two hours later ran into fog and dived again to get more hydrophone bearings. Kassel, the *U-99*'s Chief Radio Operator, nearly leapt from his seat in alarm as a loud jumbled noise crackled through his headphones. "There are propeller noises all around us, sir," he shouted. Despite the risk of collision, Kretschmer decided to surface. They came up into a thinning fog which had cleared enough for them to see they were inside the escort screen on the starboard side of the convoy with two destroyers to port and one coming towards them from ahead. Kretschmer dived, heading underneath the convoy itself and then slowed down to drop back gradually until he could surface well astern in safety. When they came up again the fog had lifted completely, and they sighted four trawlers ploughing their way northwards from the convoy, presumably to keep a rendezvous with another.

Although Lemp was the official shadower, Kretschmer

stayed astern, reasoning that it was as good a position as any to spend the daylight hours. He would work up to the convoy at dusk. Throughout the afternoon he had to dive at intervals to avoid detection by a Sunderland aircraft flying round the convoy in wide sweeps. But he was by now a good five miles behind the stern escorts and hidden in a light mist. At one time he was mystified by the appearance behind him of a curious-looking ship that seemed to be an old-time lugger. Rightly he diagnosed this as a rescue ship attached to the convoy to pick up survivors. They heard the sharp explosions of depth-charges on the convoy's starboard side, and a rapid check of U-boats in the vicinity showed it was Schepke who had been trying to work his way up to the bow for a dusk attack with a fan of torpedoes. By early evening *U-99* was working up on the port side of the convoy. The weather looked as though it had cleared for the night, so Kretschmer left his officer-of-the-watch in charge while he went down to rest.

Shortly after 3 p.m. he returned to the bridge and, to his anger and amazement, the watch had lost contact with the convoy. He altered course to bring it back in sight but they saw nothing. The convoy had altered course to starboard unnoticed by the officer-of-the-watch. Eventually, they had to dive to use the hydrophones and managed to pick up propeller noises very faint to the south. Kretschmer surfaced and made off on the bearing at full speed, making contact again shortly after dark. He wasted no time. He had been on patrol for a month, taking part in three convoy battles, and fuel was running low. He raced in between the two escorts on the port beam and attacked a large tanker in the middle of the outside port column. The range was just under a mile, and the torpedo hit just before the midships mast. There was a huge sheet of flame as high-octane petrol exploded and burned internally. Lying between the tanker and the outside destroyer escort, *U-99* was illuminated starkly in the white glare of the flames.

Surprised and not too happy at being exposed so unexpectedly and so nakedly, Kretschmer dived and made his way past the escort again into the protective blackness well beyond the escort screen. He worked his way round to the stern on the surface and, passing the burning tanker wallowing and hissing in the water, crept between the stern escort and entered the convoy inside the centre lane of ships. He was only a few yards from ships on either side of him as he slipped past them looking for a suitable target. He saw another large tanker two columns away and manœuvred through gaps in the lines of merchant ships until he was running alongside his target. He fired one torpedo, and *U-99* rocked in the blast as the tanker erupted and he was revealed again in the glare of the blaze to every merchant ship around him. The flames gave way to a huge cloud of smoke which fell slowly like fog on the sea. Kretschmer steered into the middle of it and crossed into another lane.

Meanwhile, the escorts were firing starshell and snowflakes out to sea on either side, making it difficult for U-boats who wanted to fire fans outside the escort screen. For nearly fifteen minutes, Kretschmer travelled along with the convoy as though part of it before he spotted a third tanker. He conned *U-99* alongside it and scored another hit with one torpedo. This tanker stopped and black smoke poured from beneath its decks, while angry red flames licked its decks. He repeated his previous performance, seeking shelter in the smoke-clouds and then emerging in another lane to travel along as part of the convoy. By now depth-charges were being dropped several miles astern and on the starboard side. He saw two small escort vessels hovering round his three sinking tankers picking up survivors. Moving up to the front ships of the lanes, he saw two large freighters. He fired one torpedo at each and scored direct hits amidships on both almost simultaneously. The smaller one sank immediately, but the larger only settled slightly down into the water. He stopped alongside it and fired another torpedo while other ships were still passing by. At little more than 200 yards, this shot missed – another torpedo failure. The third torpedo at this target hit astern and she subsided below the surface on an even keel.

By this time the convoy had drawn ahead and Kretschmer put on speed to catch up again. Meanwhile, Kassel had been intercepting distress signals and noting the names of each ship. The first tanker had been the *Ferm*, the second *Bedouin* (both Norwegian) and the third a British tanker, *Franche Comte*. The small freighter was the *Venetia* and the larger one the *J. B. White*. *U-99* got into the convoy from astern again, sighted another tanker and, in confident mood, went alongside the target and turned outwards for a shot from the stern tubes. The hit had extraordinary results. The ship broke in two, sagging inwards, but somehow she turned towards *U-99* and came at them in an attempt to ram. They had no torpedoes left and there was nothing to do but get away. Kassel reported this ship's wireless as giving her name as the *Karsham*. As they raced away, they saw the two clinging halves vanish into the sea with tremendous hissing and clouds of steam.

They left the convoy passing between two escorts firing starshell and vanished into the darkness, heading for home. Behind lay the wreckage-strewn battlefield of the fourth attack of the patrol, but they needed rest badly, having been at action stations on this last attack for nearly forty-eight hours.

It was quiet, the convoy had disappeared and Kretschmer decided to set course north of Lousy Bank. It was exactly 3 a.m. when Peterson wrote in the War Diary: "Have taken over watch from Captain, who has gone below." In the control-room, Kretschmer sat with Kassel totalling up tonnages, checking names and composing signals of success . . . Suddenly, alarm-bells rang with a startling clang throughout the relaxed U-boat.

For three days the convoy escort group under the command of Commander Donald Macintyre in the destroyer *Walker* had been aware they were being shadowed. It was Macintyre's first trip as captain of *Walker* and senior officer of the escort. The days of escort group teams, efficiently worked up during weeks of exercises before taking over convoy duty, had yet to come. At present the urgent need of any sort of escort made these groups a loosely-knit bunch of ill-assorted ships relying more on the decisions of individual commanders than on any practised plan to defend their convoys. Commander Macintyre hardly knew his way round *Walker*'s bridge as the main attack developed, after dusk on the 16th, when the first tanker burst into flame. The determined but ill-equipped escorts fought back desperately in an effort to tear the U-boats from the blackness of the Atlantic night – turning, twisting and firing starshell and snowflake rockets to illuminate the sea. But the hours dragged by with no trace of the enemy. It soon became apparent that the full weight of the attack was coming from the port side, where, unknown to the escorts, were gathered the U-boats of Schepke, Kretschmer, Lemp and Schultz. On *Walker*'s bridge Macintyre felt the despair of so many commanders in those days who could see convoys being ripped to pieces and had neither the ships nor the equipment to make the attackers reveal their whereabouts. Radar was still a primitive instrument, and the Asdic was of little use against a surfaced U-boat.

With the only other destroyer in the escort, *Vanoc*, in company, they turned to cover a large circle out to port, and on this sweep their luck broke. *Walker* sighted the fluorescent wash of a U-boat retreating on the surface. They gave chase at full speed and dropped a pattern of ten depth-charges over the whirlpool of water left by their quarry. It was Schepke. They lost contact and Macintyre decided to steam southwards to pick up the survivors of the tanker. When this had been done both destroyers set off to carry out another sweep to port of the convoy. Meanwhile Schepke had been damaged by the depth-charging and felt unable to stay below for a long period. He decided to risk inspecting the damage and probably escape on the surface. As *U-100* came up, the radar operator in *Vanoc* reported to the bridge that his screen showed a dark green blob that might be a U-boat. This radar sighting made naval history, for it was the first known time that those primitive and crude sets had led to a night attack on a U-boat. *Vanoc* reported by R/T to *Walker*, and on Macintyre's orders the two destroyers made a tight turn at high speed and raced along the bearing given by *Vanoc*'s radar operator.

After covering little more than a mile they sighted the tiny hull-down silhouette of *U-100* on the surface. A brief order from Macintyre sent *Vanoc* racing into the attack. The sleek destroyer headed straight for the conning-tower of *U-100*. Cries of alarm sounded thinly in the night air as the U-boat crew saw the knife-edge bows of the destroyer coming at them in a cloud of spray. Some of them jumped overboard and tried desperately to swim out of the way. On *Vanoc*'s bridge they heard the roar of Schepke's voice as he shouted in German: "Don't panic. They are going to miss us. They will pass astern." Then came the rending, grinding crash as *Vanoc* struck *U-100* amidships by the conning-tower, throwing the remainder of the crew into the water. Carried forward by her speed, *Vanoc* ran right over the stricken U-boat before coming to a halt, straining to release herself with both engines pulling astern. Eventually they came clear with a sharp jolt. *U-100* rose high in the air, and then sank beneath the heavy swell.

Vanoc played her searchlight over the scene. Schepke had died like an "ace" – on his bridge. Only five swimmers could be seen, and with *Walker* keeping an Asdic sweep around her, she picked up these survivors from a crew of nearly fifty and inspected the damage to her bows. On the bridge of *Walker*,

Commander Macintyre waited impatiently for *Vanoc* to report her damage, when suddenly his Asdic operator shouted: "Echo to starboard, sir." He gave range and bearing and, to Macintyre's astonishment, it put the target almost directly under the stern of *Vanoc*; a most unlikely position. Macintyre and his officers felt inclined to believe that this echo was caused by the disturbance of the water after the collision or by *Vanoc*'s own wake. But the Asdic operator stubbornly resisted all suggestions that the echo was caused by something other than a U-boat. It was too firm, too strong to be anything else. Still doubting, but not willing to risk missing a chance of attack, Macintyre increased speed and went into attack, while *Vanoc* drew out of the way. Could there be two U-boats where there had been only one a few seconds before?

As Petersen fell down through the conning-tower hatch, he reported a destroyer less than half a mile to starboard.

"Well, why have you dived?" Kretschmer asked. "Are you sure we could not have got away on the surface?"

Petersen looked worried. "I don't think so, sir. They must have seen us."

Kretschmer did not bother to reply. Kassel reported propeller noises approaching and they dived to 300 feet. As they waited for the attack to develop, Kretschmer extracted the story of what had happened from Petersen. The starboard forward look-out – a petty officer – had been dreaming and not properly alert. He had failed to spot the destroyer and report it to Petersen, who had sighted it by chance when they were so close that he had to look up at camouflaged cream-and-green sides gleaming in the moonlight. Astonished at being so close, his first thought had been to dive. Despite the suddenness with which events had happened, Kretschmer still had doubts if Petersen had been wise, and told him so forcibly. All this time Kassell was reporting propeller noises, and now estimated there were two destroyers near them. He shouted that one was heading their way, and the first seven depth-charges came with unexpected suddenness.

There was a long series of cracks, and *U-99* nearly rolled right over in a circle. They were the closest depth-charges Kretschmer had experienced, and an unpleasant, ominous roaring filled his ears as blast-waves rocked against the boat, sending lights out and shattering every movable object. The glass dials of the instruments were smashed, the chronometers broken and many of the gauges had gone haywire, among them the depth-gauge. They had no way now of knowing how deep they were.

It was almost certain that the depth-charges were tossing them about wildly, and they might be deeper than they thought. The patterns continued to come down, all well aimed, and a pipe split, sending a jet of water into the crew's compartment forward of the control-room. The boat was listing badly, and oil began pouring into the control-room from a leak in the after fuel-tanks. Within a few minutes they were wading ankle deep in oil and sea-water and Kretschmer decided they would have to blow themselves to the surface. The second depth-gauge in the forward torpedo compartment seemed to be in order; unbelievably, they were at 600 feet, 100 feet below the depth of no return for a submarine. By rights the pressure outside would snap their hull into shreds at any moment. The engineer reported that they were getting no thrust from the propellers and were making no speed – a danger because, without speed, a U-boat must sink. Kretschmer ordered compressed air to be blown into the ballast tanks. "Popp", the control-room Petty Officer, tugged at the air valve, but it would not budge. Petersen stood by the depth-gauge, calling out depths as they plunged downwards . . . six-fifty feet . . . seven hundred feet. . . . There was a sharp crack from the stern and a slight leak appeared in the starboard side of the after-torpedo-room. This was the beginning of the end. Their only hope lay in the valve that "Popp" could not move.

In U-boat Command Headquarters at Lorient, in German-occupied France, the radio-room was silent save for the monotonous hum of the electric generators and the regular chimes of a clock announcing each quarter of an hour. On a chart of the North Atlantic stood out the red flag of a missing U-boat, *U-47*. But, confidently, they waited for the success reports of the other "aces" – Schepke and Kretschmer.

"Popp", helped by a sailor, tugged frantically at the air valve. It gave slightly and, cautiously, he pulled it open to avoid wasting the air. "Seven hundred and twenty feet," sang out Petersen, and Kretschmer jumped over to slap "Popp" hard on the back and shout: "Open it wide and fast." The air rushed into the ballast tanks and for a moment the boat shuddered, then it rolled and Petersen sang out excitedly: "Seven hundred . . . six seventy-five . . . six-fifty . . . we are going up." At 200 feet

Kretschmer ordered the air to be turned off in an attempt to keep *U-99* at that depth, but the engines had not been repaired and there was a suspicion that the propellers were damaged. Without speed there was no way of controlling the boat, which continued its upward climb and rushed to the surface, where it arrived with such force that its nose hung out of the water before dropping back. Kretschmer rushed to the conning-tower, while the assistant radio operator, Stohrer, tapped out their final message to Lorient. It said, "Bombs . . . bombs".

At Lorient the wireless-room came to life. Alarm buzzers sounded and runners took the message to several staff officers, including Admiral Doenitz, the Commander-in-Chief. They came running to the wireless-room to listen. What was happening out there in the Atlantic?

The senior staff officer on duty ordered signals to be sent to the boats which had not reported during the night.

"All submarines accompanying *U-110* (convoy shadower) report position, conditions and success immediately."

They waited, but there was no reply.

"*U-99* and *U-100* report position at once."

Again neither ship replied. In the control-room of *U-99* First Lieutenant Knebel-Döberitz grabbed the signal Kassell was dictating and read:

"Two destroyers – depth-charges – 53,000 tons – capture – Kretschmer."

It was the curt signal of a man who was going to fight his boat to the last. But Knebel-Döberitz inserted the word "sunk", and when the full text was picked up by a nearby U-boat, *U-37*, and transmitted to Lorient, the staff there interpreted it as meaning that he had sunk two destroyers as well as 53,000 tons of shipping. Doenitz sensed the drama that was being played out on the high seas; that some of his top commanders were fighting for their lives. But he did not know how the fight was going or that his "aces" were caught in a deadly trap.

When Kretschmer reached the conning-tower, *U-99* was lying on her side to starboard and a great patch of oil was spreading over the water around her. One destroyer was lying dead ahead beam-on and stopped. Kretschmer cursed at not having a torpedo. He was sure he could have hit, despite his own crippled position. Desperately, the engineer and his men strove to get the diesel motors working, in the hope they could get some steering way, but the propellers had been blown off or damaged by the explosions. They simply wallowed in the light swell.

The crew of the slow-moving, damaged *Vanoc* were startled to see a second U-boat shoot to the surface so close under their stern. A quick report to *Walker* brought the curt reply: "Well, get out of the way." *Vanoc* increased speed, turned and the three ships set at the corner of a triangle gazed at each other speculatively for a few seconds as their commanders adjusted themselves to this swift turn of events. While Kretschmer raged impotently at having no speed to get away and no torpedoes to attack, Commander Macintyre remembered that, with his only other destroyer damaged due to ramming one U-boat, he could not afford to damage himself. The convoy had to be protected for another four days at least. He was bound also by an Admiralty order that if it was possible, escort commanders should make every attempt to capture U-boats intact.

He decided to open fire with his main and small armament in an attempt to scare the U-boat crew into abandoning their ship. The two destroyers circled the target warily, keeping *U-99* under a vicious cross-fire from four-inch turrets, machine-guns and pom-poms. Showers of tracer bullets and shell splinters splattered on the side of *U-99*'s hull and conning-tower, but as she was listing away from the fire, the crew had plenty of protection on the other side of the boat. Kretschmer thought seriously for a moment of having his own gun manned and making a duel of it, in the hope of delaying the final attack until another U-boat came to his rescue. But he realized that any man who appeared above the deck level of the listing *U-99* would be killed before he could get to the gun. Miraculously, the shells of the large guns were missing widely, making big water-spouts all around them. He ordered all his crew out on to the deck on the starboard side in readiness to abandon ship and then, unable to do anything about their predicament, sat under the protection of the conning-tower and, to his men's amazement, proceeded to light a cigar.

It seemed that the two destroyers were being over-cautious. They drew off, obviously expecting to be torpedoed at any moment. With no teeth to bite back, Kretschmer thought it likely that a boarding party would be sent over to try to capture the U-boat. He ordered scuttling charges to be set, although the boat was quite noticeably sinking, if only slowly, but the scuttling men reported that the door to the compartment where the charges were kept was jammed.

There was nothing for it but to open all the hatches. When this had been done, Kretschmer made a short speech to his crew hanging on to the guard-rails. He told them how sorry he was he would not be able to take them home this trip, and warned them that they might have a short spell in the water before being taken prisoner. He then sent them below to put on the warmest clothes they could find and return to deck to wait for his order to abandon ship. The crew returned wearing their lifebelts, and his own servant appeared carrying his peaked uniform cap, which he put on at the regulation angle. The First Lieutenant stayed below to see that all secret papers were destroyed, and the chief engineer checked if there was any more air left in the ballast tanks.

Suddenly, the stern of the boat sank with a jolt below the level of the water and the sea rushed in through the open galley hatch, and even through the conning-tower hatch itself. Kretschmer had just time to haul his First Lieutenant and engineer up to the conning-tower through the inrushing water. But the ballast tanks had been blown by the engineer and the stern rose again. Half the ship's company had been standing on the after-deck and had now been washed overboard. The tortured, straining *U-99* began drifting sideways away from the two waiting destroyers, leaving the men in the water behind. Kretschmer told Petersen to sling the portable battery flash-lamp over his shoulder so that, if left alone in the water, they could flash their whereabouts to the destroyers. He also ordered him to call up the nearest destroyer with his lamp, and dictated a signal, letter by letter, very slowly in English:

"From Captain to Captain . . . Please pick up my men drifting towards you in the water stop I am sinking."

The destroyer – it was *Walker* – flashed an acknowledgment before approaching the men in the water with her searchlight playing on the swimmers. Kretschmer and the remainder of his crew watched while the men were hauled aboard with scramble nets. As *Walker* drew close to *U-99* on the starboard beam, Kretschmer realized at once that the destroyer could drift down on top of him and send a boarding party leaping across on to his decks. He decided then that no matter what chance there might be of avoiding capture, no British sailor would ever step aboard *U-99*. He discussed the position with Knebel-Döberitz and Schroeder, the engineer, who immediately offered to go down and flood the ballast tanks.

The matter became urgent as they saw the *Walker* preparing to lower a boat. The engineer vanished below into the water-filled control-room and was never seen again. He must have operated the ballast controls, letting the air out as water rushed in with a final hiss. Kretschmer shouted to the engineer to jump up to the conning-tower, but there was no reply. The stern went down, the nose came up and the U-boat that had menaced Allied shipping so successfully for nearly a year slid stern first to her grave in the waters she had prowled for so long. Kretschmer and his men were washed off the conning-tower and decks. There was no whirlpool or backwash to drag them down. They formed a line in the water and held hands like a human chain to make sure no one was lost. *Walker* came close and they swam the short distance to safety, where Kretschmer grabbed the corner of a scramble net hanging down the destroyer's side and counted his men going up.

In addition to the engineer who had gone down with *U-99*, two of the crew were missing. One had lost his lifebelt and could not swim, and another had received concussion during the depth-charging and, when they turned to help him up the nets, went under and failed to reappear. When his officers and crew had gone up, Kretschmer tried to clamber up himself, but his sea-boots were so full of water they held him back. He could not lift his legs. Thinking all the survivors were on board, *Walker* was picking up speed and dragging him through the water at a rate that would force him to let go. Kretschmer thought this was to be his end, when his boatswain glanced over the side and shouted, "There's the Captain", and came down to help him. When he was pulled on the destroyer's deck, weak after hours of action and tension, a large Colt .45 pistol was poked into his face by a cheerful, grinning British petty officer. He had imagined all sorts of receptions if he was ever captured, but hardly this. He began laughing, but soon stopped when the petty officer glanced covetously at his binoculars. Kretschmer thought suddenly that the glasses he valued so much should not fall into the hands of an enemy sailor and tried to sling them overboard. But he was too late. Another sailor grabbed them from him as he slipped the lanyard over his head. Captain Macintyre claimed these binoculars as his personal "prize" and used them for the rest of the war. Though they have small globules of water in the lens, he says: "They were far better than ours, and still one of my most treasured acquisitions of the war."

Thomas Cochrane

The Cruise of the *Speedy*

Lord Cochrane entered the Royal Navy in 1793, and received his own command, the sloop Speedy *seven years later. It was as commander of the* Speedy *that Cochrane perpetrated the exploits – which included the seizure of fifty prizes in five months – that made him the veritable model for dashing naval commanders in fiction from Hornblower to Aubry.*

The *Speedy* was little more than a burlesque on a vessel of war, even sixty years ago. She was about the size of an average coasting brig, her burden being 158 tons. She was crowded, rather than manned, with a crew of eighty-four men and six officers, myself included. Her armament consisted of fourteen 4-*pounders!*, a species of gun little larger than a blunderbuss, and formerly known in the service under the name of "miñion", an appellation which it certainly merited.

Being dissatisfied with her armament, I applied for and obtained a couple of 12-pounders, intending them as bow and stern chasers, but was compelled to return them to the ordnance wharf, there not being room on deck to work them; besides which, the timbers of the little craft were found on trial to be too weak to withstand the concussion of anything heavier than the guns with which she was previously armed.

With her rig I was more fortunate. Having carried away her mainyard, it became necessary to apply for another to the senior officer, who, examining the list of spare spars, ordered the *foretopgallant-yard* of the *Généreux* to be hauled out *as a mainyard for the Speedy!*

The spar was accordingly sent on board and rigged, but even this appearing too large for the vessel, an order was issued to cut off the yard-arms and thus reduce it to its proper dimensions.

This order was neutralized by getting down and planing the yard-arms as though they had been cut, an evasion which, with some alteration in the rigging, passed undetected on its being again swayed up; and thus a greater spread of canvas was secured. The fact of the foretopgallant-yard of a second-rate ship being considered too large for the mainyard of my "man-of-war" will give a tolerable idea of her insignificance.

Despite her unformidable character and the personal discomfort to which all on board were subjected, I was very proud of my little vessel, caring nothing for her want of accommodation, though in this respect her cabin merits passing notice. It had not so much as room for a chair, the floor being entirely occupied by a small table surrounded with lockers, answering the double purpose of storechests and seats. The difficulty was to get seated, the ceiling being only five feet high, so that the object could only be accomplished by rolling on the locker, a movement sometimes attended with unpleasant failure. The most singular discomfort, however, was that my only practicable mode of shaving consisted in removing the skylight and putting my head through to make a toilet-table of the quarter-deck.

In the following enumeration of the various cruises in which the *Speedy* was engaged, the boarding and searching innumerable neutral vessels will be passed over, and the narrative will be strictly confined – as in most cases throughout this work – to log extracts, where captures were made, or other occurrences took place worthy of record.

> *May* 10. – Sailed from Cagliari, from which port we had been ordered to convoy fourteen sail of merchantmen to Leghorn. At 9 a.m. observed a strange sail take possession of a Danish brig under our escort. At 11:30 a.m. rescued the brig and captured the assailant. This prize – my first piece of luck – was the *Intrépide*, French privateer of six guns and forty-eight men.
>
> *May* 14. – Saw five armed boats pulling towards us from Monte Cristo. Out sweeps to protect convoy. At 4 p.m. the boats boarded and took possession of the two sternmost ships. A light breeze springing up, made all sail towards the captured vessels, ordering the remainder of the convoy to make the best of their way to Longona. The breeze freshening we came up with and recaptured the vessels

with the prize crews on board, but during the operation the armed boats escaped.

May 21. – At anchor in Leghorn Roads. Convoy all safe.
25. – Off Genoa. Joined Lord Keith's squadron of five sail of the line, four frigates and a brig.

26, 27, 28. – Ordered by his lordship to cruise in the offing, to intercept supplies destined for the French army under Massena, then in possession of Genoa.

29. – At Genoa some of the gun-boats bombarded the town for two hours.

30. – All the gun-boats bombarded the town. A partial bombardment had been going on for an hour a day, during the past fortnight, Lord Keith humanely refraining from continued bombardment, out of consideration for the inhabitants, who were in a state of absolute famine.

This was one of the *crises* of the war. The French, about a month previous, had defeated the Austrians with great slaughter in an attempt, on the part of the latter, to retake Genoa; but the Austrians, being in possession of Savona, were nevertheless able to intercept provisions on the land side, whilst the vigilance of Lord Keith rendered it impossible to obtain supplies by sea.

It having come to Lord Keith's knowledge that the French in Genoa had consumed their last horses and dogs, whilst the Genoese themselves were perishing by famine, and on the eve of revolt against the usurping force – in order to save the carnage which would ensue, his lordship caused it to be intimated to Massena that a defence so heroic would command honourable terms of capitulation. Massena was said to have replied that if the word "capitulation" were mentioned his army should perish with the city; but, as he could no longer defend himself, he had no objection to "treat". Lord Keith, therefore, proposed a treaty, viz, that the army might return to France, but that Massena himself must remain a prisoner in his hands. To this the French general demurred; but Lord Keith insisting – with the complimentary observation to Massena that "he was worth 20,000 men" – the latter reluctantly gave in, and on the 4th of June, 1800, a definite treaty to the above effect was agreed upon, and ratified on the 5th, when the Austrians took possession of the city, and Lord Keith of the harbour, the squadron anchoring within the mole.

This affair being ended, his lordship ordered the *Speedy* to cruise off the Spanish coast, and on the 14th of June we parted company with the squadron.

June 16. – Captured a tartan off Elba. Sent her to Leghorn, in the charge of an officer and four men.

22. – Off Bastia. Chased a French privateer with a prize in tow. The Frenchman abandoned the prize, a Sardinian vessel laden with oil and wool, and we took possession. Made all sail in chase of the privateer; but on our commencing to fire she ran under the fort of Caprea, where we did not think proper to pursue her. Took prize in tow, and on the following day left her at Leghorn, where we found Lord Nelson, and several ships at anchor.

25. – Quitted Leghorn, and on the 26th were again off Bastia, in chase of a ship which ran for that place, and anchored under a fort three miles to the southward. Made at and brought her away. Proved to be the Spanish letter of marque *Assuncion*, of ten guns and thirty-three men, bound from Tunis to Barcelona. On taking possession, five gun-boats left Bastia in chase of us; took the prize in tow, and kept up a running fight with the gun-boats till after midnight, when they left us.

29. – Cast off the prize in chase of a French privateer off Sardinia. On commencing our fire she set all sail and ran off. Returned and took the prize in tow; and the 4th of July anchored with her in Port Mahon.

July 9. – Off Cape Sebastian. Gave chase to two Spanish ships standing along shore. They anchored under the protection of the forts. Saw another vessel lying just within range of the forts – out boats and cut her out, the forts firing on the boats without inflicting damage.

July 19. – Off Caprea. Several French privateers in sight. Chased, and on the following morning captured one, the *Constitution*, of one gun and nineteen men. Whilst we were securing the privateer, a prize which she had taken made sail in the direction of Gorgona and escaped.

27. – Off Planosa, in chase of a privateer. On the following morning saw three others lying in a small creek. On making preparations to cut them out, a military force made its appearance, and commenced a heavy fire of musketry, to which it would have answered no purpose

to reply. Fired several broadsides at one of the privateers, and sunk her.

31. – Off Porto Ferraio in chase of a French privateer, with a prize in tow. The Frenchman abandoned his prize, of which we took possession, and whilst so doing the privateer got away.

August 3. – Anchored with our prizes in Leghorn Roads, where we found Lord Keith in the *Minotaur*.

Lord Keith received me very kindly, and directed the *Speedy* to run down the Spanish coast; pointing out the importance of harassing the enemy there as much as possible, but cautioning me against engaging anything beyond our capacity. During our stay at Leghorn, his lordship frequently invited me ashore to participate in the gaieties of the place.

Having filled up with provisions and water, we sailed on the 16th of August, and on the 21st captured a French privateer bound from Corsica to Toulon. Shortly afterwards we fell in with HMS *Mutine* and *Salamine*, which, to suit their convenience, gave into our charge a number of French prisoners, with whom and our prize we consequently returned to Leghorn.

On the 14th of September we again put to sea, the interval being occupied by a thorough overhaul of the sloop. On the 22nd, when off Caprea, fell in with a Neapolitan vessel having a French prize crew on board. Recaptured the vessel, and took the crew prisoners.

On the 5th of October, the *Speedy* anchored in Port Mahon, where information was received that the Spaniards had several armed vessels on the look-out for us, should we again appear on their coast. I therefore applied to the authorities to exchange our 4-pounders for 6-pounders, but the latter being too large for the *Speedy*'s ports, we were again compelled to forego the change as impracticable.

October 12. – Sailed from Port Mahon, cruising for some time off Cape Sebastian, Villa Nova, Oropesa, and Barcelona; occasionally visiting the enemy's coast for water, of which the *Speedy* carried only ten tons. Nothing material occurred till November 18th, when we narrowly escaped being swamped in a gale of wind, the sea breaking over our quarter, and clearing our deck, spars, &c., otherwise inflicting such damage as to compel our return to Port Mahon, where we were detained till the 12th of December.

December 15. – Off Majorca. Several strange vessels being in sight, singled out the largest and made sail in chase; shortly after which a French bombard bore up, hoisting the national colours. We now cleared for action, altering our course to meet her, when she bore up between Dragon Island and the Main. Commenced firing at the bombard, which returned our fire; but shortly afterwards getting closer in shore she drove on the rocks. Three other vessels being in the passage, we left her, and captured one of them, the *La Liza* of ten guns and thirty-three men, bound from Alicant to Marseilles. Took nineteen of our prisoners on board the *Speedy*. As it was evident that the bombard would become a wreck, we paid no further attention to her, but made all sail after the others.

December 18. – Suspecting the passage between Dragon Island and the Main to be a lurking-place for privateers, we ran in again, but found nothing. Seeing a number of troops lining the beach, we opened fire and dispersed them, afterwards engaging a tower, which fired upon us. The prisoners we had taken proving an incumbrance, we put them on shore.

December 19. – Stood off and on the harbour of Palamos, where we saw several vessels at anchor. Hoisted Danish colours and made the signal for a pilot. Our real character being evidently known, none came off, and we did not think it prudent to venture in.

It has been said that the *Speedy* had become the marked object of the Spanish naval authorities. Not that there was much danger of being caught, for they confined their search to the coast only, and that in the daytime, when we were usually away in the offing; it being our practice to keep out of sight during the day, and run in before dawn on the next morning.

On the 21st, however, when off Plane Island, we were very near "catching a Tartar." Seeing a large ship in shore, having all the appearance of a well-laden merchantman, we forthwith gave chase. On nearing her she raised her ports, which had been closed to deceive us, the act discovering a heavy broadside, a clear demonstration that we had fallen into the jaws of a formidable Spanish frigate, now crowded with men, who had before remained concealed below.

That the frigate was in search of us there could be no doubt, from the deception practised. To have encountered her with our insignificant armament would have been exceedingly imprudent, whilst escape was out of the question, for she would have outsailed us, and could have run us down by her mere weight. There was, therefore, nothing left but to try the effect of a *ruse*, prepared beforehand for such an emergency. After receiving at Mahon information that unusual measures were about to be taken by the Spaniards for our capture, I had the *Speedy* painted in imitation of the Danish brig *Clomer*; the appearance of this vessel being well known on the Spanish coast. We also shipped a Danish quartermaster, taking the further precaution of providing him with the uniform of an officer of that nation.

On discovering the real character of our neighbour, the *Speedy* hoisted Danish colours, and spoke her. At first this failed to satisfy the Spaniard, who sent a boat to board us. It was now time to bring the Danish quartermaster into play in his officer's uniform; and to add force to his explanations, we ran the quarantine flag up to the fore, calculating on the Spanish horror of the plague, then prevalent along the Barbary coast.

On the boat coming within hail – for the yellow flag effectually repressed the enemy's desire to board us – our mock officer informed the Spaniards that we were two days from Algiers, where at the time the plague was violently raging. This was enough. The boat returned to the frigate, which, wishing us a good voyage, filled, and made sail, whilst we did the same.

I have noted this circumstance more minutely than it merits, because it has been misrepresented. By some of my officers blame was cast on me for not attacking the frigate after she had been put off her guard by our false colours, as her hands – being then employed at their ordinary avocations in the rigging and elsewhere – presented a prominent mark for our shot. There is no doubt but that we might have poured in a murderous fire before the crew could have recovered from their confusion, and perhaps have taken her, but feeling averse to so cruel a destruction of human life, I chose to refrain from an attack, which might not, even with that advantage in our favour, have been successful.

It has been stated by some naval writers that this frigate was the *Gamo*, which we subsequently captured. To the best of my knowledge this is an error.

December 24. – Off Carthagena. At daylight fell in with a convoy in charge of two Spanish privateers, which came up and fired at us; but being to windward we ran for the convoy, and singling out two, captured the nearest, laden with wine. The other ran in shore under the fort of Port Genoese, where we left her.

25. – Stood for Cape St Martin, in hope of intercepting the privateers. At 8 a.m. saw a privateer and one of the convoy under Cape Lanar. Made sail in chase. They parted company; when, on our singling out the nearest privateer, she took refuge under a battery, on which we left off pursuit.

30. – Off Cape Oropesa. Seeing some vessels in shore, out boats in chase. At noon they returned pursued by two Spanish gunboats, which kept up a smart fire on them. Made sail to intercept the gun-boats, on which they ran in under the batteries.

January 10, 1801. – Anchored in Port Mahon, and having refitted, sailed again on the 12th.

16. – Off Barcelona. Just before daylight chased two vessels standing towards that port. Seeing themselves pursued, they made for the battery at the entrance. Bore up and set steering sails in chase. The wind falling calm, one of the chase drifted in shore and took the ground under Castel De Ferro. On commencing our fire, the crew abandoned her, and we sent boats with anchors and hawsers to warp her off, in which they succeeded. She proved to be the Genoese ship *Ns. Señora de Gratia*, of ten guns.

22. – Before daylight, stood in again for Barcelona. Saw several sail close in with the land. Out boats and boarded one, which turned out a Dane. Cruising off the port till 3 a.m., we saw two strange vessels coming from the westward. Made sail to cut them off. At 6 p.m. one of them hoisted Spanish colours and the other French. At 9 p.m. came up with them, when after an engagement of half an hour both struck. The Spaniard was the *Ecce Homo*, of eight guns and nineteen men, the Frenchman, *L'Amitié*, of one gun and thirty-one men. Took all the prisoners on board the *Speedy*.

23. – Still off Barcelona. Having sent most of our crew to man the prizes, the number of prisoners on board the

> *Speedy* became dangerous; we therefore put twenty-five of the Frenchmen into one of their own launches and told them to make the best of their way to Barcelona. As the prizes were a good deal cut up about the rigging, repaired their damages and made sail for Port Mahon, where we arrived on the 24th, with our convoy in company.
>
> 28th. – Quitted Port Mahon for Malta, not being able to procure at Minorca various things of which we stood in need; and on the 1st of February, came to an anchor at Valetta, where we obtained anchors and sweeps.

An absurd affair took place during our short stay at Malta, which would not have been worthy of notice, had it not been made the subject of comment.

The officers of a French royalist regiment, then at Malta, patronized a fancy ball, for which I amongst others purchased a ticket. The dress chosen was that of a sailor – in fact, my costume was a tolerable imitation of that of my worthy friend, Jack Larmour, in one of his relaxing moods, and personated in my estimation as honourable a character as were Greek, Turkish, or other kinds of Oriental disguises in vogue at such reunions. My costume was, however, too much to the life to please French royalist taste, not even the marlinspike and the lump of grease in the hat being omitted.

On entering the ball-room, further passage was immediately barred, with an intimation that my presence could not be permitted in such a dress. Good-humouredly expostulating that, as the choice of costume was left to the wearer, my own taste – which was decidedly nautical – had selected that of a British seaman, a character which, though by no means imaginary, was quite as picturesque as were the habiliments of an Arcadian shepherd; further insisting that as no rule had been infringed, I must be permitted to exercise my discretion. Expostulation being of no avail, a brusque answer was returned that such a dress was not admissible, whereupon I as brusquely replied that having purchased my ticket, and chosen my own costume in accordance with the regulations, no one had any right to prevent me from sustaining the character assumed.

Upon this a French officer, who appeared to act as master of the ceremonies, came up, and without waiting for further explanation, rudely seized me by the collar with the intention of putting me out; in return for which insult he received a substantial mark of British indignation, and at the same time an uncomplimentary remark in his own language. In an instant all was uproar; a French picket was called, which in a short time overpowered and carried me off to the guard-house of the regiment.

I was, however, promptly freed from detention on announcing my name, but the officer who had collared me demanded an apology for the portion of the *fracas* concerning him personally. This being of course refused, a challenge was the consequence; and on the following morning we met behind the ramparts and exchanged shots, my ball passing through the poor fellow's thigh, and dropping him. My escape, too, was a narrow one – his ball perforating my coat, waistcoat, and shirt, and bruising my side. Seeing my adversary fall, I stepped up to him – imagining his wound to be serious – and expressed a hope that he had not been hit in a vital part. His reply – uttered with all the politeness of his nation – was, that "he was not materially hurt." I, however, was not at ease, for it was impossible not to regret this, to him, serious *dénouement* of a trumpery affair, though arising from his own intemperate conduct. It was a lesson to me in future never to do anything in frolic which might give even unintentional offence.

On the 3rd of February we sailed under orders for Tripoli, to make arrangements for fresh provisions for the fleet. This being effected, the *Speedy* returned to Malta, and on the 20th again left port in charge of a convoy for Tunis.

24th. – At the entrance of Tunis Bay we gave chase to a strange sail, which wore and stood in towards the town, anchoring at about the distance of three miles. Suspecting some reason for this movement, I dispatched an officer to examine her, when the suspicion was confirmed by his ascertaining her to be *La Belle Caroline*, French brig of four guns, bound for Alexandria with field-pieces, ammunition, and wine for the use of the French army in Egypt.

Our position was one of delicacy, the vessel being in a neutral port, where, if we remained to watch her, she might prolong our stay for an indefinite period or escape in the night; whilst, from the warlike nature of the cargo, it was an object of national importance to effect her capture. The latter appearing the most beneficial course under all circumstances, we neared her so as to prevent escape, and soon after midnight boarded her, and having weighed her anchor, brought her close to the *Speedy*, before she had an opportunity of holding any communication with the shore.

The following day was employed in examining her stores, a portion of her ammunition being transferred to our magazine, to replace some damaged by leakage. Her crew, now on board the *Speedy* as prisoners, becoming clamorous at what they considered an illegal seizure, and being, moreover, in our way, an expedient was adopted to get rid of them, by purposely leaving their own launch within reach during the following night, with a caution to the watch not to prevent their desertion should they attempt it. The hint was taken, for before daylight on the 27th they seized the boat, and pulled out of the bay without molestation, not venturing to go to Tunis lest they should be retaken. We thus got rid of the prisoners, and at the same time of what might have turned out their reasonable complaint to the Tunisian authorities, for that we had exceeded the bounds of neutrality there could be no doubt.

On the 28th we weighed anchor, and proceeded to sea with our prize. After cruising for some days off Cape Bon, we made sail for Cagliari, where we arrived on the 8th of March, and put to sea on the 11th with the prize in tow. On the 16th, anchored in Port Mahon.

On the 18th we again put to sea, and towards evening observed a large frigate in chase of us. As she did not answer the private signal, it was evident that the stranger was one of our Spanish friends on the look-out. To cope with a vessel of her size and armament would have been folly, so we made all sail away from her, but she gave instant chase, and evidently gained upon us. To add to our embarrassment, the *Speedy* sprung her main-topgallant-yard, and lost ground whilst fishing it.

At daylight the following morning the strange frigate was still in chase, though by crowding all sail during the night we had gained a little upon her; but during the day she again recovered her advantage, the more so, as the breeze freshening, we were compelled to take in our royals, whilst she was still carrying on with everything set. After dark, we lowered a tub overboard with a light in it, and altering our course thus fortunately evaded her. On the 1st of April we returned to Port Mahon, and again put to sea on the 6th.

April 11. – Observing a vessel near the shoal of Tortosa, gave chase. On the following morning her crew deserted her, and we took possession. In the evening anchored under the land.

13. – Saw three vessels at anchor in a bay to the westward of Oropesa. Made sail up to them and anchored on the flank of a ten-gun fort. Whilst the firing was going on, the boats were sent in to board and bring out the vessels, which immediately weighed and got under the fort. At 5:30 p.m. the boats returned with one of them; the other two being hauled close in shore, we did not make any further attempt to capture them. As the prize, the *Ave Maria*, of four guns, was in ballast, we took the sails and spars out of her, and set her on fire.

On the following morning at daybreak, several vessels appeared to the eastward. Made all sail to intercept them, but before we could come up, they succeeded in anchoring under a fort. On standing towards them, they turned out to be Spanish gun-boats, which commenced firing at us. At 10 a.m. anchored within musket-shot, so as to keep an angle of the tower on our beam, thus neutralising its effect. Commenced firing broadsides alternately at the tower and the gunboats, with visible advantage. Shortly before noon made preparation to cut out the gun-boats, but a fresh breeze setting in dead on shore, rendered it impossible to get at them without placing ourselves in peril. We thereupon worked out of the bay.

15. – Two strange sail in sight. Gave chase, and in a couple of hours came up with and captured them. Made sail after a convoy in the offing, but the wind falling light at dusk, lost sight of them.

On the 26th we anchored in Mahon, remaining a week to refit and procure fresh hands, many having been sent away in prizes. On the 2nd of May put to sea with a reduced crew, some of whom had to be taken out of HM's prison.

We again ran along the Spanish coast, and on the 4th of May were off Barcelona, where the *Speedy* captured a vessel which reported herself as Ragusan, though in reality a Spanish four-gun tartan. Soon after detaining her we heard firing in the WN-W and steering for that quarter fell in with a Spanish privateer, which we also captured, the *San Carlos*, of seven guns. On this a swarm of gun-boats came out of Barcelona, seven of them giving chase to us and the prizes, with which we made off shore, the gun-boats returning to Barcelona.

On the following morning the prizes were sent to Port Mahon, and keeping out of sight for the rest of the day, the *Speedy* returned at midnight off Barcelona, where we found the gun-boats on the watch; but on our approach they ran in shore, firing at us occasionally. Suspecting that the object was to decoy us within reach of some larger vessel, we singled out one of them and made at her, the others, however, supporting her so well that some of our rigging being shot away, we made off shore to repair, the gun-boats following. Having thus got them to some distance, and repaired damages, we set all sail, and again ran in shore, in the hope of getting between them and the land, so as to cut off some of their number. Perceiving our intention, they all made for the port as before, keeping up a smart fight, in which our foretopgallant-yard was so much injured, that we had to shift it, and were thus left astern. The remainder of the day was employed in repairing damages, and the gun-boats not venturing out again, at 9 p.m. we again made off shore.

Convinced that something more than ordinary had actuated the gun-boats to decoy us – just before daylight on the 6th we again ran in for Barcelona, when the trap manifested itself in the form of a large ship, running under the land, and bearing ES-E. On hauling towards her, she changed her course in chase of us, and was shortly made out to be a Spanish xebec frigate.

As some of my officers had expressed dissatisfaction at not having been permitted to attack the frigate fallen in with on the 21st of December, after her suspicions had been lulled by our device of hoisting Danish colours, &c., I told them they should now have a fair fight, notwithstanding that, by manning the two prizes sent to Mahon, our numbers had been reduced to fifty-four, officers and boys included. Orders were then given to pipe all hands, and prepare for action.

Accordingly we made towards the frigate, which was now coming down under steering sails. At 9:30 a.m., she fired a gun and hoisted Spanish colours, which the *Speedy* acknowledged by hoisting American colours, our object being, as we were now exposed to her full broadside, to puzzle her, till we got on the other tack, when we ran up the English ensign, and immediately afterwards encountered her broadside without damage.

Shortly afterwards she gave us another broadside, also without effect. My orders were not to fire a gun till we were close to her; when, running under her lee, we locked our yards amongst her rigging, and in this position returned our broadside, such as it was.

To have fired our popgun 4-pounders at a distance would have been to throw away the ammunition; but the guns being doubly, and, as I afterwards learned, trebly, shotted, and being elevated, they told admirably upon her main deck; the first discharge, as was subsequently ascertained, killing the Spanish captain and the boatswain.

My reason for locking our small craft in the enemy's rigging was the one upon which I mainly relied for victory, viz that from the height of the frigate out of the water, the whole of her shot must necessarily go over our heads, whilst our guns, being elevated, would blow up her main-deck.

The Spaniards speedily found out the disadvantage under which they were fighting, and gave the order to board the *Speedy*; but as this order was as distinctly heard by us as by them, we avoided it at the moment of execution by sheering off sufficiently to prevent the movement, giving them a volley of musketry and a broadside before they could recover themselves.

Twice was this manoeuvre repeated, and twice thus averted. The Spaniards finding that they were only punishing themselves, gave up further attempts to board and stood to their guns, which were cutting up our rigging from stem to stern, but doing little farther damage; for after the lapse of an hour the loss to the *Speedy* was only two men killed and four wounded.

This kind of combat, however, could not last. Our rigging being cut up and the *Speedy*'s sails riddled with shot, I told the men that they must either take the frigate or be themselves taken, in which case the Spaniards would give no quarter – whilst a few minutes energetically employed on their part would decide the matter in their own favour.

The doctor, Mr Guthrie, who, I am happy to say, is still living to peruse this record of his gallantry, volunteered to take the helm; leaving him therefore for the time both commander and crew of the *Speedy*, the order was given to board, and in a few seconds every man was on the enemy's deck – a feat rendered the more easy as the doctor placed the *Speedy* close alongside with admirable skill.

For a moment the Spaniards seemed taken by surprise, as though unwilling to believe that so small a crew would have the audacity to board them; but soon recovering themselves, they made a rush to the waist of the frigate, where the fight was for some minutes gallantly carried on. Observing the enemy's colours still flying, I directed one of our men immediately to haul them down, when the Spanish crew, without pausing to consider by whose orders the colours had been struck, and naturally believing it the act of their own officers, gave in, and we were in possession of the *Gamo* frigate, of thirty-two heavy guns and 319 men, who an hour and a half before had looked upon us as a certain if not an easy prey.

Our loss in boarding was Lieutenant Parker, severely wounded in several places, one seaman killed and three wounded, which with those previously killed and wounded gave a total of three seamen killed, and one officer and seventeen men wounded.

The *Gamo*'s loss was Captain de Torres – the boatswain – and thirteen seamen killed, together with forty-one wounded; her casualties thus exceeding the whole number of officers and crew on board the *Speedy*.

Some time after the surrender of the *Gamo*, and when we were in quiet possession, the officer who had succeeded the deceased Captain Don Francisco de Torres, not in command, but in rank, applied to me for a certificate that he had done his duty during the action; whereupon he received from me a certificate that he had "conducted himself like a true Spaniard", with which document he appeared highly gratified, and I had afterwards the satisfaction of learning that it procured him further promotion in the Spanish service.

Shortly before boarding, an incident occurred which, by those who have never been placed in similar circumstances, may be thought too absurd for notice. Knowing that the final struggle would be a desperate one, and calculating on the superstitious wonder which forms an element in the Spanish character, a portion of our crew were ordered to blacken their faces, and what with this and the excitement of combat, more ferocious looking objects could scarcely be imagined. The fellows thus disguised were directed to board by the head, and the effect produced was precisely that calculated on. The greater portion of the Spaniard's crew was prepared to repel boarders in that direction, but stood for a few moments as it were transfixed to the deck by the apparition of so many diabolical looking figures emerging from the white smoke of the bow guns; whilst our other men, who boarded by the waist, rushed on them from behind, before they could recover from their surprise at the unexpected phenomenon.

In difficult or doubtful attacks by sea – and the odds of 50 men to 320 comes within this description – no device can be too minute, even if apparently absurd, provided it have the effect of diverting the enemy's attention whilst you are concentrating your own. In this, and other successes against odds, I have no hesitation in saying that success in no slight degree depended on out-of-the-way devices, which the enemy not suspecting, were in some measure thrown off their guard.

The subjoined tabular view of the respective force of the two vessels will best show the nature of the contest.

Gamo	**Speedy**
Main-deck guns. – Twenty-two long 12-pounders.	*Fourteen 4-pounders.*
Quarter-deck. – Eight long 5-pounders, and two 24-pounder carronades.	*None.*
No. of crew, 319.	*No. of crew, 54.*
Broadside weight of shot, 190 lbs.	*Broadside weight of shot, 28 lbs.*
Tonnage, 600 and upwards.	*Tonnage, 158.*

It became a puzzle what to do with 263 unhurt prisoners now we had taken them, the *Speedy* having only forty-two men left. Promptness was however necessary; so driving the prisoners into the hold, with guns pointing down the hatchway, and leaving thirty of our men on board the prize – which was placed under the command of my brother, the Hon. Archibald Cochrane, then a midshipman – we shaped our course to Port Mahon – not Gibraltar, as has been recorded – and arrived there in safety; the Barcelona gun-boats, though spectators of the action, not venturing to rescue the frigate. Had they made the attempt, we should have had some difficulty in evading them and securing the prize, the prisoners manifesting every disposition to rescue themselves, and only being deterred by their own main deck guns loaded with cannister, and pointing down the hatchways, whilst our men stood over them with lighted matches.

Our success hitherto had procured us some prize-money, notwithstanding the peculations of the Mediterranean Admiralty Courts, by which the greater portion of our captures was absorbed.

Despite this drawback, which generally disinclined officers and crews from making extraordinary exertions, my own share of the twelvemonth's zealous endeavours in our little sloop was considerable, and even the crew were in receipt of larger sums than those constituting the ordinary pay of officers; a result chiefly owing to our nocturnal mode of warfare, together with our refraining from meddling with vessels ascertained to be loading in the Spanish ports, and then lying in wait for them as they proceeded on their voyage.

One effect of our success was no slight amount of ill-concealed jealousy on the part of officers senior to myself, though there were some amongst these who, being in command of small squadrons instead of single vessels, might, had they adopted the same means, have effected far more than the *Speedy*, with an armament so insignificant, was calculated to accomplish.

O.D. Gallagher

The Loss of the *Repulse*

HMS Repulse *and HMS* Prince of Wales *were lost to Japanese fighter-bombers within a single hour on 10 December 1941.*

THIS is the simple story of a naval force which went into north-eastern Malayan waters on Monday. *Prince of Wales* and *Repulse* were the backbone of this force. I was in *Repulse*. The aim of the force was, in the words of the signal C-in-C Admiral Sir Tom Phillips sent to all ships: "The enemy has made several landings on the north coast of Malaya and has made local progress. Meanwhile fast transports lie off the coast. This is our opportunity before the enemy can establish himself.

"We have made a wide circuit to avoid air reconnaissance and hope to surprise the enemy shortly after sunrise to-morrow (Wednesday). We may have the luck to try our metal against the old Jap battle cruiser *Kongo* or against some Jap cruisers or destroyers in the Gulf of Siam.

"We are sure to get some useful practices with our high-angle armament, but whatever we meet I want to finish quickly and get well clear to eastward before the Japanese can mass a too formidable scale of air attack against us. So shoot to sink."

But at 5.20 that same evening a bugle sounded throughout my ship *Repulse* over the ship's loud-speakers, giving immediate orders to the whole ship's company and filling every space of engine room and wardroom with its urgent bugle notes, followed by the order: "Action stations. Enemy aircraft!"

I rushed on to the flag deck which was my action station. It was a single Nakajama Naka 93 twin-floated Jap reconnaissance plane. She kept almost on the horizon, too far for engagement, for a couple of hours.

A voice from the bridge came out of the loud-speakers: "We

are being shadowed by enemy aircraft. Keep ready for immediate action to repel aircraft."

Two more Nakajama Naka 93s appeared. They kept a long relay watch on us. What an admiral most wishes to avoid has happened.

His ships were out at sea, sufficiently distant from shore to prevent him receiving air support before dawn the following morning, when a mass enemy air attack now seemed certain. We had not yet sighted any enemy naval force or received reports of an enemy transport convoy.

For dinner in our wardroom that night we had hot soup, cold beef, ham, meat pie, oranges, bananas, pineapples, and coffee. We discussed this unfortunate happening. We had travelled all day in good visibility without being spotted. Now, as the last hour of darkness fell, a lucky Jap had found us.

One of the *Repulse*'s Fleet Air Arm pilots who fly the ship's aircraft – this one a young New Zealander with a ginger beard – came in cursing: "My God! Someone's blacked the right eye of my air-gunner – the one he shoots with." The laughter ended. Everyone was fitted with a tight-fitting asbestos helmet which makes you look like a Disney drawing. We were all expecting action at dawn to-morrow, hoping to meet a Jap cruiser. At 9.5 p.m. came a voice from the loudspeakers: "Stand by for the Captain to speak to you."

Captain: "A signal has just been received from the Commander-in-Chief. We were shadowed by three planes. We were spotted after dodging them all day. Their troop convoy will now have dispersed. We will find enemy aircraft waiting for us now. We are now returning to Singapore."

Then followed a babble of voices and groans. A voice said: "This ship will never get into action. It's too lucky."

So it was. In the message from the Captain the previous day, in which he said: "We're going looking for trouble and I expect we shall find it," he noted that *Repulse* had travelled 53,000 miles in this war without action, although it trailed the *Bismarck* and was off northern Norway and has convoyed throughout the war.

I slept in the wardroom fully clothed that night and awoke to the call "Action stations" at 5 a.m. on Wednesday. It was a thin oriental dawn, when a cool breeze swept through the fuggy ship, which had been battened down all night as a result of the order to "darken ship".

The sky was luminous as pearl. We saw from the flag deck a string of black objects on the port bow. They turned out to be a line of landing barges, "like railway trucks", as a young signaller said. At 6.30 a.m. the loud-speaker voice announced: "Just received message saying enemy is making landing north of Singapore. We're going in."

We all rushed off to breakfast, which consisted of coffee, cold ham, bread, and marmalade. Back at action stations all the ship's company kept a look-out. We cruised in line-ahead formation, *Prince of Wales* leading, the *Repulse* second, and with our destroyer screen out.

Down the Malayan coast, examining with the help of terrier-like destroyers all coves for enemy landing parties.

At 7.55 a.m. *Prince of Wales* catapulted one of her planes on reconnaissance, with instructions not to return to the ship, but to land ashore after making a report to us on what she found.

We watched her become midget-size and drop out of sight behind two hummock-back islands, behind which was a beach invisible to us. We all thought that was where the enemy lay. But it reappeared and went on, still searching.

Meanwhile all the ship's company on deck had put on anti-flash helmets, elbow-length asbestos gloves, goggles and tin hats.

Prince of Wales looked magnificent. White-tipped waves rippled over her plunging bows. The waves shrouded them with watery lace, then they rose high again and once again dipped. She rose and fell so methodically that the effect of staring at her was hypnotic. The fresh breeze blew her White Ensign out stiff as a board.

I felt a surge of excited anticipation rise within me at the prospect of her and the rest of the force sailing into enemy landing parties and their escorting warships.

A young Royal Marines lieutenant who was my escort when first I went aboard the *Repulse* told me: "We've not had any action but we're a perfect team – the whole twelve hundred and sixty of us. We've been working together so long. We claim to have the Navy's best gunners."

My anticipatory reverie was broken by the voice from the loudspeakers again: "Hello, there. Well, we've sighted nothing yet, but we'll go down the coast having a look for them."

More exclamations of disappointment. The yeoman of signals said: "Don't say this one's off, too."

As we sped down Malaya's changing coastline the wag of the flag-deck said travel-talkwise: "On the starboard beam, dear listeners, you see the beauty spots of Malaya, land of the orang-outang."

Again the loud-speaker announces: "Nothing sighted."

The *Repulse* sends off one of her aircraft. The pilot is not the ginger-bearded New Zealander, as he tossed a coin with another pilot and lost the toss, which means that he stays behind.

We drift to the wardroom again until 10.20 a.m. We are spotted again by a twin-engined snooper of the same type as attacked Singapore the first night of this new war.

We can do nothing about it, as she keeps well beyond range while her crew presumably studies our outlines and compares them with silhouettes in the Jap equivalent of *Jane*'s *Fighting Ships*.

At 11 a.m. a twin-masted single funnel ship is sighted on the starboard bow. The force goes to investigate her. She carries no flag.

I was looking at her through my telescope when the shock of an explosion made me jump so that I nearly poked my right eye out. It was 11.15 a.m. The explosion came from the *Prince of Wales*'s portside secondary armament. She was firing at a single aircraft.

We open fire. There are about six aircraft.

A three-quarter-inch screw falls on my tin hat from the bridge deck above from the shock of explosion of the guns. "The old tub's falling to bits," observes the yeoman of signals.

That was the beginning of a superb air attack by the Japanese, whose air force was an unknown quantity.

Officers in the *Prince of Wales* whom I met in their wardroom when she arrived here last week said they expected some unorthodox flying from the Japs. "The great danger will be the possibility of these chaps flying their whole aircraft into a ship and committing hara-kiri."

It was nothing like that. It was most orthodox. They even came at us in formation, flying low and close.

Aboard the *Repulse*, I found observers as qualified as anyone to estimate Jap flying abilities. They know from first-hand experience what the RAF and the Luftwaffe are like. Their verdict was: "The Germans have never done anything like this in the North Sea, Atlantic or anywhere else we have been."

They concentrated on the two capital ships, taking the *Prince of Wales* first and the *Repulse* second. The destroyer screen they left completely alone except for damaged planes forced to fly low over them when they dropped bombs defensively.

At 11.18 the *Prince of Wales* opened a shattering barrage with all her multiple pom-poms, or Chicago Pianos as they call them. Red and blue flames poured from the eight-gun muzzles of each battery. I saw glowing tracer shells describe shallow curves as they went soaring skyward surrounding the enemy planes. Our Chicago Pianos opened fire; also our triple-gun four-inch high-angle turrets. The uproar was so tremendous I seemed to feel it.

From the starboard side of the flag-deck I could see two torpedo planes. No, they were bombers. Flying straight at us.

All our guns pour high-explosives at them, including shells so delicately fused that they explode if they merely graze cloth fabric.

But they swing away, carrying out a high-powered evasive action without dropping anything at all. I realize now what the purpose of the action was. It was a diversion to occupy all our guns and observers on the air defence platform at the summit of the mainmast.

There is a heavy explosion and the *Repulse* rocks. Great patches of paint fall from the funnel on to the flag-deck. We all gaze above our heads to see planes which during the action against the low fliers were unnoticed.

They are high-level bombers. Seventeen thousand feet. The first bomb, the one that rocked us a moment ago, scored a direct hit on the catapult deck through the one hangar on the port side.

I am standing behind a multiple Vickers gun, one which fires 2,000 half-inch bullets per minute. It is at the after end of the flag-deck.

I see a cloud of smoke rising from the place where the final bomb hit. Another comes down bang again from 17,000 feet. It explodes in the sea, making a creamy blue and green patch ten feet across. The *Repulse* rocks again. It was three fathoms from the port side. It was a miss, so no one bothers.

Cooling fluid is spouting from one of the barrels of a Chicago Piano. I can see black paint on the funnel-shaped covers at the muzzles of the eight barrels actually rising in blisters big as fists.

The boys manning them – there are ten to each – are sweating, saturating their asbestos anti-flash helmets. The whole gun swings this way and that as spotters pick planes to be fired at.

Two planes can be seen coming at us. A spotter sees another at a different angle, but much closer.

He leans forward, his face tight with excitement, urgently pounding the back of the gun swiveller in front of him. He hits that back with his right hand and points with the left a stabbing forefinger at a single sneaker plane. Still blazing two-pounders the whole gun platform turns in a hail of death at the single plane. It is some 1,000 yards away.

I saw tracers rip into its fuselage dead in the centre. Its fabric opened up like a rapidly spreading sore with red edges. Fire . . .

It swept to the tail, and in a moment stabilizer and rudder became a framework skeleton. Her nose dipped down and she went waterward.

We cheered like madmen. I felt my larynx tearing in the effort to make myself heard above the hellish uproar of guns.

A plane smacked the sea on its belly and was immediately transformed into a gigantic shapeless mass of fire which shot over the waves fast as a snake's tongue. The *Repulse* had got the first raider.

For the first time since the action began we can hear a sound from the loud-speakers, which are on every deck at every action station. It is the sound of a bugle.

Its first notes are somewhat tortured. The young bugler's lips and throat are obviously dry with excitement. It is that most sinister alarm of all for seamen: "Fire!"

Smoke from our catapult deck is thick now. Men in overalls, their faces hidden by a coat of soot, man-handle hoses along decks. Water fountains delicately from a rough patch made in one section by binding it with a white shirt.

It sprays on the Vickers gunners, who, in a momentary lull, lift faces, open mouths and put out tongues to catch the cooling jets. They quickly avert faces to spit – the water is salt and it is warm. It is sea water.

The Chicago Piano opens up again with a suddenness that I am unable to refrain from flinching at, though once they get going with their erratic shell-pumping it is most reassuring.

All aboard have said the safest place in any battleship or cruiser or destroyer is behind a Chicago Piano. I believe them.

Empty brass cordite cases are tumbling out of the gun's scuttle-like exit so fast and so excitedly it reminds me of the forbidden fruit machine in Gibraltar on which I once played. It went amok on one occasion and ejected £8 in shillings in a frantic rush.

The cases bounce off the steel C deck, roll and dance down the sloping base into a channel for easy picking up later.

At 11.25 we see an enormous splash on the very edge of the horizon. The splash vanishes and a whitish cloud takes its place.

A damaged enemy plane jettisoning its bombs or another enemy destroyed? A rapid Gallup poll on the flag deck says: "Another duck down." Duck is a word they have rapidly taken from the Aussie Navy. It means enemy plane.

Hopping about the flag-deck from port to starboard, whichever side is being attacked, is the plump figure of the naval photographer named Tubby Abrahams.

He was a Fleet Street agency pictureman now in the Navy. But all his pictures are lost. He had to throw them into the sea with his camera. He was saved. So was United States broadcaster Cecil Brown, of Columbia System.

Fire parties are still fighting the hangar outbreak, oblivious of any air attack used so far. Bomb splinters have torn three holes in the starboard side of the funnel on our flag-deck.

Gazing impotently with no more than fountain pen and notebook in my hands while gunners, signallers, surgeons and range-finders worked, I found emotional release in shouting rather stupidly, I suppose, at the Japanese.

I discovered depths of obscenity previously unknown, even to me.

One young signaller keeps passing me pieces of information in between running up flats. He has just said: "A couple of blokes are caught in the lift from galley to servery. They're trying to get them out."

The yeoman of signals interjected: "How the bloody hell they got there, God knows."

There is a short lull. The boys dig inside their overalls and pull out cigarettes. Then the loud-speaker voice: "Enemy aircraft ahead." Lighted ends are nipped off cigarettes. The ship's company goes into action again. "Twelve of them." The flag-deck boys whistle. Someone counts them aloud: "One, two, three, four, five, six, seven, eight, nine – yes, nine." The flag-deck wag, as he levels a signalling lamp at the *Prince of Wales*: "Any advance on nine? Anybody? No? Well, here they come."

It is 12.10 p.m. They are all concentrating on the *Prince of Wales*. They are after the big ships all right. A mass of water and smoke rises in a tree-like column from the *Prince of Wales*'s stern. They've got her with a torpedo.

A ragged-edged mass of flame from her Chicago Piano does not stop them, nor the heavy instant flashes from her high-angle secondary armament.

She is listing to port – a bad list. We are about six cables from her.

A snottie, or midshipman, runs past, calls as he goes: "*Prince of Wales*'s steering gear gone." It doesn't seem possible that those slight-looking planes could do that to her.

The planes leave us, having apparently dropped all their bombs and torpedoes. I don't believe it is over, though. "Look, look!" shouts someone, "there's a line in the water right under our bows, growing longer on the starboard side. A torpedo that missed us. Wonder where it'll stop."

The *Prince of Wales* signals us again asking if we've been torpedoed. Our Captain Tennant replies: "Not yet. We've dodged nineteen."

Six stokers arrive on the flag-deck. They are black with smoke and oil and are in need of first aid. They are ushered down to the armoured citadel at the base of the mainmast.

The *Prince of Wales*'s list is increasing. There is a great rattle of empty two-pounder cordite cases as Chicago Piano boys gather up the empties and stow them away and clear for further action.

12.20 p.m. . . . The end is near, although I didn't know it.

A new wave of planes appears, flying around us in formation and gradually coming nearer. The *Prince of Wales* lies about ten cables astern of our port side. She is helpless.

They are making for her. I don't know how many. They are splitting up our guns as they realize they are after her, knowing she can't dodge their torpedoes. So we fire at them to defend the *Prince of Wales* rather than attend to our own safety.

The only analogy I can think of to give an impression of the *Prince of Wales* in those last moments is of a mortally wounded tiger trying to beat off the *coup de grâce*.

Her outline is hardly distinguishable in smoke and flame from all her guns except the fourteen-inchers. I can see one plane release a torpedo. It drops nose heavy into the sea and churns up a small wake as it drives straight at the *Prince of Wales*. It explodes against her bows.

A couple of seconds later another explodes amidships and another astern. Gazing at her turning over on the port side with her stern going under and with dots of men leaping from her, I was thrown against the bulkhead by a tremendous shock as the *Repulse* takes a torpedo on her portside stern.

With all others on the flag-deck I am wondering where it came from, when the *Repulse* shudders gigantically. Another torpedo.

Now men are cheering with more abandon than at a Cup Final. What the heck is this? I wonder. Then see it is another plane down. It hits the sea in flames also. There have been six so far as I know.

My notebook, which I have got before me, is stained with oil and is ink-blurred. It says: "Third torp."

The *Repulse* now listing badly to starboard. The loud-speakers speak for the last time: "Everybody on main deck."

We all troop down ladders, most orderly except for one lad who climbs the rail and is about to jump when an officer says: "Now then – come back – we are all going your way." The boy came back and joined the line.

It seemed slow going. Like all the others I suppose I was tempted to leap to the lower deck, but the calmness was catching. When we got to the main deck the list was so bad our shoes and feet could not grip the steel deck. I kicked off mine, and my damp stockinged feet made for sure movement.

Nervously opening my cigarette case I found I hadn't a match. I offered a cigarette to a man beside me. He said: "Ta. Want a match?" We both lit up and puffed once or twice. He said: "I'll be seeing you, mate." To which I replied: "Hope so, cheerio."

We were all able to walk down the ship's starboard side, she lay so much over to port.

We all formed a line along a big protruding anti-torpedo blister, from where we had to jump some twelve feet into a sea which was black – I discovered it was oil.

I remember jamming my cap on my head, drawing a breath and leaping.

Oh, I forgot – the last entry in my notebook was: "Sank about 12.20 p.m." I made it before leaving the flag-deck. In the water I glimpsed the *Prince of Wales*'s bows disappearing.

Kicking with all my strength, I with hundreds of others tried to get away from the *Repulse* before she went under, being afraid of getting drawn under in the whirlpool.

I went in the wrong direction, straight into the still spreading oil patch, which felt almost as thick as velvet. A wave hit me and swung me round so that I saw the last of the *Repulse*.

Her underwater plates were painted a bright, light red. Her bows rose high as the air trapped inside tried to escape from underwater forward regions, and there she hung for a second or two and easily slid out of sight.

I had a tremendous feeling of loneliness, and could see nothing capable of carrying me. I kicked, lying on my back, and felt my eyes burning as the oil crept over me, in mouth, nostrils, and hair.

When swamped by the waves, I remember seeing the water I spurted from my mouth was black. I came across two men hanging on to a round lifebelt. They were black, and I told them they looked like a couple of Al Jolsons: They said: "Well, we must be a trio, 'cos you're the same."

We were joined by another, so we had an Al Jolson quartet on one lifebelt. It was too much for it and in the struggle to keep it lying flat on the sea we lost it.

We broke up, with the possibility of meeting again, but none of us would know the other, owing to the complete mask of oil.

I kicked, I must confess somewhat panicky, to escape from the oil, but all I achieved was a bumping into a floating paravane. Once again there were four black faces with red eyes gathered together in the sea.

Then we saw a small motor boat with two men in it. The engine was broken. I tried to organize our individual strength into a concerted drive to reach the idly floating boat. We tried to push or pull ourselves by hanging on the paravane, kicking our legs, but it was too awkward, and it overturned.

I lost my grip and went under. My underwater struggles happily took me nearer to the boat.

After about two hours in the water, two hours of oil-fuel poisoning, I reached a thin wire rope which hung from the boat's bows.

My fingers were numb and I was generally weak as the result of the poisoning, but I managed to hold on to the wire by clamping my arms around it. I called to the men aboard to help me climb the four feet to the deck.

They tried with a boat hook, but finally said: "You know, we are pretty done in, too. You've got to try to help yourself. We can't do it alone."

I said I could not hold anything. They put the boathook in my shirt collar, but it tore and finally they said: "Sorry pal, we can't lift you. Have you got that wire?"

"Yes," I said. They let me go and there I hung. Another man arrived and caught the wire. He was smaller than I was. I am thirteen stone. The men aboard said they would try to get him up. "He's lighter than you," they said.

They got him aboard during which operation I went under again when he put his foot on my shoulder. The mouth of one black face aboard opened and showed black-slimed teeth, red gums and tongue. It said: "To hell with this."

He dived through the oil into the sea, popped up beside me with a round lifebelt which he put over my head, saying: "Okay. Now let go the wire."

But I'm sorry to say I couldn't. I couldn't bear to part with it. It had kept me on the surface about fifteen minutes.

They separated us, however, and the next thing I was draped through the lifebelt like a dummy being hauled aboard at a rope's end, which they could grip as it was not oily or shiny.

Another oil casualty was dragged aboard, and later thirty of us were lifted aboard a destroyer. We were stripped, bathed and left naked on the fo'c'sle benches and tables to sweat the oil out of the pores in the great heat.

Yoshida Mitsuru

Requiem for *Yamato*

The Japanese Yamato, *along with its sister ship* Musashi, *was the largest battleship in service during the second World War.* Yamato *measured 263 metres in length and displaced 67,000 tons. On April 6–7 1945* Yamato *was committed to Operation* Ten'ichigo *– a suicidal sortie against American forces off Okinawa. Yoshida Mitsuru was an ensign aboard* Yamato *in her final battle.*

Escape from the Bridge

The bridge is already just a dark chamber lying on its side.

Two volumes of operation documents, fallen from somewhere: without thinking, I pick them up and put them away in the chart stand.

Around me, all at once, I see no one.

The command "All hands on deck," manipulating the exhausted survivors like puppets, has lured them to leave the bridge.

My post: should I leave it? The bridge: a capital place to die.

Nothing left for me to do here?

These twenty square meters of space, to which I have entrusted my fate, to live or to die, for these two hours.

For a moment, an involuntary restlessness.

As if possessed by something, I stick my fingers in the grating of the deck and clamber up onto the lookout stand.

Some guy ahead of me kicks me, and I roll off onto the bottom of the bridge; but grumbling "Here I go again," I crawl up onto it once more.

Behind me, a cheerful voice: "Okay, I'm bringing up the rear." Ensign Watanabe, communications officer.

He was stationed on the bridge in place of the wounded

Ensign Nishio; as a result he is lucky and survives. The last to escape from the bridge, he reports that when he had climbed halfway out the port he and the ship were engulfed together by the water.

He went through the window, he says, as if blown out by the pressure of air or water; he was thrown into the sea.

Wriggling through the port, I look back almost longingly: poor bridge, on its side and completely dark. Surprisingly narrow, burrowlike.

Their bodies lashed together, the navigation officer and the assistant navigation officer reject a second and third time our exhortations to escape; they shrug off their shoulders the hands of their fellow officers.

I watch until the end: both have their eyes wide open and stare fixedly at the water rising toward them.

Thus the end of Commander Mogi and Lieutenant (jg.) Hanada.

Is the responsibility of running the ship so great?

Even now vivid in my memory: the voice of Ensign Mori, an aide attached to the captain, continuing to shout encouragement; and the sight of him, thumping the sailors on the shoulders with a swagger stick, right up to the moment of sinking.

For him who never did take off his steel helmet and flak jacket, it is an admirable end.

Because we were posted on the enclosed part of the bridge, we had no such protective gear. To go into the water still wearing such gear makes it impossible to stay afloat for any length of time.

The Great Whale Sinks

I crawl out the lookout port and stand on the starboard bulkhead of the bridge. The survivors are lined up on the brownish belly of the ship, a distant thirty or forty meters away, all with hands raised in unison. They must just have finished three shouts of *banzai!*

In a small cluster and moving as one, they look like toy soldiers; my heart goes out to them.

The last moments of Rear Admiral Ariga Kōsaku, captain of *Yamato:*

In the antiaircraft command post on the very top of the bridge, still wearing his helmet and flak jacket, he binds himself to the binnacle.

Having completed the final dispositions – the code-books, the command for all hands to come on deck, and so on – he shouts *banzai* three times. On finishing, he turns to look at the four surviving lookouts standing by his side.

They are too devoted to this resolute ruddy-faced captain to be able to leave him. Seeing a resolve to die together crystallizing among them, he slaps each on the shoulder, encourages them to keep their spirits up, and pushes them off into the water.

The final sailor presses his last four biscuits into the captain's hands, as if to show his innermost feelings. The captain takes them with a grin. As he has the second one in his mouth, he is engulfed along with the ship.

To eat biscuits at such a time! Iron nerves without equal.

Thus the words of the chief lookout. He too was unable to leave the captain's side. In the end he was thrown into the water while standing right next to the captain; but not having lashed himself down, he floated to the surface.

Fluttering atop the main mast, the great battle ensign is about to touch the water.

As I watch, a young sailor comes forward and clambers up to the base of the mast. Would he serve the battle ensign, soul of this sinking giant?

No one could have ordered him to do so.

So he has chosen this glorious duty. How proud his death!

It seems foolish to think such thoughts now, but when I drop my glance to the hull of the ship towering above the water and to its exposed undersides, it looks like a great whale.

That this vast piece of metal, 270 meters long and 40 meters wide, is about to plunge beneath the waves!

I recognize near me many shipmates. That fellow, and that one.

This one's eyebrows are very dark, that one's ears very pale. All of them have childlike expressions on their faces; better, they are all completely without expression.

For each of them, it must be a moment of absolute innocence, an instant of complete obliviousness.

For all I know, I too am in the same condition.

At what do they gaze with ecstatic eyes?

The eddies, extending as far as they can see. The boiling waves, interlocking in a vast pattern.

Pure white and transparent, like ice congealing around this giant ship and propping her up.

And the sound of the waves, deafening our ears, induces still deeper rapture.

We see a sheet of white; we hear only the thundering of the turbulent waters.

"Are we sinking?" For the first time, as if on fire, I ask myself that question. The spectacle is so mysterious, so resplendent, that I am overcome with the premonition that something extraordinary is about to happen.

The water already begins to creep up on the starboard half of the ship.

Bodies flying in all directions. It is not simply a matter of being swallowed by the waves. The pressure of the water boiling up sends bodies flying like projectiles.

The bodies become mere gray dots and scatter in all directions, effortlessly, happily. Even as I watch, a whirlpool runs fifty meters in one swoop.

And spray springs up at my feet; water contorted as in a funhouse mirror, gleaming in countless angles, countless formations, glitters before my nose.

In multiple mirrors the water engulfs human figures. Some pop back up; some hang upside down in the water.

This exquisite glass design colors the uniform white of the foam, as do stripes of pure blue scattered all over this blanket of bubbles. The effect of the churning created by the many eddies?

Just as my heart delights for a moment in this beauty, this gracefulness, I am swept into a large whirlpool.

Without thinking, I draw as deep a breath as I can.

Grabbing my feet and rolling up into a ball, like a baby in the womb, I brace myself and do my utmost to avoid being injured; but the snarling whirlpool is so strong it almost wrenches off my arms and legs.

Tossed up, thrown down, beaten, torn limb from limb, I think: o world, seen with half an eye at the last moment. Even twisted and upended, how alluring your form! how exquisite your colors!

This mental image, flitting past, is welcome solace for my suffocating breast.

Not one person managed to swim far enough away in time to escape this whirlpool.

They say that with a great ship like this one, the danger zone has a radius of 300 meters.

The decision to save the men came too late and robbed us of the margin of time needed to swim that distance.

All hands dead in battle – this has become our fate.

Explosion

Now *Yamato*'s list is virtually 90 degrees.

Such instances are rare. Most ships sink when the list reaches 30 degrees.

Because of the 90-degree list, the shells for the main batteries fall over in the magazines, slide in the direction of their pointed ends, knock their fuses on the overhead, and explode.

This reconstruction of events was agreed upon at a staff conference of the officers after our return. The staff officers recognized that at the time no fire could have reached the magazines.

The ship is already completely under water; I am in the whirlpool.

There is a full load of shells for the main batteries: armor-piercing shells, one round of which can sink a ship, and type-3 shells, one round of which can knock out a squadron of planes.

First, the fore magazine explodes. Twenty seconds after the ship went under?

Had it happened before the sinking, with the ship still on the surface, the explosion would have turned us outright into shrapnel and scattered us in all directions.

But the water, even while toying with us, deadened its force.

Had there been no explosion, I would have sunk rapidly in the whirlpool, to the bottom of the sea.

At the instant *Yamato*, rolling over, turns belly up and plunges beneath the waves, she emits one great flash of light and sends a gigantic pillar of flame high into the dark sky. Armor plate, equipment, turrets, guns – all the pieces of the ship go flying off.

Moreover, thick smoke, dark brown and bubbling up from the ocean depths, soon engulfs everything, covers everything.

The navigation officer on one of the destroyers calculated that the pillar of fire reached a height of 2,000 meters, that the mushroom-shaped cloud rose to a height of 6,000 meters.

Newspapers reported afterward that the flash of light could be seen easily from Kagoshima.

Opening out like an umbrella, the top of the pillar of light engulfs and destroys several American planes circling to observe the end.

In the general explosion, the shells for the main batteries, carried below decks but of no use because of the bad weather, do get their shot at the enemy.

The pressure generated by the explosion of the fore magazine alone is not equal to that of the whirlpool.

While being tossed about in the whirlpool, my whole body absorbs the extraordinary concussion of the shock wave from the first explosion; I am thrust back, around, and up, crashing into a thick yet undulating wall overhead.

This wall: the corpses of comrades who surfaced quickly and are now being baptized in the fiery rain.

Did they shield us with their bodies from the arrows of fire?

Meanwhile, the whirlpool still has enough force to pull us back down again, away from the surface.

Then, about twenty seconds later, the second explosion. Perhaps part of the aft magazine?

This blast finally hurls my body up to the surface.

The repeated explosions send countless pieces of shrapnel flying. Did the shrapnel turn all the men into living targets, save only those few of us on the aft side of the main tower?

Again, we alone were able to avoid fearsome injury from the underwater explosions.

To think that we who stayed put on the bridge and stuck close to the ship's superstructure should be protected time and again and be safest of all!

As for those who left before us and neared the deck, the nearer they were, the more exposed they were to the blasts.

What an irony!

Even so, every one among us is at least slightly wounded. Most have injuries to head and feet.

Only those whose wounds were slight were able to survive the hardships to follow.

⋆ ⋆ ⋆

I receive a long burn wound and a cut on the top left of my head. According to the examination carried out later by the medical officer, the fragment of shrapnel must have been pretty large, and chances were that the injury would have been a fatal one; but because it hit my head at a tangent, I narrowly escaped death.

It hit me while I was being tossed about as if by a hurricane. This being so, how tiny the probability that the piece of shrapnel and I should collide at a tangent!

To be born a human being and yet to owe my life to the fact that something hit me tangentially! Should I laugh?

A great number of men must have been sucked under by the funnel.

Fearsome, its suction. A great cavity sucks up a vast amount of water and, with it, any solid object.

After we got back the survivors were asked to mark on a diagram where they were when they entered the water. In the vicinity of the funnel: a large blank space.

Had I been five paces to the right, I would have been in danger.

The pillar of fire blows straight back down. The sky is filled with red-hot shrapnel and pieces of wood falling with a roar.

The debris kills or wounds most of the men who struggled to float to the surface but got there too soon.

Only those of us who float up at the last, after having taken the roundabout route through the whirlpool, escape that and do not see a scorching hot sky. Instead, we see only dense smoke.

Those who came to the surface a few moments earlier than we looked up as in a daze at countless pieces of metal falling out of a blazing orange sky.

Caught up in the whirlpool, my body suffered great torment. By comparison, how trivial the thoughts that flooded my mind!

. . . There was still a good bit of soda left in that bottle . . . and I've still got five packets of candy . . . late, isn't it? . . . from a quick look at the charts I know the ocean in this area is 430 meters deep; at this speed, how long before I sink to the bottom? what does it feel like to drop 430 meters? and, and then? . . .

Finally, still with only the one deep breath, I approach my limit.

Perhaps ten seconds after the shock from the second explosion, the agonizing pressure on my chest rises sharply; at last my throat seizes, and suddenly I start to swallow sea water.

Through my nose, through my mouth, I breathe in sea water as if poured in by a pump – unconsciously my body registers the movements of my jaw . . . seven . . . ten . . . fifteen . . . seventeen.

Still, I have no sense of suffocating. Am I unable to die until water fills all the nooks and crannies of my body and spills out my mouth?

Is the peace of death still far off?

Kill me . . . Death, take me.

The edges of my eyes register a dim light. The backs of my eyelids are yellowish. In my nostrils a burnt smell surges up. My feet feel light. Everything is hazy as in a dream, and my body floats in space. No sooner do I think these thoughts than I break through the surface of the water.

Had the second explosion come even five seconds later, my lungs would have burst. It would have been all over with me.

Only those survived who were delayed on a circuitous route just long enough to avoid the falling pillar of fire and yet were thrust to the surface before their lungs burst.

Lieutenant Watanabe. "The darkness turned bright, so I heaved a sigh of relief – 'Hooray! It's the next world!' "

Cadet Sako: "I'm not sure, but I think I said two prayers to Amida. Strange – I've never been one to pray."

To think that if even one element of this repeated good fortune had been lacking, we would not have seen the light of day again!

Cut off by the water, the smoke gradually clears away only to leave behind waves covered with bubbly heavy oil.

"*Yamato* sinks. 1423 hours." Friend and foe flash the message simultaneously.

Two hours of uninterrupted battle against airplanes. It's over now.

William of Poitiers

Hastings

On the death of Edward the Confessor, Harold II was chosen as King of England; however, his claim was disputed by Duke William of Normandy, who landed in England on 29 September 1066. William of Poitiers was the Duke's chaplain.

Rejoicing greatly at having secured a safe landing, the Normans seized and fortified first Pevensey and then Hastings, intending that these should serve as a stronghold for themselves and as a refuge for their ships. Marius and Pompey the Great, both of whom earned their victories by courage and ability (since the one brought Jugurtha in chains to Rome while the other forced Mithridates to take poison), were so cautious when they were in enemy territory that they feared to expose themselves to danger even by separating themselves with a legion from their main army: their custom was (like that of most generals) to direct patrols and not to lead them. But William, with twenty-five knights and no more, himself went out to gain information about the neighbourhood and its inhabitants. Because of the roughness of the ground he had to return on foot, a matter doubtless for laughter, but if the episode is not devoid of humour it none the less deserves serious praise. For the duke came back carrying on his shoulder, besides his own hauberk, that of William fitz Osbern, one of his companions. This man was famed for his bodily strength and courage, but it was the duke who relieved him in his necessity of the weight of his armour.

A rich inhabitant of the country who was a Norman by race, Robert, son of Wimarc, a noble lady, sent a messenger to Hastings to the duke who was his relative and his lord. "King Harold," he said, "has just given battle to his brother and to the king of Norway, who is reputed to be the greatest warrior under heaven, and he has killed both of them in one fight, and has destroyed their mighty armies. Heartened by this success he now hastens towards you at the head of innumerable troops all well equipped for war, and against them your own warriors will prove of no more account than a pack of curs. You are accounted a wise man, and at home you have hitherto acted prudently both in peace and war. Now therefore take care for your safety lest your boldness lead you into a peril from which you will not escape. My advice to you is to remain within your entrenchments and not at present to offer battle." But the duke replied to the messenger thus: "Although it would have been better for your master not to have mingled insults with his message, nevertheless I thank him for his advice. But say this also to him: I have no desire to protect myself behind any rampart, but I intend to give battle to Harold as soon as possible. With the aid of God I would not hesitate to oppose him with my own brave men even if I had only ten thousand of these instead of the sixty thousand I now command."

One day when the duke was visiting the guards of his fleet, and was walking about near the ships, he was told that a monk had arrived sent to him by Harold. He at once accosted him and discreetly said: "I am the steward of William, duke of Normandy, and very intimate with him. It is only through me you will have an opportunity of delivering your message. Say therefore what you have to say to me, and I will deliver a faithful report of your message, for no one is dearer to him that I am. Afterwards through my good offices you may in person say to him whatever you wish." The monk then delivered his message without further delay, and the duke at once caused him to be well housed and kindly entertained. At the same time he carefully considered with his followers what reply he should make to the message.

The next day, seated in the midst of his magnates, he summoned the monk to his presence and said: "I am William, by the grace of God, prince of the Normans. Repeat now therefore in the presence of these men what you said to me yesterday." The envoy then spoke: "This is what Harold bids you know. You have come into his land with he knows not what temerity. He recalls that King Edward at first appointed you as his heir to the kingdom of England, and he remembers that he was himself sent by the king to Normandy to give you an assurance of the succession. But he knows also that the same king, his lord, acting within his rights, bestowed on him the kingdom of England when dying. Moreover, ever since the time when blessed Augustine came to these shores it has been the unbroken custom of the English to treat death-bed bequests as inviolable. It is therefore with justice that he bids you return with your followers to your own country. Otherwise he will break the friendship and the pacts he made with you in Normandy. And he leaves the choice entirely to you."

When the duke had heard this message he asked the monk whether he would conduct his messenger safely into Harold's presence, and the monk promised that he would take as much care for his safety as for his own. Then the duke ordered a certain monk of Fécamp to carry this message forthwith to Harold: "It is not with temerity nor unjustly but after deliberation and in defence of right that I have crossed the sea into this country. My lord and kinsman, King Edward, made me the heir of this kingdom even as Harold himself has testified; and he did so because of the great honours and rich benefits conferred upon him and his brother and followers by me and my magnates. He acted thus because among all his acquaintance he held me to be the best capable of supporting him during his life and of giving just rule to the kingdom after his death. Moreover his choice was not made without the consent of his magnates since Archbishop Stigand, Earl Godwine, Earl Leo-fric and Earl Siward confirmed it, swearing in his hands that after King Edward's death they would serve me as lord, and that during his lifetime they would not seek to have the country in any way occupied so as to hinder my coming. He gave me the son and the nephew of Godwine as hostages. And finally he sent me Harold himself to Normandy that in my presence he might personally take the oath which his father and the others had sworn in my absence. While he was on his way to me Harold fell into a perilous captivity from which he was rescued by my firmness and prudence. He made himself my man by a solemn act of homage, and with his hands in mine he pledged to me the security of the English kingdom. I am ready to submit my case against his for judgment either by the law of Normandy or better still by the law of England, whichever he may choose; and if according to truth and equity either the Normans or the English decide that the kingdom is his by right, let him possess it in peace. But if it be decided that in justice the kingdom should be mine, let him yield it up. Moreover, if he refuses these conditions, I do not think it right that either my men or his should perish in conflict over a quarrel that is none of their making. I am therefore ready to risk my life against his in single combat to decide whether the kingdom of England should by right be his or mine."

We have been careful to record all this speech in the duke's own words rather than our own, for we wish posterity to regard him with favour. Anyone may easily judge that he showed himself wise and just, pious and brave. On reflection it will be considered that the strength of his argument was such that it could not have been shaken by Tully himself, the glory of Roman eloquence; and it brought to nought the claims of Harold. The duke (it will be seen) was ready to accept the judgment prescribed by the law of nations, since he did not desire that his enemies, the English, should perish because of his quarrel, but rather he wanted to decide the issue by means of a single combat and at the peril of his own life. When Harold advanced to meet the duke's envoy and heard this message he grew pale and for a long while remained as if dumb. And when the monk had asked more than once for a reply he first said: "We march at once," and then added, "We march to battle." The envoy besought him to reconsider this reply, urging that what the duke desired was a single combat and not the double slaughter of two armies. (For that good and brave man was willing to renounce something that was just and agreeable to him in order to prevent the death of many: he wished for Harold's head, knowing that it was defended by less fortitude than his own, and that it was not protected by justice.) Then Harold, lifting up his face to heaven, exclaimed: "May the Lord decide this day between William and me, and may he pronounce which of us has the right." Thus, blinded by his lust for dominion, and in his fear unmindful of the wrongs he had committed, Harold made his conscience his judge and that to his own ruin.

In the meantime trusty knights who had been sent out by the duke on patrol came back in haste to report the approach of the enemy. The king was the more furious because he had heard that the Normans had laid waste the neighbourhood of their camp, and he planned to take them unawares by a surprise or night attack. Further, in order to prevent their escape, he sent out a fleet of seven hundred armed vessels to block their passage home. Immediately the duke summoned to arms all those within the camp, for the greater part of his host had gone out foraging. He himself attended mass with the greatest devotion, and fortified both his body and soul by partaking of the Body and Blood of our Lord. With great humility he hung round his neck the relics on which Harold had sworn the oath he had now broken, and whose protection he had therefore lost. The duke had with him two bishops from Normandy, Odo, bishop of Bayeux, and Geoffrey, bishop of Coutances; and there were also with him many secular clergy and not a few monks. This company made ready to fight for him with their prayers. Anyone but the duke would have been alarmed at seeing his hauberk turn to the left when he put it on, but he merely laughed and did not allow the unlucky omen to disturb him.

Although no one has reported to us in detail the short harangue with which on this occasion he increased the courage of his troops, we doubt not it was excellent. He reminded the Normans that with him for their leader they had always proved victorious in many perilous battles. He reminded them also of their fatherland, of its noble history, and of its great renown. "Now is the time," he said, "for you to show your strength, and the courage that is yours." "You fight," he added, "not merely for victory but also for survival. If you bear yourselves valiantly you will obtain victory, honour and riches. If not you will be ruthlessly butchered, or else led ignominiously captive into the hands of pitiless enemies. Further, you will incur abiding disgrace. There is no road for retreat. In front, your advance is blocked by an army and a hostile countryside; behind you, there is the sea where an enemy fleet bars your flight. Men worthy of the name do not allow themselves to be dismayed by the number of their foes. The English have again and again fallen to the sword of an enemy; often, being vanquished, they have submitted to a foreign yoke; nor have they ever been famed as soldiers. The vigorous courage of a few men armed in a just cause and specially protected by heaven must prevail against a host of men unskilled in combat. Only be bold so that nothing shall make you yield, and victory will gladden your hearts."

He then advanced in good order with the papal banner which had been granted to him borne aloft at the head of his troops. In the van he placed foot-soldiers equipped with arrows and crossbows; in the second rank came the more heavily armed infantry clad in hauberks; and finally came the squadrons of knights in the midst of whom he rode himself, showing invincible courage and in such a position that he could give his orders by hand or by voice. If any ancient writer had described the host of Harold, he would have said that at its passage the rivers became dry and the forests were turned into plains. From all the provinces of the English a vast host had gathered together. Some were moved by their zeal for Harold, but all were inspired by the love of their country which they desired, however unjustly, to defend against foreigners. The land of the Danes who were allied to them had also sent copious reinforcements. But fearing William more than the king of Norway and not daring to fight with him on equal terms, they took up their position on higher ground, on a hill abutting the forest through which they had just come. There, at once dismounting from their horses, they drew themselves up on foot and in very close order. The duke and his men in no way dismayed by the difficulty of the ground came slowly up the hill, and the terrible sound of trumpets on both sides signalled the beginning of the battle. The eager boldness of the Normans gave them the advantage of attack, even as in a trial for theft it is the prosecuting counsel who speaks first. In such wise the Norman foot drawing nearer provoked the English by raining death and wounds upon them with their missiles. But the English resisted valiantly, each man according to his strength, and they hurled back spears and javelins and weapons of all kinds together with axes and stones fastened to pieces of wood. You would have thought to see our men overwhelmed by this death-dealing weight of projectiles. The knights came after the chief, being in the rearmost rank, and all disdaining to fight at long range were eager to use their swords. The shouts both of the Normans and of the barbarians were drowned in the clash of arms and by the cries of the dying, and for a long time the battle raged with the utmost fury. The English, however, had the advantage of the ground and profited by remaining within their position in close order. They gained further superiority from their numbers, from the impregnable front which they preserved, and most of all from the manner in which their weapons found easy passage through the shields and armour of their enemies. Thus they bravely withstood and successfully repulsed those who were engaging them at close quarters, and inflicted losses upon the men who were shooting missiles at them from a distance. Then the foot-soldiers and the Breton knights, panic-stricken by the violence of the assault, broke in flight before the English and also the auxiliary troops on the left wing, and the whole army of the duke was in danger of retreat. This may be said without disparagement to the unconquerable Norman race. The army of the Roman emperor, containing the soldiers of kings accustomed to victory on sea and land, sometimes fled on the report, true or false, that their leader was dead. And in this case the Normans believed that their duke and lord was killed. Their flight was thus not so much shameful as sad, for their leader was their greatest solace.

Seeing a large part of the hostile host pursuing his own troops, the prince thrust himself in front of those in flight, shouting at them and threatening them with his spear. Staying their retreat, he took off his helmet, and standing before them bareheaded he cried: "Look at me well. I am still alive and by the grace of God I shall yet prove victor. What is this madness which makes you fly, and what way is open for your retreat? You are allowing yourselves to be pursued and killed by men whom you could slaughter like cattle. You are throwing away victory and lasting glory, rushing into ruin and incurring abiding disgrace. And all for naught since by flight none of you can escape destruction." With these words he restored their courage, and, leaping to the front and wielding his death-dealing sword, he defied the enemy who merited death for their disloyalty to him their prince. Inflamed by his ardour the Normans then surrounded several thousands of their pursuers and rapidly cut them down so that not one escaped. Heartened by this success, they then furiously carried their attack on to the main body of the English host, which even after their losses scarcely seemed diminished in number. The English fought confidently with all their strength, striving in particular to prevent the attackers from penetrating within their ranks, which indeed were so closely massed together that even the dead had not space in which to fall. The swords of the bravest warriors hewed a gap in some places, and there they were followed by the men of Maine, by the French, by the Bretons and the men of Aquitaine, and by the Normans who showed the greatest valour.

A certain Norman, Robert, son of Roger of Beaumont, being nephew and heir to Henry, count of Meulan, through Henry's sister, Adeline, found himself that day in battle for the first time: he was as yet but a young man and he performed feats of valour worthy of perpetual remembrance. At the head of the troop which he commanded on the right wing, he attacked with the utmost bravery and success. It is not, however, our purpose, or within our capacity, to describe as they deserve the exploits of individuals. Even a master of narrative who had actually been present that day would find it very difficult to narrate them all in detail. For our part we shall hasten to the point at which, having ended our praise of William the count, we shall begin to describe the glory of William the king.

Realising that they could not without severe loss overcome an army massed so strongly in close formation, the Normans and their allies feigned flight and simulated a retreat, for they recalled that only a short while ago their flight had given them an advantage. The barbarians thinking victory within their grasp shouted with triumph, and heaping insults upon our men, threatened utterly to destroy them. Several thousand of them, as before, gave rapid pursuit to those whom they thought to be in flight; but the Normans suddenly wheeling their horses surrounded them and cut down their pursuers so that not one was left alive. Twice was this ruse employed with the utmost success, and then they attacked those that remained with redoubled fury. This army was still formidable and very difficult to overwhelm. Indeed this was a battle of a new type: one side vigorously attacking; the other resisting as if rooted to the ground. At last the English began to weary, and as if confessing their crime in their defeat they submitted to their punishment. The Normans threw and struck and pierced. The movements of those who were cut down to death appeared greater than that of the living; and those who were lightly wounded could not escape because of the density of their formation but were crushed in the throng. Thus fortune crowned the triumph of William.

There were present in this battle: Eustace, count of Boulogne; William, son of Richard, count of Evreux; Geoffrey, son of Rotrou, count of Mortagne; William fitz Osbern; Haimo, *vicomte* of Thouars; Walter Giffard; Hughe of Montfort-sur-Risle; Rodulf of Tosny; Hugh of Grantmesnil; William of Warenne; and many other most renowned warriors whose names are worthy to be commemorated in histories among the bravest soldiers of all time. But Duke William excelled them all both in bravery and soldier-craft, so that one might esteem him as at least the equal of the most praised generals of ancient Greece and Rome. He dominated this battle, checking his own men in flight, strengthening their spirit, and sharing their dangers. He bade them come with him, more often than he ordered them to go in front of him. Thus it may be understood how he led them by his valour and gave them courage. At the mere sight of this wonderful and redoubtable knight, many of his enemies lost heart even before they received a scratch. Thrice his horse fell under him; thrice he leapt upon the ground; and thrice he quickly avenged the death of his steed. It was here that one could see his prowess, and mark at once the strength of his arm and the height of his spirit. His sharp sword pierced shields, helmets and armour, and not a few felt the weight of his shield. His knights seeing him thus fight on foot were filled with wonder, and although many were wounded they took new heart. Some weakened by loss of blood went on resisting, supported by their shields, and others unable themselves to carry on the struggle, urged on their comrades by voice and gesture to follow the duke. "Surely," they cried, "you will not let victory slip from your hands." William himself came to the rescue of many . . .

Evening was now falling, and the English saw that they could not hold out much longer against the Normans. They knew they had lost a great part of their army, and they knew also that their king with two of his brothers and many of their greatest men had fallen. Those who remained were almost exhausted, and they realised that they could expect no more help. They saw the Normans, whose numbers had not been much diminished, attack them with even greater fury than at the beginning of the battle, as if the day's fighting had actually increased their vigour. Dismayed at the implacable bearing of the duke who spared none who came against him and whose prowess could not rest until victory was won, they began to fly as swiftly as they could, some on horseback, some on foot, some along the roads, but most over the trackless country. Many lay on the ground bathed in blood, others who struggled to their feet found themselves too weak to escape, while a few, although disabled, were given strength to move by fear. Many left their corpses in the depths of the forest, and others were found by their pursuers lying by the roadside. Although ignorant of the countryside the Normans eagerly carried on the pursuit, and striking the rebels in the back brought a happy end to this famous victory. Many fallen to the ground were trampled to death under the hooves of runaway horses.

But some of those who retreated took courage to renew the struggle on more favourable ground. This was a steep valley intersected with ditches. These people, descended from the ancient Saxons (the fiercest of men), are always by nature eager for battle, and they could only be brought down by the greatest valour. Had they not recently defeated with ease the king of Norway at the head of a fine army?

The duke who was following the victorious standards did not turn from his course when he saw these enemy troops rallying. Although he thought that reinforcements had joined his foes he stood firm. Armed only with a broken lance he was more formidable than others who brandished long javelins. With a harsh voice he called to Eustace of Boulogne, who with fifty knights was turning in flight, and was about to give the signal for retreat. This man came up to the duke and said in his ear that he ought to retire since he would court death if he went forward. But at the very moment when he uttered the words Eustace was struck between the shoulders with such force that blood gushed out from his mouth and nose, and half dead he only made his escape with the aid of his followers. The duke, however, who was superior to all fear and dishonour, attacked and beat back his enemies. In this dangerous phase of the battle many Norman nobles were killed since the nature of the ground did not permit them to display their prowess to full advantage.

Having thus regained his superiority, the duke returned to the main battlefield, and he could not gaze without pity on the carnage, although the slain were evil men, and although it is good and glorious in a just war to kill a tyrant. The bloodstained battle-ground was covered with the flower of the youth and nobility of England. The two brothers of the king were found near him, and Harold himself stripped of all badges of honour could not be identified by his face, but only by certain marks on his body. His corpse was brought into the duke's camp, and William gave it for burial to William, surnamed Malet, and not to Harold's mother, who offered for the body of her beloved son its weight in gold. For the duke thought it unseemly to receive money for such merchandise, and equally he considered it wrong that Harold should be buried as his mother wished, since so many men lay unburied because of his avarice. They said in jest that he who had guarded the coast with such insensate zeal should be buried by the seashore . . .

Julius Caesar

The Defeat of the Nervii

The Belgic Nervii were amongst the most powerful of the tribes rallied against Caesar as he tried to extend Roman rule in the West in 57BC. Caesar writes in the third person.

The Ambiani were neighbours of the Nervii, about whose character and habits Caesar made enquiries. He learned that they did not admit traders into their country and would not allow the importation of wine or other luxuries, because they thought such things made men soft and took the edge off their courage; that they were a fierce, warlike people, who bitterly reproached the other Belgae for throwing away their inheritance of bravery by submitting to the Romans, and vowed that they would never ask for peace or accept it on any terms. After three days' march through Nervian territory, Caesar learned from prisoners that the river Sambre was not more than ten miles from the place where he was encamped, and that all the Nervian troops were posted on the farther side of it, awaiting the arrival of the Romans. Already with them in the field, he was told, were their neighbours the Atrebates and the Viromandui, whom they had persuaded to try the fortune of war along with them; and they were expecting to be joined by the forces of the Atuatuci, which were already on the way. They had hastily thrust their women, and all who were thought too young or too old to fight, into a place which marshes made inaccessible to an army.

On receiving this information Caesar sent forward a reconnoitring party, accompanied by some centurions, to choose a good site for a camp. A large number of Gauls, including some of the Belgae who had surrendered, had attached themselves to Caesar and were marching with the troops. Some of these, as was afterwards ascertained from prisoners, had observed the order in which our army marched during the previous days, and at night made their way to the Nervii and explained to them that each legion was separated from the following one by a long baggage-train, so that when the first reached camp the others would be far away; it would be quite easy to attack it while the men were still burdened with their packs, and when it was routed and its baggage plundered, the others would not dare to make a stand. There was one thing that favoured the execution of the plan suggested by these deserters. The Nervii, having virtually no cavalry,[1] long ago devised a method of hindering their neighbours' cavalry when it made plundering raids into their territory. They cut off the tops of saplings, bent them over, and let a thick growth of side branches shoot out; in between them they planted briars and thorns, and thus made hedges like walls, which gave such protection that no one could even see through them, much less penetrate them. As these obstacles hindered the march of our column, the Nervii thought the proposed plan too good to leave untried.

At the place that the Romans had chosen for their camp a hill sloped down evenly from its summit to the Sambre. Opposite it, on the other side of the river, rose another hill with a similar gradient, on the lower slopes of which were some three hundred yards of open ground, while the upper part was covered by a wood which it was not easy to see into. In this wood the main part of the enemy's forces lay concealed, while on the open ground along the river bank a few pickets of cavalry were visible. The depth of the river was about three feet.

Caesar had sent his cavalry a little in advance and was following with the rest of his forces. But the column was formed up in a different manner from that which the Belgic deserters had described to the Nervii. In accordance with his usual practice when approaching an enemy, Caesar marched at the head of the column with six legions, unencumbered by heavy baggage; then came the transport of the entire army, protected by the two newly-enrolled legions, which brought up the rear. First of all, our cavalry crossed the river with the slingers and archers and engaged the enemy's horsemen. These kept on retiring into the wood where their comrades were and then reappearing to charge our troops, who dared not pursue them beyond the end of the open ground. Meanwhile the six legions

1 To this day they pay no attention to that arm, their whole strength consisting of infantry.

that were the first to arrive measured out the ground and began to construct the camp. The Gauls concealed in the wood had already formed up in battle-order and were waiting full of confidence. As soon as they caught sight of the head of the baggage-train – the moment which they had agreed upon for starting the battle – they suddenly dashed out in full force and swooped down on our cavalry, which they easily routed. Then they ran down to the river at such an incredible speed that almost at the same instant they seemed to be at the edge of the wood, in the water, and already upon us. With equal rapidity they climbed the hill towards our camp to attack the men who were busy entrenching it.

Caesar had everything to do at once – hoist the flag which was the signal for running to arms, recall the men from their work on the camp, fetch back those who had gone far afield in search of material for the rampart, form the battle-line, address the men, and sound the trumpet-signal for going into action. Much of this could not be done in the short time left available by the enemy's swift onset. But the situation was saved by two things – first, the knowledge and experience of the soldiers, whose training in earlier battles enabled them to decide for themselves what needed doing, without waiting to be told; secondly, the order which Caesar had issued to all his generals, not to leave the work, but to stay each with his own legion until the camp fortifications were completed. As the enemy was so close and advancing so swiftly, the generals did not wait for further orders but on their own responsibility took the measures they thought proper.

After giving the minimum of essential orders, Caesar hastened down to the battlefield to address the troops and happened to come first upon the 10th legion, to which he made only a short speech, urging them to live up to their tradition of bravery, to keep their nerve, and to meet the enemy's attack boldly. Then, as the Nervii were within range, he gave the signal for battle. On going to the other side of the field to address the troops there, he found them already in action. The soldiers were so pushed for time by the enemy's eagerness to fight, that they could not even take the covers off their shields or put on helmets – not to speak of fixing on crests or decorations. Each man, on coming down from his work at the camp, went into action under the first standard he happened to see, so as not to waste time searching for his own unit. The battle-front was not formed according to the rules of military theory, but as necessitated by the emergency and the sloping ground of the hill-side. The legions were facing different ways and fighting separate actions, and the thick hedges obstructed their view. The result was that Caesar could not fix upon definite points for stationing reserves or foresee what would be needed in each part of the field, and unity of command was impossible. In such adverse circumstances there were naturally ups and downs of fortune.

The 9th and 10th legions were on the left, and discharged a volley of spears at the Atrebates, who happened to be facing them. Breathless and exhausted with running, and many of them now wounded, the Atrebates were quickly driven down to the river, and when they tried to cross it our soldiers with their swords attacked them at a disadvantage and destroyed a large number. Crossing the river themselves without hesitation and pushing forward up the steep slope, they renewed the fight when the enemy began to resist once more, and again put them to flight. Meanwhile in another part of the field, on a front facing in a slightly different direction, the 11th and 8th legions engaged the Viromandui, drove them down the hill, and were now fighting right on the river banks. By this time, however, the Roman camp was almost entirely exposed in front and on the left, and the 12th and 7th legions, which were posted fairly close together on the right, were attacked by the whole force of the Nervii, led in a compact mass by their commander-in-chief Boduognatus. Some of them began to surround the legions on their right flank, while the rest made for the hill-top where the camp stood.

At the same time, the Roman cavalry and light-armed troops, routed by the first attack, were in the act of retreating into the camp, when they found themselves face to face with the Nervii and took to flight again in a different direction. The servants, too, who from the back gate on the summit of the hill had seen our victorious troops cross the river, and had gone out to plunder, on looking back and seeing the enemy in the camp immediately ran for their lives. Meanwhile shouting and din arose from the drivers coming up with the baggage, who rushed panic-stricken in every direction. With the army were some auxiliary cavalry sent by the Treveri, a people with a unique reputation for courage among the Gauls. When these horsemen saw the Roman camp full of the enemy, the legions hard pressed and almost surrounded, and the non-combatants, cavalry, slingers, and Numidians scattered and stampeding in every direction, they decided that our case was desperate, and, riding off home in terror, reported that the Romans were utterly defeated and their camp and baggage captured.

After addressing the 10th legion Caesar had gone to the right wing, where he found the troops in difficulties. The cohorts of the 12th legion were packed together so closely that the men were in one another's way and could not fight properly. All the centurions of the 4th cohort, as well as a standard-bearer, were killed, and the standard was lost; nearly all the centurions of the other cohorts were either killed or wounded, including the chief centurion Publius Sextius Baculus, a man of very great courage, who was so disabled by a number of severe wounds that he could no longer stand. The men's movements were slow, and some in the rear, feeling themselves abandoned, were retiring from the fight and trying to get out of range. Meanwhile the enemy maintained unceasing pressure up the hill in front, and were also closing in on both flanks. As the situation was critical and no reserves were available, Caesar snatched a shield from a soldier in the rear (he had not his own shield with him), made his way into the front line, addressed each centurion by name, and shouted encouragement to the rest of the troops, ordering them to push forward and open out their ranks, so that they could use their swords more easily. His coming gave them fresh heart and hope; each man wanted to do his best under the eyes of his commander-in-chief, however desperate the peril, and the enemy's assault was slowed down a little.

Noticing that the 7th legion, which stood close by, was likewise hard put to it, Caesar told the military tribunes to join the two legions gradually together and adopt a square formation, so that they could advance against the enemy in any direction. By this manoeuvre the soldiers were enabled to support one another, and were no longer afraid of being surrounded from behind, which encouraged them to put up a bolder resistance. Meanwhile the two legions which had acted as a guard to the baggage at the rear of the column, having received news of the battle, had quickened their pace, and now appeared on the hill-top, where the enemy could see them; and Labienus, who had captured the enemy's camp, and from the high ground on which it stood could see what was going on in ours, sent the 10th legion to the rescue. The men of the 10th, who could tell from the flight of the cavalry and the non-combatants how serious things were, and what peril threatened the camp, the legions, and their commander-in-chief, strained every nerve to make the utmost speed.

Their arrival so completely changed the situation that even some of the Roman soldiers who had lain down, exhausted by wounds, got up and began to fight again, leaning on their shields. The non-combatants, observing the enemy's alarm, stood up to their attack, unarmed as they were; and the cavalry, anxious to wipe out the disgrace of their flight, scoured the whole battlefield and tried to outdo the legionaries in gallantry. But the enemy, even in their desperate plight, showed such bravery that when their front ranks had fallen those immediately behind stood on their prostrate bodies to fight; and when these too fell and the corpses were piled high, the survivors still kept hurling javelins as though from the top of a mound, and flung back the spears intercepted by their shields. Such courage accounted for the extraordinary feats they had performed already. Only heroes could have made light of crossing a wide river, clambering up the steep banks, and launching themselves on such a difficult position.

So ended this battle, by which the tribe of the Nervii was almost annihilated and their name almost blotted out from the face of the earth.

John Froissart

English Archers at Crécy

The Battle of Crécy was fought on September 1346 in pursuit of Edward II's claim to the throne of France. Although the English numbered a mere 9,000 against a French army of 30,000 they emerged triumphant because of their superior technology – the longbow.

The English, who were drawn up in three divisions, and seated on the ground, on seeing their enemies advance rose undauntedly up, and fell into their ranks. That of the prince was the first to do so, whose archers were formed in the manner of a portcullis, or harrow, and the men-at-arms in the rear. The earls of Northampton and Arundel, who commanded the second division, had posted themselves in good order on his wing, to assist and succour the prince if necessary.

You must know that these kings, earls, barons, and lords of France did not advance in any regular order, but one after the other, or any way most pleasing to themselves. As soon as the king of France came in sight of the English, his blood began to boil, and he cried out to his marshals, "Order the Genoese forward, and begin the battle, in the name of God and St. Denis." There were about fifteen thousand Genoese crossbowmen; but they were quite fatigued, having marched on foot that day six leagues, completely armed, and with their crossbows. They told the constable they were not in a fit condition to do any great things that day in battle. The Earl of Alençon, hearing this, said, "This is what one gets by employing such scoundrels, who fall off when there is any need for them." During this time a heavy rain fell, accompanied by thunder and a very terrible eclipse of the sun; and before this rain a great flight of crows hovered in the air over all those battalions, making a loud noise. Shortly afterwards it cleared up, and the sun shone very bright; but the Frenchmen had it in their faces, and the English in their backs. When the Genoese were somewhat in order, and approached the English, they set up a loud shout, in order to frighten them; but they remained quite still, and did not seem to attend to it. They then set up a second shout, and advanced a little forward; but the English never moved. They hooted a third time, advancing with their cross-bows presented, and began to shoot. The English archers then advanced one step forward, and shot their arrows with such force and quickness that it seemed as if it snowed. When the Genoese felt these arrows, which pierced their arms, heads, and through their armour, some of them cut the strings of their cross-bows, others flung them on the ground, and all turned about and retreated, quite discomfited. The French had a large body of men-at-arms on horseback, richly dressed, to support the Genoese. The king of France, seeing them thus fall back, cried out, "Kill me those scoundrels; for they stop up our road, without any reason." You would then have seen the above-mentioned men-at-arms lay about them, killing all they could of these runaways.

The English continued shooting as vigorously and quickly as before; some of their arrows fell among the horsemen, who were sumptuously equipped, and, killing and wounding many, made them caper and fall among the Genoese, so that they were in such confusion they could never rally again. In the English army there were some Cornish and Welshmen on foot, who had armed themselves with large knives; these, advancing through the ranks of the men-at-arms and archers, who made way for them, came upon the French when they were in this danger, and, falling upon earls, barons, knights, and squires, slew many, at which the king of England was afterwards much exasperated. The valiant king of Bohemia was slain there. He was called Charles of Luxembourg; for he was the son of the gallant king and emperor, Henry of Luxembourg: having heard the order of the battle, he inquired where his son, the lord Charles, was; his attendants answered that they did not know, but believed he was fighting. The king said to them, "Gentlemen, you are all my people, my friends and brethren-at-arms this day; therefore, as I am blind, I request of you to lead me so far into the engagement that I may strike one stroke with my sword." The knights replied they would directly lead him forward; and in order that they might not lose him in the crowd, they fastened all the reins of their horses together, and put the king at their head, that he might gratify his wish, and advanced towards the enemy. The lord Charles of Bohemia, who already signed his name as king of Germany, and bore the arms, had come in good order to the engagement; but when he perceived that it was likely to turn out against the French he departed, and I do not well know what road he took. The king, his father, had ridden in among the enemy, and made good use of his sword; for he and his companions had fought most gallantly. They had advanced so far that they were all slain; and on the morrow they were found on the ground, with their horses all tied together.

The earl of Alençon advanced in regular order upon the English, to fight with them; as did the earl of Flanders, in another part. These two lords, with their detachments, coasting as it were the archers, came to the prince's battalion, where they fought valiantly for a length of time. The king of France was eager to march to the place where he saw their banners displayed, but there was a hedge of archers before him. He had that day made a present of a handsome black horse to sir John of Hainault, who had mounted on it a knight of his called sir John de Fuselles, that bore his banner; which horse ran off with him, and forced his way through the English army, and, when about to return, stumbled and fell into a ditch and severely wounded him; he would have been dead if his page had not followed him round the battalions, and found him unable to rise; he had not, however, any other hindrance than from his horse, for the English did not quit the ranks that day to make prisoners. The page alighted, and raised him up; but he did not return the way he came, as he would have found it difficult from the crowd. This battle, which was fought on the Saturday between la Broyes and Creçy, was very murderous and cruel; and many gallant deeds of arms were performed that were never known. Towards evening, many knights and squires of the French had lost their masters; they wandered up and down the plain, attacking the English in small parties; they were soon destroyed, for the English had determined that day to give no quarter or hear of ransom from any one.

Early in the day some French, Germans, and Savoyards had broken through the archers of the prince's battalion, and had engaged with the men-at-arms; upon which the second battalion came to his aid, and it was time, for otherwise he would have been hard pressed. The first division, seeing the danger they were in, sent a knight in great haste to the king of England, who was posted upon an eminence, near a windmill. On the knight's arrival, he said, "Sir, the earl of Warwick, the lord Stafford, the lord Reginald Cobham, and the others who are about your son, are vigorously attacked by the French; and they entreat that you would come to their assistance with your battalion, for, if their numbers should increase, they fear he will have too much to do." The king replied, "Is my son dead, unhorsed, or so badly wounded that he cannot support himself?" "Nothing of the sort, thank God," rejoined the knight; "but he is in so hot an engagement that he has great need of your help." The king answered, "Now, sir Thomas, return back to those that sent you and tell them from me, not to send again for me this day, or expect that I shall come, let what will happen, as long as my son has life; and say that I command them to let the boy win his spurs; for I am determined, if it please God, that all the glory and honour of this day shall be given to him, and to those into whose care I have entrusted him." The knight returned to his lords, and related the king's answer, which mightily encouraged them, and made them repent they had ever sent such a message.

It is a certain fact, that sir Godfrey de Harcourt, who was in the prince's battalion, having been told by some of the English that they had seen the banner of his brother engaged in the battle against him, was exceedingly anxious to save him; but he was too late, for he was left dead on the field and so was the earl of Aumarle, his nephew. On the other hand, the earls of Alençon and Flanders were fighting lustily under their banners, and with their own people; but they could not resist the force of the English, and were there slain, as well as many other knights and squires that were attending on or accompanying them. The earl of Blois, nephew to the king of France, and the duke of Lorraine, his brother-in-law, with their troops, made a gallant defence; but they were surrounded by a troop of English and Welsh, and slain in spite of their prowess. The earl of St. Pol and the earl of Auxerre were also killed, as well as many others. Late after vespers, the king of France had not more about him than sixty men, every one included. Sir John of Hainault, who was of the number, had once remounted the king; for his horse had been killed under him by an arrow; he said to the king, "Sir, retreat whilst you have an opportunity, and do not expose yourself so simply; if you have lost this battle, another time you will be the conqueror." After he had said this, he took the bridle of the king's horse, and led him off by force; for he had before entreated of him to retire. The king rode on until he came to the castle of la Broyes, where he found the gates shut, for it was very dark. The king ordered the governor of it to be summoned; he came upon the battlements, and asked who it was that called at such an hour? The king answered, "Open, open, governor; it is the fortune of France." The governor, hearing the king's voice, immediately descended, opened the gate, and let down the bridge. The king and his company entered the castle; but he had only with him five barons, sir John of Hainault, the lord Charles of Montmorency, the lord of Beaujeu, the lord of Aubigny, and the lord of Montfort. The king would not bury himself in such a place as that, but having taken some refreshments, set out again with his attendants about midnight, and rode on, under the direction of guides who were well acquainted with the country, until, about day-break, he came to Amiens, where he halted. This Saturday the English never quitted their ranks in pursuit of any one, but remained on the field, guarding their position, and defending themselves against all who attacked them. The battle was ended at the hour of vespers.

When, on this Saturday night, the English heard no more hooting, or shouting, nor any more crying out to particular lords or their banners, they looked upon the field as their own, and their enemies as beaten. They made great fires, and lighted torches because of the obscurity of the night. King Edward then came down from his post, who all that day had not put on his helmet, and, with his whole battalion, advanced to the prince of Wales, whom he embraced in his arms and kissed, and said, "Sweet son, God give you good perseverance: you are my son, for most loyally have you acquitted yourself this day; you are worthy to be a sovereign." The prince bowed down very low, and humbled himself, giving all honour to the king his father. The English, during the night, made frequent thanksgivings to the Lord, for the happy issue of the day, and without rioting; for the king had forbidden all riot or noise. On the Sunday morning, there was so great a fog that one could scarcely see the distance of half an acre. The king ordered a detachment from the army, under the command of the two marshals, consisting of about five hundred lances and two thousand archers, to make an excursion and see if there were any bodies of French collected together. The quota of troops, from Rouen and Beauvais, had, this Sunday morning, left Abbeville and St. Ricquier in Ponthieu, to join the French army, and were ignorant of the defeat of the preceding evening: they met this detachment, and, thinking they must be French, hastened to join them.

As soon as the English found who they were, they fell upon them; and there was a sharp engagement; but the French soon turned their backs, and fled in great disorder. There were slain in this fight in the open fields, under hedges and bushes, upwards of seven thousand; and had it been clear weather, not one soul would have escaped.

A little time afterwards, this same party fell in with the archbishop of Rouen, and the great prior of France, who were also ignorant of the discomfiture of the French; for they had been informed that the king was not to fight before Sunday. Here began a fresh battle; for those two lords were well attended by good men-at-arms; however, they could not withstand the English, but were almost all slain, with the two chiefs who commanded them; very few escaping. In the course of the morning, the English found many Frenchmen, who had lost their road on the Saturday, and had lain in the open fields, not knowing what was become of the king, or their own leaders. The English put to the sword all they met; and it has been assured to me for fact, that of foot soldiers, sent from the cities, towns, and municipalities, there were slain, this Sunday morning, four times as many as in the battle of the Saturday.

This detachment, which had been sent to look after the French, returned as the king was coming from mass, and related to him all that they had seen and met with. After he had been assured by them that there was not any appearance of the French collecting another army, he sent to have the numbers and condition of the dead examined.

He ordered on this business, lord Reginald Cobham, lord Stafford, and three heralds to examine their arms, and two secretaries to write down all the names. They took much pains to examine all the dead, and were the whole day in the field of battle, not returning but just as the king was sitting down to supper. They made to him a very circumstantial report of all they had observed, and said they had found eighty banners, the bodies of eleven princes, twelve hundred knights, and about thirty thousand common men.

The English halted there that day, and on the Monday morning prepared to march off. The king ordered the bodies of the principal knights to be taken from the ground, and carried to the monastery of Mountenay, which was hard by, there to be interred in consecrated ground. He had it proclaimed in the neighbourhood that he should grant a truce for three days, in order that the dead might be buried.

Alex Bowlby

Officer Selection

For the Rifle Brigade in Britain during the Second World War.

I had volunteered for the Army – I hadn't fancied being called up – and this, plus the fact of my having been to one of the public schools which the regiment preferred its officers from, automatically ear-marked me as a potential officer. This upset my platoon sergeant even more than my arms-drill. One bleak November morning he could stand it no longer. The squad was practising gas-drill. I had hidden myself in the back rank but the Sergeant had turned the squad round. When everyone else had replaced their respirators I was still wrestling with the head-piece. The eye of the Sergeant was upon me. Desperately I rammed home the head-piece. When I buttoned up the respirator it bulged like a pregnant serpent. The Sergeant moved in for the kill. Unbuttoning the respirator he replaced it correctly. Then he thrust his face into mine.

"If you ever get a commission my prick's a bloater!"

A week later I was sent to a War Office Selection Board. Its highlight was an interview with a psychiatrist. I thought this would be fun. When I entered his room I had to stop myself giggling. He motioned me to sit down, and continued to correct papers (we had all answered a word-association test). After five minutes' silence I no longer found anything funny about the interview. After ten minutes I felt like screaming.

The psychiatrist suddenly looked up from the papers. He stared at me until I had to look down.

"You were unhappy at school, are extremely self-conscious, and find it difficult to concentrate. Correct?"

I nodded dumbly, wondering how on earth he did it.

"Both your parents are neurotic, aren't they?"

"I – I don't know."

"H'm. Have you ever had a woman?"

"No."

"Do you want to?"

"Of course!"

The psychiatrist gave me another long stare. I ended up looking at the floor.

"What do you like most in life?"

"Poetry, I suppose."

"Why?"

"Because it's part of my ideals."

"What ideals?"

"I don't quite know how to explain. I suppose my ideals are what I believe in."

"What do you believe in?"

"Helping other people. Doing what I feel is right."

The psychiatrist leant across the table.

"What would your feelings be if you bayoneted a German?"

This was much better.

"I'd feel sorry for him. I don't think he would have caused the war any more than I did."

The psychiatrist frowned.

"Well, what would you feel if *you* were bayoneted by a German?"

"A great deal of pain."

"Yes, but *what else?*"

I couldn't think of anything else.

"Nothing."

The psychiatrist glared at me. I stared back. We looked at each other until I felt dizzy.

"You should avoid going out alone at nights," he said finally.

I nearly burst out laughing. But he hadn't quite finished.

"And if you don't give up these so-called ideals of yours you'll go mad within eighteen months."

I was so shaken I couldn't speak. Finally I said: "But what shall I do?"

"That's up to you."

When I got out of the room I fainted.

For some weeks afterwards my nerves were all to bits. I lived for letters from a friend who slowly convinced me that the psychiatrist was talking through his hat.

Roger Hall

Finest Hour

Hall flew with the RAF's 152 Squadron during the Battle of Britain, summer 1940.

Beneath us, as we reached thirty-thousand feet and levelled out, there was a flat carpet of cloud, pure white in the bright sunlight. Above us, apart from a few delicate and remote wisps of cirrus, the sky was an intense blue, the sort of blue you find on an artist's palette. Behind us and slightly to our starboard the sun was still high in the sky and was dazzling to look into. To the east the stratus cloud was beginning to disperse and we could see across the North Sea to the Dutch Islands. Visibility was limited only by the curvature of the Earth.

The entire firmament, the vault of the heavens, was revealed to us. It stretched from Lille and St. Omer in the rolling plains of the Pas-de-Calais to the south, eastwards down the sandy coastline of Northern France, past Dunkirk to the Belgian frontier, beyond that to the Dutch Islands and past them to the faint line of the German coast, and up as far north as the Norfolk coast of our own country. Such was a panorama that confronted us as we levelled out five miles above the earth and higher than the highest mountain.

"Hallo Mandrake," Maida Leader called. "Maida Squadron now at angels three-zero – Bandits in sight – Tally Ho." "Well done, Maida Leader – Good luck – Good luck – over to you," Mandrake replied.

Yes, there they were all right. Very many bandits, too. The sky was full of black dots, which, from where we were at the moment, might have been anything; but we knew only too well what they were. They were coming from the south; squadron upon squadron, fleet upon fleet, an aerial Armada the size of which I don't suppose Jules Verne or even Wells had envisaged.

The main body of them was below us by quite ten thousand feet, but above them as escort, winged the protective fighter screen proudly trailing their long white plumes of vapour.

Our position was somewhere over Surrey at the moment, and as we approached the enemy formations which were still some miles away, we saw our own fighters – the eleven group squadrons and some from twelve group in the Midlands – coming up from the north. There seemed to be quite a number of us. They too were black dots, climbing in groups of twelve or thirty-six in wing formation. Most of them were Hurricanes.

The enemy appeared to be disposed in three distinct and separate groups each comprising a hundred or more bombers. Above each group were about fifty fighters – M.E. 109's, and M.E. 110's. The bombers were Heinkels, Dorniers and Junkers 88's.

"Line astern formation – Maida squadron," ordered Maida Leader. We took up our battle formations at once, with "A" Flight in the order of Red, Yellow and White. There were two machines behind me and three in front. "Come up into line abreast 'B' Flight" came the next order from Red one. When we had completed this change the squadron was disposed in two lines of six machines flying abreast and at a distance of about fifty yards between each Flight.

We were ready to attack. We were now in the battle area and three-quarters of an hour had elapsed since we had taken off.

The two bomber formations furthest from us were already being attacked by a considerable number of our fighters. Spitfires and Hurricanes appeared to be in equal numbers at the time. Some of the German machines were already falling out of their hitherto ordered ranks and floundering towards the earth. There was a little ack-ack fire coming from up somewhere on the ground although its paucity seemed pathetic and its effect was little more than that of a defiant gesture.

We approached the westernmost bomber formation from the front port quarter, but we were some ten thousand feet higher than they were and we hadn't started to dive yet. Immediately above the bombers were some twine-engined fighters, M.E. 110's. Maida Leader let the formation get a little in front of us then he gave the order "Going down now Maida aircraft," turning his machine upside-down as he gave it. The whole of "A" Flight, one after the other, peeled off after him, upside-down at first and then into a vertical dive.

When they had gone "B" Flight followed suit. Ferdie and I turned over with a hard leftward pressure to the stick to bring the starboard wing up to right angles to the horizon, and some application to the port or bottom rudder pedal to keep the nose from rising. Keeping the controls like this, the starboard wing fell over until it was parallel to the horizon again, but upside-down. Pulling the stick back from this position the nose of my machine fell towards the ground and followed White one in front, now going vertically down on to the bombers almost directly below us. Our speed started to build up immediately. It went from three hundred miles per hour to four and more. White one in front, his tail wheel some distance below me but visible through the upper part of my windscreen, was turning his machine in the vertical plane from one side to the other by the use of his ailerons. Red Section had reached the formation and had formed into a loosened echelon to starboard as they attacked. They were coming straight down on top of the bombers, having gone slap through the protective M.E. 110 fighter screen, ignoring them completely.

Now it was our turn. With one eye on our own machines I slipped out slightly to the right of Ferdie and placed the red dot of my sight firmly in front and in line with the starboard engine of a Dornier vertically below me and about three hundred yards off. I felt apprehensive lest I should collide with our own machines in the mêlée that was to ensue. I seemed to see one move ahead what the positions of our machines would be, and where I should be in relation to them if I wasn't careful. I pressed my trigger and through my inch thick windscreen I saw the tracers spiralling away hitting free air in front of the bomber's engine. I was allowing too much deflection. I must correct. I pushed the stick further forward. My machine was past the vertical and I was feeling the effect of the negative gravity trying to throw me out of the machine, forcing my body up into the perspex hood of the cockpit. My Sutton harness was biting into my shoulders and blood was forcing its way to my head, turning everything red. My tracers were hitting the bomber's engine and bits of metal were beginning to fly off it. I was getting too close to it, much too close. I knew I must pull away but I seemed hypnotised and went still closer, fascinated by what was happening. I was oblivious to everything else. I pulled away just in time to miss hitting the Dornier's starboard wing-tip. I turned my machine to the right on ailerons and heaved back on the stick, inflicting a terrific amount of gravity on to the machine. I was pressed down into the cockpit again and a black veil came over my eyes and I could see nothing.

I eased the stick a little to regain my vision and to look for Ferdie. I saw a machine, a single Spitfire, climbing up after a dive about five hundred yards in front of me and flew after it for all I was worth. I was going faster than it was and I soon caught up with it – in fact I overshot it. It was Ferdie all right. I could see the "C" Charlie alongside our squadron letters on his fuselage. I pulled out to one side and back again hurling my machine at the air without any finesse, just to absorb some speed so that Ferdie could catch up with me. "C" Charlie went past me and I thrust my throttle forward lest I should lose him. I got in behind him again and called him up to tell him so. He said: "Keep an eye out behind and don't stop weaving." I acknowledged his message and started to fall back a bit to get some room. Ferdie had turned out to the flank of the enemy formation and had taken a wide sweeping orbit to port, climbing fast as he did so. I threw my aircraft first on to its port wing-tip to pull it round, then fully over to the other tip for another steep turn, and round again and again, blacking out on each turn. We were vulnerable on the climb, intensely so, for we were so slow.

I saw them coming quite suddenly on a left turn; red tracers coming towards us from the centre of a large black twin-engined M.E. 110 which wasn't quite far enough in the sun from us to be totally obscured, though I had to squint to identify it. I shouted to Ferdie but he had already seen the tracers flash past him and had discontinued his port climbing turn and had started to turn over on his back and to dive. I followed, doing the same thing, but the M.E. 110 must have done so too for the tracers were still following us. We dived for about a thousand feet, I should think, and I kept wondering why my machine had not been hit.

Ferdie started to ease his dive a bit. I watched him turn his machine on to its side and stay there for a second, then its nose came up, still on its side, and the whole aircraft seemed to come round in a barrel-roll as if clinging to the inside of some revolving drum. I tried to imitate this manoeuvre but I didn't know how to, so I just thrust open the throttle and aimed my machine in Ferdie's direction and eventually caught him up.

The M.E. 110 had gone off somewhere. I got up to Ferdie and slid once more under the doubtful protection of his tail and told him that I was there. I continued to weave like a pilot inspired, but my inspiration was the result of sheer terror and nothing more.

All the time we were moving towards the bombers; but we moved indirectly by turns, and that was the only way we could move with any degree of immunity now. Four Spitfires flashed past in front of us, they weren't ours, though, for I noticed the markings. There was a lot of talking going on on the ether and we seemed to be on the same frequency as a lot of other squadrons. "Hallo Firefly Yellow Section – 110 behind you" – "Hallo Cushing Control – Knockout Red leader returning to base to refuel." "Close up Knockout 'N' for Nellie and watch for those 109's on your left" – "All right Landsdown Squadron – control answering – your message received – many more bandits coming from the east – over" – "Talker White two where the bloody hell are you?" – "Going down now Sheldrake Squadron – loosen up a bit" – "You clumsy clot – Hurricane 'Y' Yoke – what the flaming hades do you think you are doing?" – "I don't know Blue one but there are some bastards up there on the left – nine o'clock above" – Even the Germans came in intermittently: "Achtung, Achtung – drei Spitfeuer unter, unter Achtung, Spitfeuer, Spitfeuer." "Tally Ho – Tally Ho – Homer Red leader attacking now." "Get off the bastard air Homer leader" – "Yes I can see Rimmer leader – Red two answering – Glycol leak I think – he's getting out – yes he's baled out he's o.k."

And so it went on incessantly, disjointed bits of conversation coming from different units all revealing some private little episode in the great battle of which each story was a small part of the integral whole.

Two 109's were coming up behind the four Spitfires and instinctively I found myself thrusting forward my two-way radio switch to the transmitting position and calling out "Look out those four Spitfires – 109's behind you – look out." I felt that my message could hardly be of less importance than some that I had heard, but no heed was taken of it. The two 109's had now settled themselves on the tail of the rear Spitfire and were pumping cannon shells into it. We were some way off but Ferdie too saw them and changed direction to starboard, opening up his throttle as we closed. The fourth Spitfire, or "tail-end Charlie", had broken away, black smoke pouring from its engine, and the third in line came under fire now from the same 109. We approached the two 109's from above their starboard rear quarter and, taking a long deflection shot from what must have been still out of range, Ferdie opened fire on the leader. The 109 didn't see us for he still continued to fire at number three until it too started to trail Glycol from its radiator and turned over on its back breaking away from the remaining two. "Look out Black one – look out Black Section Apple Squadron – 109's – 109's" came the belated warning, possibly from number three as he went down. At last number one turned steeply to port, with the two 109's still hanging on to their tails now firing at number two. They were presenting a relatively stationary target for us now for we were directly behind them. Ferdie's bullets were hitting the second 109 now and pieces of its tail unit were coming away and floating past underneath us. The 109 jinked to the starboard. The leading Spitfire followed by its number two had now turned full circle in a very tight turn and as yet it didn't seem that either of them had been hit. The 109 leader was vainly trying to keep into the same turn but couldn't hold it tight enough so I think his bullets were skidding past the starboard of the Spitfires. The rear 109's tail unit disintegrated under Ferdie's fire and a large chunk of it slithered across the top surface of my starboard wing, denting the panels but making no noise. I put my hand up to my face for a second.

The fuselage of the 109 fell away below us and we came into the leader. I hadn't fired at it yet but now I slipped out to port of Ferdie as the leader turned right steeply and over on to its back to show its duck-egg blue belly to us. I came up almost to line abreast of Ferdie on his port side and fired at the under surface of the German machine, turning upside-down with it. The earth was now above my perspex hood and I was trying to keep my sights on the 109 in this attitude, pushing my stick forward to do so. Pieces of refuse rose up from the floor of my machine and the engine spluttered and coughed as the carburettor became temporarily starved of fuel. My propeller idled helplessly for a second and my harness straps bit into my shoulders again. Flames leapt from the engine of the 109 but at the same time there was a loud bang from somewhere behind me and I heard "Look out Roger" as a large hole appeared near my starboard wing-tip throwing up the matt green metal into a ragged rent to show the naked aluminium beneath.

I broke from the 109 and turned steeply to starboard throwing the stick over to the right and then pulling it back into me and blacking out at once. Easing out I saw three 110's go past my tail in "V" formation but they made no attempt to follow me round. "Hallo Roger – Are you O.K.?" I heard Ferdie calling. "I think so – where are you?" I called back.

"I'm on your tail – keep turning" came Ferdie's reply. Thank God, I thought. Ferdie and I seemed to be alone in the sky. It was often like this. At one moment the air seemed to be full of aircraft and the next there was nothing except you. Ferdie came up in "V" on my port side telling me at the same time that he thought we had better try to find the rest of the squadron.

The battle had gone to the north. We at this moment were somewhere over the western part of Kent, and a little less than a quarter of an hour had elapsed since we had delivered our first attack on the bombers. Ferdie set course to the north where we could see in the distance the main body of aircraft. London with its barrage balloons floating unconcernedly, like a flock of grazing sheep, ten thousand feet above it, was now feeling the full impact of the enemy bombers. Those that had got through – and the majority of them had – were letting their bombs go. I recalled for an instant Mr Baldwin's prophecy, not a sanguine one, made to the House of Commons some five years before when he said that the bomber will always get through.

Now it was doing just that. I wondered if it need have done. As we approached South London the ground beneath us became obscured by smoke from the bomb explosions which appeared suddenly from the most unlikely sort of places – an open field, a house, a row of houses, a factory, railway sidings, all sorts of things. Suddenly there would be a flash, then a cloud of reddish dust obscuring whatever was there before and then drifting away horizontally to reveal once more what was left of the target.

I saw a whole stick of bombs in a straight line advancing like a creeping barrage such as you would see on the films in pictures like "Journey's End" or "All Quiet on the Western Front", but this time they were not over the muddy desolation of No-Man's Land but over Croydon, Surbiton and Earl's Court. I wondered what the people were like who were fighting the Battle of Britain just as surely as we were doing but in a less spectacular fashion. I thought of the air raid wardens shepherding their flocks to the air raid trenches without a thought of their own safety; the Auxiliary Firemen and the regular fire brigades who were clambering about the newly settled rubble strewn with white-hot and flaming girders and charred wood shiny black with heat, to pull out the victims buried beneath; the nurses, both the professional ones and the V.A.D.'s in their scarlet cloaks and immaculate white caps and cuffs, who were also clambering about the shambles to administer first aid to the wounded and give morphine to the badly hurt; the St John's Ambulance brigade who always were on the spot somehow no matter where or under that circumstances an accident or emergency occurred, helping, encouraging and uplifting the victims without thought for themselves; the Red Cross and all the civilian volunteers who, when an emergency arises, always go to assist. Not least I thought of the priests and clergy who would also be there, not only to administer the final rites to the dying but to provide an inspiration to those who had lost faith or through shock seemed temporarily lost. The clergy were there all right and showed that their job was not just a once-a-week affair at the Church, but that religion was as much a part of everyday living as was eating and sleeping.

I felt humble when I thought of what was going on down there on the ground. We weren't the only people fighting the Battle of Britain. There were the ordinary people, besides these I've mentioned, all going about their jobs quietly yet heroically and without any fuss or complaint. They had no mention in the press or news bulletins, their jobs were routine and hum-drum and they got no medals.

We were now in the battle area once again and the fighting had increased its tempo. The British fighters were becoming more audacious, had abandoned any restraint that they might have had at the outset, and were allowing the bombers no respite at all. If they weren't able to prevent them from reaching their target they were trying desperately to prevent them from getting back to their bases in Northern France. The air was full of machines, the fighters, British and German, performing the most fantastic and incredibly beautiful evolutions. Dark oily brown streams of smoke and fire hung vertically in the sky from each floundering aircraft, friend or foe, as it plunged to its own funeral pyre miles below on the English countryside. The sky, high up aloft, was an integrated medley of white tracery, delicately woven and interwoven by the fighters as they searched for their opponents. White puffs of ack-ack fire hung limply in mid-air and parachute canopies drifted slowly towards the ground.

It was an English summer's evening. It was about a quarter to six. We had been in the air now for about an hour and a quarter and our fuel would not last much longer. We had failed to join up with the rest of the flight, but this was understandable and almost inevitable under the circumstances. I don't suppose the others were in any formation other than sections now.

Beneath us at about sixteen thousand feet, while we were at twenty-three, there were four Dorniers by themselves still going north and I presumed, for that reason, they hadn't yet dropped their bombs. Ferdie had seen them and was making for them. Three Hurricanes in line astern had seen the same target, had overtaken them, turned, and were delivering a head-on attack in a slightly echeloned formation. It was an inspiring sight, but the Dorniers appeared unshaken as the Hurricanes flew towards them firing all the time. Then the one on the port flank turned sharply to the left, jettisoning its bomb load as it went. The leading Hurricane got on to its tail and I saw a sheet of flame spring out from somewhere near its centre section and billow back over the top surface of its wing, increasing in size until it had enveloped the entire machine except the extreme tips of its two wings. I didn't look at it any more.

We were now approaching the remaining three Dorniers and we came up directly behind them in line astern. "Get out to port Roger" cried Ferdie "and take the left one." I slid outside Ferdie and settled my sight on the Dornier's starboard engine nacelle. We were not within range yet but not far off. The Dorniers saw us coming all right and their rear-gunners were opening fire on us, tracer bullets coming perilously close to our machines. I jinked out to port in a lightning steep turn and then came back to my original position and fired immediately at the gunner and not the engine. The tracers stopped coming from that Dornier. I changed my aim to the port engine and fired again, one longish burst and my "De-Wilde" ammunition ran up the trailing edge of the Dornier's port wing in little dancing sparks of fire until they reached the engine. The engine exploded and the machine lurched violently for a second as if a ton weight had landed on the wing and then fallen off again for, as soon as the port wing had dropped it picked up again and the bomber still kept formation despite the damage to its engine. The engine was now totally obscured by thick black smoke which was being swept back on to my windscreen. I was too close to the bomber now to do anything but break off my attack and pull away. I didn't see what had happened to the Dornier that Ferdie had attacked and what's more I could no longer see Ferdie.

I broke off in a steep climbing turn to port scanning the sky for a single Spitfire – "C" Charlie. There were lots of lone Spitfires, there were lots of lone Hurricanes and there were lots of lone bombers but it was impossible now and I thought improvident to attempt to find Ferdie in all this mêlée. I began to get concerned about my petrol reserves as we had been in the air almost an hour and a half now and it was a long way back to base.

I pressed my petrol indicator buttons and one tank was completely empty, the other registering twenty-two gallons. I began to make some hasty calculations concerning speed, time and distance and decided that if I set course for base now and travelled fairly slowly I could make it. I could put down at another airfield of course and get refuelled, but it might be had policy, especially for a new pilot.

I called up Ferdie, thinking, not very hopefully, that he might hear me, and told him what I was doing. Surprisingly he came back on the air at once in reply and said that he was also returning to base and asked me if I thought I had got enough fuel. He said that he thought it ought to be enough and added as an after-thought that I should make certain that my wheels and flaps were working satisfactorily before coming too low, for they could be damaged. I thanked him for his advice and listened out. I was by myself now and still in the battle area and I was weaving madly for I realised how vulnerable I was. I was easy meat to German fighters, just their cup of tea, particularly if there should be more than one of them, for the Germans always seemed to fancy themselves when the odds were in their favour, particularly numerical odds. It was past six o'clock now and the sun was getting lower in the west, the direction I was travelling in. If I were going to be attacked from the sun, then it would be a head on attack. I felt fairly secure from behind, provided I kept doing steep turns.

I could see a single Spitfire in front of me and a little lower. It must be Ferdie, I thought at once, and chased after it to catch it up. It would be nice to go back to base together. When I got closer to it I noticed a white stream of Glycol coming away from the underneath. There wasn't very much but it was enough to tell me that the machine had been hit in the radiator. It seemed to be going down on a straight course in a shallow dive. I got to within about three hundred yards of it and called up Ferdie to ask his position, feeling that he would be sure to tell me if he had been hit in the radiator, although he might not have wanted me to know in the first instance. I got no reply and for a second I became convinced that he had been attacked since I had last spoken to him. I opened up my throttle, although I ought to have been conserving my fuel. From the direct rear all Spitfires look exactly the same and I had to get up close to read the lettering. I came up on its port side and at a distance of about twenty yards. It wasn't Ferdie. I felt relief. It didn't belong to Maida Squadron at all. It was "G" for George and belonged to some totally different squadron. I made a mental note of the lettering for "Brains's" benefit. I closed in a bit to see what it was all about. The Glycol leak wasn't severe. I couldn't think what to make of it at all. Perhaps the pilot wasn't aware of the leak. Perhaps he had baled out already and the machine, as they have been known to, was carrying on alone, like the "Marie Celeste". Perhaps it was my imagination, an hallucination after the excitement and strain of the past hour. I came in very close to it as though I were in squadron formation and it no longer presented a mystery to me. The pilot was there, his head resting motionless against the side of the perspex hood. Where it was resting, and behind where it was resting, the perspex was coloured crimson. Now and then as the aircraft encountered a disturbance and bumped a little, the pilot's head moved forward and back again. The hood was slightly open at the front which gave me the impression that he had made an instinctive last minute bid to get out before he had died. The wind had blown into the cockpit and had blown the blood which must have gushed from his head, back along the entire length of the cockpit like scarlet rain. I became suddenly and painfully aware that I was being foolhardy to stay so close as this for a sudden reflex from the pilot, dead though he was, a sudden thrust of the rudder bar or a movement from the stick could hurl the aircraft at me. I swung out and left it. I didn't look back any more. Before I left it, it had started to dive more steeply, and the Glycol flowed more freely as the nose dipped and the speed increased.

I thanked God for many things as I flew back away from the din and noise of the battle through the cool and the peace of the evening across the New Forest and above Netley to base. I landed my machine at six-thirty, stepped out and went to the hut.

Brains was very much in evidence and busy collecting reports from different people. Most of the pilots had landed and Ferdie, I was glad to see, was among them. I gave my report to Brains and Ferdie checked it. I was granted two damaged aircraft and Ferdie got one confirmed and two damaged. There were still three of our pilots unaccounted for. P/O Watty was not down and Red two and Blue two were overdue. We were allowed up to the mess in parties of six at a time, for we were still on readiness until nine o'clock. Ferdie and I went together and discussed the events of the last hour or so. We had some supper and then went down to dispersal again to relieve the others. It was unlikely, I was told, that we should be scrambled again in any strength for it was getting late now and the Germans would hardly be likely to mount another large offensive as late as this.

Brains was still down in the hut and was spending most of his time at the telephone answering calls from Group Intelligence and making enquiries from other stations as to the possible fate of our own missing pilots. Eventually news came through that Watty was safe but had been shot down near Southampton on his way back to base. He had been attacked by two M.E. 109's in this area and his machine had been hit in the Glycol tank but he had managed to force-land. He was taken to the hospital there because the Medical Officer had found a rip in his tunic which, upon further investigation, had revealed that he had got some shrapnel of some sort into his arm. We heard later that Red two and Blue two had both been shot down and both of them had been killed. Blue two had gone down in flames in front of a M.E. 110 and Red two had pressed his attack too closely to a Heinkel 111 and had gone into it. Both of these were sergeant pilots.

The squadron, according to Brains's assessment, had accounted for eight confirmed aircraft, three probables and seven damaged. There was no further flying that day and we were released at nine o'clock. We went up to the mess as usual and after some drinks we got into our cars and left the camp. We were to rendezvous at The Sunray.

We got to The Sunray after five minutes or so. It wasn't far from the aerodrome and was tucked away at the end of a lane leading from the main Weymouth–Wareham road.

The Sunray was blacked out and it was pitch dark outside when we switched off our lights. We groped our way to the door which Chumley seemed able to find in some instinctive manner. He opened the front door calling to me "Switch your radar on Roger" and pulled aside a blanket which had been rigged up to act as a further precaution to prevent the light from escaping as the main door was opened. We got inside to find the others already drinking. Cocky seemed to be in the chair as Chumley and I came in and he called out "Lost again White Section – biggies coming up for both of you."

The Sunray was an old pub and full of atmosphere. The ceilings were low and oak beams ran the entire length of them. In between the beams, the ceiling itself was made of wood of the same colour. It seemed dark at first but there was a liberal amount of lamps, not on the ceiling itself but on the walls, and these gave a soft light that was distinctly cosy. There were tables of heavy oak around which were chairs made out of barrels, highly polished and each containing soft plushy cushions. Around the walls ran an almost continuous cushion-covered bench, and the windows, from what I could see of them, for they were heavily curtained, were made of bottle-glass and were only translucent. The serving bar in the middle of the room was round and from it hung a varied assortment of brilliantly polished copper and brass ornaments. There were roses in copper vases standing on some of the tables and a bowl or two on the bar itself. There were sandwiches beneath glass cases and sausage-rolls as well. The visible atmosphere in the room was cloudy with tobacco smoke which seemed to reach its optimum height a foot or so from the ceiling where it appeared to flatten out and drift in horizontal layers until someone passed through it and then it appeared to follow whoever did so for a moment. There was a wireless somewhere in the room, for I could hear music coming from near where I was standing.

I was by the bar with the others and I had finished my third pint of bitter and was talking to Cocky. The night was quite early yet and Bottle was standing up at the bar with Dimmy, Chumley and Pete; they were all laughing at the top of their voices and a bit further along was Ferdie listening to what might, I think, have been a rather long-drawn-out story from one of the sergeant pilots, while two others seemed impatiently trying to get him to the point. Ferdie seemed to be quite amused at the process. There were two of our Polish pilots here too, both non-commissioned and their names were so difficult to pronounce that we simply called them "Zig" and "Zag". They didn't seem to take any offence at this abbreviation. They were excellent pilots, both of them.

The wireless now started to play the theme of Tchaikovsky's "Swan Lake" ballet and when I'd got my sixth pint I mentally detached myself from the rest for a moment.

"Wotcher Roger, mine's a pint of black and tan – have one yourself." I was jolted back to reality by this, accompanied by a hearty slap on the back from Ferdie, who had wormed his way across to me.

I had my seventh pint with Ferdie and we both edged up closer to the bar where the main body of the squadron seemed to have congregated. It was Cocky who, high spirited and irrepressible as ever, said "Come on boys, we've had this – next stop The Crown." We picked up our caps and made for the door. "Mind the light," someone shouted as the protective blanket was thrust aside for a moment. The air outside was cold and it hit me like a cold shower for a brief second while I gathered my wits. Chumley piled into the passenger seat. I was feeling perhaps a little too self-confident after the drinks but I felt sure I would make it somehow.

We got on to the main road again and Chumley directed us to The Crown in Weymouth. The road was fairly free of traffic and I gave the little car full rein for a while. It was dark and just in front of me there seemed to be an even darker but obscure sort of shape which I found difficulty in identifying for a moment. "For Christ's sake, man," Chumley shouted. Cocky's large Humber had pulled up on the verge and its occupants were busy relieving themselves by the roadside, but one of them was standing in front of the rear light and obscuring it. We were travelling at not much less than seventy-five m.p.h. when Chumley shouted at me and the Humber was only about thirty yards from us when I recognised it. My slow-wittedness only now became evident but I felt quite confident and in complete control of my faculties as I faced the emergency. I pulled the wheel over to the right, not abruptly but absolutely surely and with a calculated pressure to allow me only inches, inches enough to guide the left mud-guard past the Humber's off rear bumper. At the time I was in full control and thinking how fine and assured were my reactions, how much finer they were now than they ever were when I had had no drink. The sense of complete infallibility and the consequent denial of any risk had overtaken me and the feeling, if anything, became accentuated when the little car had passed Cocky's large Humber, which it did by the barest fraction of an inch, to the accompanying shout of "Look out, 109's behind" from those who were standing by the verge and otherwise engaged. "No road sense, those boys," Chumley remarked.

The Crown was quite a different sort of place from The Sunray. From the outside it was distinctly unpretentious in appearance, just a flat-sided building flanking the back street down by the harbour. It had four windows, two top and bottom and a door in the middle. We went in, and as I had rather expected, it was an ordinary working-man's pub. There were no furnishings to speak of, the floor was just plain wooden boards and the few tables were round with marble tops and the conventional china ash-trays advertising some type of lager or whisky. The bar occupied the whole one side of the room and the barman greeted us warmly as we arrived. Chumley ordered two pints of bitter. Apparently the squadron were well-known and held in high esteem.

The others arrived soon after we got there and the drinks were on me this time. There was a dart-board in the corner of the room and, not surprisingly, we threw badly. What did it matter how we played, I thought, as long as we let off some steam.

When we left The Crown at closing time I was drunk, but we didn't return to the aerodrome. Bottle had some friends in Bournemouth and it was to Bournemouth that he'd decided to go. I was too drunk to drive and so was Chumley, who had left The Crown before closing time and taken up his position in the passenger seat of my car where he was now fast asleep. Dimmy and I lifted him out, still asleep, into the back of Cocky's Humber. Dimmy, who, so he claimed, was more sober than I, said he would drive my car. I made no protest. I relapsed into the passenger seat and fell asleep as the car gathered speed towards Bournemouth. I woke up as soon as the car came to a standstill, feeling a lot more sober. It was about half-past eleven when we went through the door of this quite large private house. Bottle's and Cocky's car had already arrived and the occupants had apparently gone inside. The door opened and a girl greeted us. "I'm Pam, come on in, the others are here," she said. Everyone was seated in or on some sort of chair or stool and all had a glass of some sort in their hand. There were two other girls there besides Pam.

I was beginning to feel rather tired about this time and I

would have been glad to get back to camp, especially as I had to be on dawn readiness again. The atmosphere here didn't seem conducive to any sort of rowdery like The Crown or The Compass and the girls didn't somehow seem to fit into the picture. They weren't on the same wave-length. It was about two-thirty in the morning when we finally left.

We arrived back at the mess just after four o'clock, having stopped at an all-night café for eggs and bacon and coffee. I had to be on readiness at five-thirty and it seemed hardly worthwhile going to bed, so I decided to go straight down to dispersal, to find I was the only one there. I had just an hour and a half's sleep before I was due to take-off on dawn patrol.

Cornelius Tacitus

Boudicca's Revolt

The uprising against Roman rule led by Britain's most famous female warrior occurred in AD61.

Prasutagus, the late king of the Icenians, in the course of a long reign had amassed considerable wealth. By his will he left the whole of his two daughters and the emperor in equal shares, conceiving, by that stroke of policy, that he should provide at once for the tranquility of his kingdom and his family.

The event was otherwise. His dominions were ravaged by the centurions; the slaves pillaged his house, and his effects were seized as lawful plunder. His wife, Boudicca, was disgraced with cruel stripes; her daughters were ravished, and the most illustrious of the Icenians were, by force, deprived of the positions which had been transmitted to them by their ancestors. The whole country was considered as a legacy bequeathed to the plunderers. The relations of the deceased king were reduced to slavery.

Exasperated by their acts of violence, and dreading worse calamities, the Icenians had recourse to arms. The Trinobantians joined in the revolt. The neighboring states, not as yet taught to crouch in bondage, pledged themselves, in secret councils, to stand forth in the cause of liberty. What chiefly fired their indignation was the conduct of the veterans, lately planted as a colony at Camulodunum. These men treated the Britons with cruelty and oppression; they drove the natives from their habitations, and calling them by the (shameful) names of slaves and captives, added insult to their tyranny. In these acts of oppression, the veterans were supported by the common soldiers; a set of men, by their habits of life, trained to licentiousness, and, in their turn, expecting to reap the same advantages. The temple built in honour of Claudius was another cause of discontent. In the eye of the Britons it seemed the citadel of eternal slavery. The priests, appointed to officiate at the altars, with a pretended zeal for religion, devoured the whole substance of the country. To over-run a colony, which lay quite naked and exposed, without a single fortification to defend it, did not appear to the incensed and angry Britons an enterprise that threatened either danger or difficulty. The fact was, the Roman generals attended to improvements to taste and elegance, but neglected the useful. They embellished the province, and took no care to defend it.

While the Britons were preparing to throw off the yoke, the statue of victory, erected at Camulodunum, fell from its base, without any apparent cause, and lay extended on the ground with its face averted, as if the goddess yielded to the enemies of Rome. Women in restless ecstasy rushed among the people, and with frantic screams denounced impending ruin. In the council-chamber of the Romans hideous clamours were heard in a foreign accent; savage howlings filled the theatre, and near the mouth of the Thames the image of a colony in ruins was seen in the transparent water; the sea was purpled with blood, and, at the tide of ebb, the figures of human bodies were traced in the sand. By these appearances the Romans were sunk in despair, while the Britons anticipated a glorious victory. Suetonius, in the meantime, was detained in the isle of Mona. In this alarming crisis, the veterans sent to Catus Decianus, the procurator of the province, for a reinforcement. Two hundred men, and those not completely armed, were all that officer could spare. The colony had but a handful of soldiers. Their temple was strongly fortified, and there they hoped to make a stand. But even for the defense of that place no measures were concerted. Secret enemies mixed in all their deliberations. No fosse was made; no palisade thrown up; nor were the women, and such as were disabled by age or infirmity, sent out of the garrison. Unguarded and unprepared, they were taken by surprise, and, in the moment of profound peace, overpowered by the Barbarians in one general assault. The colony was laid waste with fire and sword.

The temple held out, but, after a siege of two days, was taken by storm. Petilius Cerealis, who commanded the ninth legion, marched to the relief of the place. The Britons, flushed with success, advanced to give him battle. The legion was put to the rout, and the infantry cut to pieces. Cerealis escaped with the cavalry to his entrenchments. Catus Decianus, the procurator of the province, alarmed at the scene of carnage which he beheld on every side, and further dreading the indignation of a people, whom by rapine and oppression he had driven to despair, betook himself to flight, and crossed over into Gaul.

Suetonius, undismayed by this disaster, marched through the heart of the country as far as London; a place not dignified with the name of a colony, but the chief residence of merchants, and the great mart of trade and commerce. At that place he meant to fix the feat of war; but reflecting on the scanty numbers of his little army, and the fatal rashness of Cerealis, he resolved to quit the station, and, by giving up one post, secure the rest of the province. Neither supplications, nor the tears of the inhabitants could induce him to change his plan. The signal for the march was given. All who chose to follow his banners were taken under his protection. Of all who, on account of their advanced age, the weakness of their sex, of the attractions of the situation, thought proper to remain behind, not one escaped the rage of the Barbarians. The inhabitants of Verulamium, a municipal town, were in like manner put to the sword. The genius of a savage people leads them always in quest of plunder; and, accordingly, the Britons left behind them all places of strength. Wherever they expected feeble resistance, and considerable booty, there they were sure to attack with the fiercest rage. Military skill was not the talent of Barbarians. The number massacred in the places which have been mentioned, amounted to no less than seventy thousand, all citizens or allies of Rome. To make prisoners, and reserve them for slavery, or to exchange them, was not in the idea of a people, who despised all the laws of war. The halter and the gibbet, slaughter and defoliation, fire and sword, were the marks of savage valour. Aware that vengeance would overtake them, they were resolved to make sure of their revenge, and glut themselves with the blood of their enemies.

The fourteenth legion, with the veterans of the twentieth, and the auxiliaries from the adjacent stations, having joined Suetonius, his army amounted to little less than ten thousand men. Thus reinforced, he resolved, without loss of time, to bring on a decisive action. For this purpose he chose a spot encircled with woods, narrow at the entrance, and sheltered in the rear by a thick forest. In that situation he had no fear of an ambush. The enemy, he knew, had no approach but in front. An open plain lay before him. He drew up his men in the following order: the legions in close array formed the center, the light armed troops were stationed at hand to serve as occasion might require: the cavalry took post in the wings. The Britons brought into the field an incredible multitude. They formed no regular line of battle. Detached parties and loose battalions displayed their numbers, in frantic transport bounding with exultation, and so sure of victory, that they placed their wives in wagons at the extremity of the plain, where they might survey the scene of action, and behold the wonders of British valour.

Boudicca, in a chariot, with her two daughters before her, drove through the ranks. She harangued the different nations in their turn: "This," she said, "is not the first time that the Britons have been led to battle by a woman." But now she did not come to boast the pride of a long line of ancestry, nor even to recover her kingdom and the plundered wealth of her family. She took the field, like the meanest among them, to assert the cause of public liberty, and to seek revenge for her body seamed with ignominious stripes, and her two daughters infamously ravished. "From the pride and arrogance of the Romans nothing is sacred; all are subject to violation; the old endure the scourge, and the virgins are deflowered. But the vindictive gods are now at hand. A Roman legion dared to face the warlike Britons: with their lives they paid for their rashness; those who survived the carnage of that day, lie poorly hid behind their entrenchments, meditating nothing but how to save themselves by an ignominious flight. From the din of preparation, and the shouts of the British army, the Romans, even now, shrink back with terror. What will be their case when the assault begins? Look round, and view your numbers. Behold the proud display of warlike spirits, and consider the motives for which we draw the avenging sword. On this spot we must either conquer, or die with glory. There is no alternative. Though a woman, my resolution is fixed: the men, if they please, may survive with infamy, and live in bondage."

Suetonius, in a moment of such importance, did not remain silent. He expected every thing from the valour of his men, and yet urged every topic that could inspire and animate them to the attack. "Despise," he said, "the savage uproar, the yells and shouts of undisciplined Barbarians. In that mixed multitude, the women out-number the men. Void of spirit, unprovided with arms, they are not soldiers who come to offer battle; they are bastards, runaways, the refuse of your swords, who have often fled before you, and will again betake themselves to flight when they see the conqueror flaming in the ranks of war. In all engagements it is the valour of a few that turns the fortune of the day. It will be your immortal glory, that with a scanty number you can equal the exploits of a great and powerful army. Keep your ranks; discharge your javelins; rush forward to a close attack; bear down all with your bucklers, and hew a passage with your swords. Pursue the vanquished, and never think of spoil and plunder. Conquer, and victory gives you everything."

This speech was received with warlike acclamations. The soldiers burned with impatience for the onset, the veterans brandished their javelins, and the ranks displayed such an intrepid countenance, that Suetonius, anticipating the victory, gave the signal for the charge.

The engagement began. The Roman legion presented a close embodied line. The narrow defile gave them the shelter of a rampart. The Britons advanced with ferocity, and discharged their darts at random. In that instant, the Romans rushed forward in the form of a wedge. The auxiliaries followed with equal ardour. The cavalry, at the same time, bore down upon the enemy, and, with their pikes, overpowered all who dared to make a stand. The Britons betook themselves to flight, but their waggons in the rear obstructed their passage. A dreadful slaughter followed. Neither sex nor age was spared. The cattle, falling in one promiscuous carnage, added to the heaps of slain. The glory of the day was equal to the most splendid victory of ancient times. According to some writers, not less than eighty thousand Britons were put to the sword. The Romans lost about four hundred men, and the wounded did not exceed that number. Boudicca, by a dose of poison, ended her life. Poenius Postumius, the Prefect in the camp of the second legion, as soon as he heard of the brave exploits of the fourteenth and twentieth legions, felt the disgrace of having, in disobedience to the orders of his general, robbed the soldiers under his command of their share in so complete a victory. Stung with remorse, he fell upon his sword, and expired on the spot.

Plutarch

Alexander's Expedition Against Darius

Inheriting the throne of Macedonia in 336, Alexander set out on a great trail of conquest that would reach to India. Darius III, the King of Persia, was amongst the earliest of Alexander's hapless opponents.

ALEXANDER was but twenty years old when his father was murdered, and succeeded to a kingdom, beset on all sides with great dangers and rancorous enemies. For not only the barbarous nations that bordered on Macedonia were impatient of being governed by any but their own native princes, but Philip like-wise, though he had been víctorious over the Grecians, yet, as the time had not been sufficient for him to complete his conquest and accustom them to his sway, had simply left all things in a general disorder and confusion. It seemed to the Macedonians a very critical time; and some would have persuaded Alexander to give up all thought of retaining the Grecians in subjection by force of arms, and rather to apply himself to win back by gentle means the allegiance of the tribes who were designing revolt, and try the effect of indulgence in arresting the first motions towards revolution. But he rejected this counsel as weak and timorous, and looked upon it to be more prudent to secure himself by resolution and magnanimity, than, by seeming to truckle to any, to encourage all to trample on him. In pursuit of this opinion, he reduced the barbarians to tranquillity, and put an end to all fear of war from them, by a rapid expedition into their country as far as the river Danube, where he gave Syrmus, King of the Triballians, an entire overthrow. And hearing the Thebans were in revolt, and the Athenians in correspondence with them, he immediately marched through the pass of Thermopylæ, saying that to Demosthenes, who had called him a child while he was in Illyria and in the country of the Triballians, and a youth when he was in Thessaly, he would appear a man before the walls of Athens.

When he came to Thebes, to show how willing he was to accept of their repentance for what was past, he only demanded of them Phœnix and Prothytes, the authors of the rebellion, and proclaimed a general pardon to those who would come over to him. But when the Thebans merely retorted by demanding Philotas and Antipater to be delivered into their hands, and by a proclamation on their part invited all who would assert the liberty of Greece to come over to them, he presently applied himself to make them feel the last extremities of war. The Thebans indeed defended themselves with a zeal and courage beyond their strength, being much outnumbered by their enemies. But when the Macedonian garrison sallied out upon them from the citadel, they were so hemmed in on all sides that the greater part of them fell in the battle; the city itself being taken by storm, was sacked and razed. Alexander's hope being that so severe an example might terrify the rest of Greece into obedience, and also in order to gratify the hostility of his confederates, the Phocians and Platæans. So that, except the priests, and some few who had heretofore been the friends and connections of the Mace-donians, the family of the poet Pindar, and those who were known to have opposed the public vote for the war, all the rest, to the number of thirty thousand, were publicly sold for slaves; and it is computed that upwards of six thousand were put to the sword.

Among the other calamities that befell the city, it happened that some Thracian soldiers, having broken into the house of a matron of high character and repute, named Timoclea, their captain, after he had used violence with her, to satisfy his avarice as well as lust, asked her, if she knew of any money concealed; to which she readily answered she did, and bade him follow her into a garden, where she showed him a well, into which, she told him, upon the taking of the city, she had thrown what she had of most value. The greedy Thracian presently stooping down to view the place where the thought the treasure lay, she came behind him and pushed him into the well, and then flung great stones in upon him, till she had killed him. After which, when the soldiers led her away bound to Alexander, her very mien and gait showed her to be a woman of dignity, and of a mind no less elevated, not betraying the least sign of fear or astonishment. And when the king asked her who she was, "I am," said she, "the sister of Theagenes, who fought the battle of Chæronea with your father Philip, and fell there in command for the liberty of Greece." Alexander was so surprised, both at what she had done and what she said, that he could not choose but give her and her children their freedom to go whither they pleased.

After this he received the Athenians into favour, although they had shown themselves so much concerned at the calamity of Thebes that out of sorrow they omitted the celebration of the Mysteries, and entertained those who escaped with all possible humanity. Whether it were, like the lion, that his passion was now satisfied, or that, after an example of extreme cruelty, he had a mind to appear merciful, it happened well for the Athenians; for he not only forgave them all past offences, but bade them look to their affairs with vigilance, remembering that if he should miscarry, they were likely to be the arbiters of Greece. Certain it is, too, that in aftertime he often repented of his severity to the Thebans, and his remorse had such influence on his temper as to make him ever after less rigorous to all others. He imputed also the murder of Clitus, which he committed in his wine, and the unwillingness of the Macedonians to follow him against the Indians, by which his enterprise and glory was left imperfect to the wrath and vengeance of Bacchus, the protector of Thebes. And it was observed that whatsoever any Theban, who had the good fortune to survive this victory, asked of him, he was sure to grant without the least difficulty.

Soon after, the Grecians, being assembled at the Isthmus, declared their resolution of joining with Alexander in the war against the Persians, and proclaimed him their general. While he stayed here, many public ministers and philosophers came from all parts to visit him and congratulated him on his election, but contrary to his expectation, Diogenes of Sinope, who then was living at Corinth, thought so little of him, that instead of coming to compliment him, he never so much as stirred out of the suburb called the Cranium, where Alexander found him lying alone in the sun. When he saw so much company near him, he raised himself a little, and vouchsafed to look upon Alexander; and when he kindly asked him whether he wanted anything, "Yes," said he, "I would have you stand from between me and the sun." Alexander was so struck at this answer, and surprised at the greatness of the man, who had taken so little notice of him, that as he went away he told his followers, who were laughing at the moroseness of the philosopher, that if he were not Alexander, he would choose to be Diogenes.

Then he went to Delphi, to consult Appolo concerning the success of the war he had undertaken, and happening to come on one of the forbidden days, when it was esteemed improper to give any answer from the oracle, he sent messengers to desire the priestess to do her office; and when she refused, on the plea of a law to the contrary, he went up himself, and began to draw her by force into the temple, until tired and overcome with his importunity, "My son," said she, "thou art invincible." Alexander taking hold of what she spoke, declared he had received such an answer as he wished for, and that it was needless to consult the god any further. Among other prodigies that attended the departure of his army, the image of Orpheus at Libethra, made of cypress-wood, was seen to sweat in great abundance, to the discouragement of many. But Aristander told him that, far from presaging any ill to him, it signified he should perform acts so important and glorious as would make the poets and musicians of future ages labour and sweat to describe and celebrate them.

His army, by their computation who make the smallest amount, consisted of thirty thousand foot and four thousand horse; and those who make the most of it, speak but of forty-three thousand foot and three thousand horse. Aristobulus says, he had not a fund of above seventy talents for their pay, nor had he more than thirty days' provision, if we may believe Duris; Onesicritus tells us he was two hundred talents in debt. However narrow and disproportionable the beginnings of so vast an undertaking might seem to be, yet he would not embark his army until he had informed himself particularly what means his friends had to enable them to follow him, and supplied what they wanted, by giving good farms to some, a village to one, and the revenue of some hamlet or harbour-town to another. So that at last he had portioned out or engaged almost all the royal property; which giving Perdiccas an occasion to ask him what he would leave himself, he replied, his hopes. "Your soldiers," replied Perdiccas, "will be your partners in those," and refused to accept of the estate he had assigned him. Some others of his friends did the like, but to those who willingly received or desired assistance of him, he liberally granted it, as far as his patrimony in Macedonia would reach, the most part of which was spent in these donations.

With such vigorous resolutions, and his mind thus disposed, he passed the Hellespont and at Troy sacrificed to Minerva, and honoured the memory of the heroes who were buried there, with solemn libations; especially Achilles, whose gravestone he anointed, and with his friends, as the ancient custom is, ran naked about his sepulchre, and crowned it with garlands, declaring how happy he esteemed him, in having while he lived so faithful a friend, and when he was dead, so famous a poet to proclaim his actions. While he was viewing the rest of the antiquities and curiosities of the place, being told he might see Paris's harp, if he pleased, he said he thought it not worth looking on but he should be glad to see that of Achilles, to which he used to sing the glories and great actions of brave men.

In the meantime, Darius's captains, having collected large forces, were encamped on the further bank of the river Granicus, and it was necessary to fight, as it were, in the gate of Asia for an entrance into it. The depth of the river, with the unevenness and difficult ascent of the opposite bank, which was to be gained by main force, was apprehended by most, and some pronounced it an improper time to engage, because it was unusual for the kings of Macedonia to march with their forces in the month called Dæsius. But Alexander broke through these scruples, telling them they should call it a second Artemisius. And when Parmenio advised him not to attempt anything that day, because it was late, he told him that he should disgrace the Hellespont should he fear the Granicus. And so, without more saying, he immediately took the river with thirteen troops of horse, and advanced against whole showers of darts thrown from the steep opposite side, which was covered with armed multitudes of the enemy's horse and foot, notwithstanding the disadvantage of the ground and the rapidity of the stream; so that the action seemed to have more frenzy and desperation in it, than of prudent conduct. However, he persisted obstinately to gain the passage, and at last with much ado making his way up the banks, which were extremely muddy and slippery, he had instantly to join in a mere confused hand-to-hand combat with the enemy, before he could draw up his men, who were still passing over, into any order. For the enemy pressed upon him with loud and warlike outcries; and charging horse against horse, with their lances, after they had broken and spent these, they fell to it with their swords. And Alexander, being easily known by his buckler, and a large plume of white feathers on each side of his helmet, was attacked on all sides, yet escaped wounding, though his cuirass was pierced by a javelin in one of the joinings. And Rhœsaces and Spithridates, two Persian commanders, falling upon him at once, he avoided one of them, and struck at Rhœsaces, who had a good cuirass on, with such force that, his spear breaking in his hand, he was glad to betake himself to his dagger. While they were thus engaged, Spithridates came up on one side of him, and raising himself upon his horse, gave him such a blow with his battle-axe on the helmet that he cut off the crest of it, with one of his plumes, and the helmet was only just so far strong enough to save him, that the edge of the weapon touched the hair of his head. But as he was about to repeat his stroke, Clitus, called the black Clitus, prevented him, by running him through the body with his spear. At the same time Alexander despatched Rhœsaces with his sword. While the horse were thus dangerously engaged, the Macedonian phalanx passed the river, and the foot on each side advanced to fight. But the enemy hardly sustaining the first onset, soon gave ground and fled, all but the mercenary Greeks, who, making a stand upon a rising ground, desired quarter, which Alexander, guided rather by passion than judgment, refused to grant, and charging them himself first, had his horse (not Bucephalus, but another) killed under him. And this obstinacy of his cut off these experienced desperate men and cost him the lives of more of his own soldiers than all the battle before, besides those who were wounded. The Persians lost in this battle twenty thousand foot and two thousand five hundred horse. On Alexander's side, Aristobulus says there were not wanting above four-and-thirty, of whom nine were foot-soldiers; and in memory of them he caused so many statues of brass, of Lysippus's making, to be erected. And that the Grecians might participate in the honour of his victory he sent a portion of the spoils home to them, particularly to the Athenians three hundred bucklers and upon all the rest he ordered this inscription to be set: "Alexander the son of Philip, and the Grecians, except the Lacedæmonians, won these from the barbarians who inhabit Asia." All the plate and purple garments, and other things of the same kind that he took from the Persians, except a very small quantity which he reserved for himself, he sent as a present to his mother.

This battle presently made a great change of affairs to Alexander's advantage. For Sardis itself, the chief seat of the barbarians, power in the maritime provinces, and many other considerable places, were surrendered to him; only Halicarnassus and Miletus stood out, which he took by force, together with the territory about them. After which he was a little unsettled in his opinion how to proceed. Sometimes he thought it best to find out Darius as soon as he could, and put all to the hazard of a battle; another while he looked up it as a more prudent course to make an entire reduction of the sea-coast, and not to seek the enemy till he had first exercised his power here and made himself secure of the resources of these provinces. While he was thus deliberating what to do, it happened that a spring of water near the city of Xanthus in Lycia, of its own accord, swelled over its banks, and threw up a copper plate, upon the margin of which was engraven in ancient characters, that the time would come when the Persian empire should be destroyed by the Grecians. Encouraged by this accident, he proceeded to reduce the maritime parts of Cilicia and Phœnicia, and passed his army along the sea-coasts of Pamphylia with such expedition that many historians have described and extolled it with that height of admiration, as if it were no less than a miracle, and an extraordinary effect of divine favour, that the waves which usually come rolling in violently from the main, and hardly ever leave so much as a narrow beach under the steep, broken cliffs at any time uncovered, should on a sudden retire to afford him passage. Menander, in one of his comedies, alludes to this marvel when he says–

> Was Alexander ever favoured more?
> Each man I wish for meets me at my door,
> And should I ask for passage through the sea,
> The sea I doubt not would retire for me.

But Alexander himself in his epistles mentions nothing unusual in this at all, but says he went from Phaselis, and passed through what they call the Ladders. At Phaselis he stayed some time, and finding the statue of Theodectes, who was a native of this town and was now dead, erected in the market-place, after he had supped, having drunk pretty plentifully, he went and danced about it and crowned it with garlands, honouring not ungracefully, in his sport, the memory of a philosopher whose conversation he had formerly enjoyed when he was Aristotle's scholar.

Then he subdued the Pisidians who made head against him, and conquered the Phyrgians, at whose chief city, Gordium, which is said to be the seat of the ancient Midas, he saw the famous chariot fastened with cords made of the rind of the cornel-tree, which whosoever should untie, the inhabitants had a tradition, that for him was reserved the empire of the world. Most authors tell the story that Alexander finding himself unable to untie the knot, the ends of which were secretly twisted round and folded up within it, cut it asunder with his sword. But Aristobulus tells us it was easy for him to undo it, by only pulling the pin out of the pole, to which the yoke was tied, and afterwards drawing off the yoke itself from below. From hence he advanced into Paphlagonia and Cappadocia, both which countries he soon reduced to obedience, and then hearing of the death of Memnon, the best commander Darius had upon the sea-coasts, who, if he had lived, might, it was supposed, have put many impediments and difficulties in the way of the progress of his arms, he was the rather encouraged to carry the war into the upper provinces of Asia.

Darius was by this time upon his march from Susa, very confident, not only in the number of his men, which amounted to six hundred thousand, but likewise in a dream, which the Persian soothsayers interpreted rather in flattery to him than according to the natural probability. He dreamed that he saw the Macedonian phalanx all on fire, and Alexander waiting on him, clad in the same dress which he himself had been used to wear when he was courier to the late king; after which, going into the temple of Belus, he vanished out of his sight. The dream would appear to have supernaturally signified to him the illustrious actions the Macedonians were to perform, and that as he, from a courier's place, had risen to the throne, so Alexander should come to be master of Asia, and not long surviving his conquests, conclude his life with glory. Darius's confidence increased the more, because Alexander spent so much time in Cilicia, which he imputed to his cowardice. But it was sickness that detained him there, which some say he contracted from his fatigues, others from bathing in the river Cydnus, whose waters were exceedingly cold. However it happened none of his physicians would venture to give him any remedies, they thought his case so desperate, and were so afraid of the suspicions and ill-will of the Macedonians if they should fail in the cure; till Philip, the Acarnanian, seeing how critical his case was, but relying on his own well-known friendship for him, resolved to try the last efforts of his art, and rather hazard his own credit and life than suffer him to perish for want of physic, which he confidently administered to him, encouraging him to take it boldly, if he desired a speedy recovery, in order to prosecute the war. At this very time, Parmenio wrote to Alexander from the camp, bidding him have a care of Philip, as one who was bribed by Darius to kill him, with great sums of money, and a promise of his daughter in marriage. When he had perused the letter, he put it under his pillow, without showing it so much as to any of his most intimate friends, and when Philip came in with the potion, he took it with great cheerfulness and assurance, giving him meantime the letter to read. This was a spectacle well worth being present at, to see Alexander take the draught and Philip read the letter at the same time, and then turn and look upon one another, but with different sentiments; for Alexander's looks were cheerful and open, to show his kindness to and confidence in his physician, while the other was full of surprise and alarm at the accusation, appealing to the gods to witness his innocence, sometimes lifting up his hands to heaven, and then throwing himself down by the bedside, and beseeching Alexander to lay aside all fear, and follow his directions without apprehension. For the medicine at first worked so strongly as to drive, so to say, the vital forces into the interior; he lost his speech, and falling into a swoon, had scarce any sense or pulse left. However, in no long time, by Philip's means, his health and strength returned, and he showed himself in public to the Macedonians, who were in continual fear and dejection until they saw him abroad again.

There was at this time in Darius's army a Macedonian refugee, named Amyntas, one who was pretty well acquainted with Alexander's character. This man, when he saw Darius intended to fall upon the enemy in the passes and defiles, advised him earnestly to keep where he was, in the open and extensive plains, it being the advantage of a numerous army to have field-room enough when it engages with a lesser force. Darius, instead of taking his counsel, told him he was afraid the enemy would endeavour to run away, and so Alexander would escape out of his hands. "That fear," replied Amyntas, "is needless, for assure yourself that far from avoiding you, he will make all the speed he can to meet you, and is now most likely on his march toward you." But Amyntas's counsel was to no purpose, for Darius immediately decamping, marched into Cilicia at the same time that Alexander advanced into Syria to meet him; and missing one another in the night, they both turned back again. Alexander, greatly pleased with the event, made all the haste he could to fight in the defiles, and Darius to recover his former ground, and draw his army out of so disadvantageous a place. For now he began to perceive his error in engaging himself too far in a country in which the sea, the mountains, and the river Pinarus running through the midst of it would necessitate him to divide his forces, render his horse almost unserviceable, and only cover and support the weakness of the enemy. Fortune was not kinder to Alexander in the choice of the ground, than he was careful to improve it to his advantage. For being much inferior in numbers, so far from allowing himself to be outflanked, he stretched his right wing much further out than the left wing of his enemies, and fighting there himself in the very foremost ranks, put the barbarians to flight. In this battle he was wounded in the thigh, Chares says, by Darius, with whom he fought hand to hand. But in the account which he gave Antipater of the battle, though indeed he owns he was wounded in the thigh with a sword, though not dangerously, yet he takes no notice who it was that wounded him.

Nothing was wanting to complete this victory, in which he overthrew above an hundred and ten thousand of his enemies, but the taking of the person of Darius, who escaped very narrowly by flight. However, having taken his chariot and his bow, he returned from pursuing him, and found his own men busy in pillaging the barbarians' camp, which (though to disburden themselves they had left most of their baggage at Damascus) was exceedingly rich. But Darius's tent, which was full of splended furniture and quantities of gold and silver, they reserved for Alexander himself, who, after he had put off his arms, went to bathe himself, saying, "Let us now cleanse ourselves from the toils of war in the bath of Darius." "Not so," replied one of his followers, "but in Alexander's rather; for the property of the conquered is and should be called the conqueror's." Here, when he beheld the bathing vessels, the water-pots, the pans, and the ointment boxes, all of gold curiously wrought, and smelt the fragrant odours with which the whole place was exquisitely perfumed, and from thence passed into a pavilion of great size and height, where the couches and tables and preparations for an entertainment were perfectly magnificent, he turned to those about him and said, "This, it seems, is royalty."

Geoffrey Evans

Admin Box

In February 1944, the Japanese launched the Ha-Go offensive against British forces in Arakan, and attacked among other objectives, the vital Corps Adminstrative Area at Sinzweya. Sinzweya – or "Admin Box" as it came to be known – was an unprepossessing place to make a stand: a 1,200 yard-square clearing ringed by jungle hills from which the Japanese could enfilade at will. Brigadier Geoffrey Evans led the defence of the Admin Box.

THE first Japanese attack was not long in coming.

At 3 p.m. a number of Zero fighters flew in low over the clearing and strafed it heavily. There were a number of casualties. Two mules, driven frenzied by the aerial attack, broke loose and were killed close to my headquarters. As there was nowhere to bury them, we suffered from this as the days went by and decomposition set in; the stench was quite frightful. Luckily there was a bulldozer at hand, and eventually the problem was partly solved by building a mound of earth over the bodies. This did not entirely remove the unpleasantness, but it reduced it. At 5 p.m. unit commanders came to Box Headquarters. There was little that I could say except that the Tanks and the West Yorkshires were there to back them in the fight. "Your job," I said, "is to stay put and keep the Japanese out. Hold your fire and conserve ammunition. Wait till you see the yellow of their eyes before you shoot. Make sure one round means one dead Jap."

At 7 p.m. it was dark and the first Japanese ground attack came in on the north-west side where the mule company was defending the perimeter. The attack was extremely noisy; tracer bullets pierced the darkness as they flew all over the place in a brilliant pyrotechnic display. There was a great deal of shouting by the Japs. But the attack was not pressed home and after about an hour it died down.

The rest of the night was fairly quiet. The second night we were not to be so fortunate. That was a ghastly affair.

The second day in the Box – February 7 – started badly.

Patrols going up the Ngakyedauk Pass were ambushed by the enemy. The Japanese had cut the pass. Now we on the east of the Mayu Range were cut off from the 5th Indian Division to the west. Not only were we cut off, we were surrounded. The first part of the Japanese plan had succeeded.

Frank Messervy had ordered 4/8 Gurkha Rifles, part of 89 Brigade, to come into the Box. To them I allotted that part of the perimeter between what became known as the Eastern Gate and Point 315. This meant that we could now hold part of the open side for which there had, before, been no troops available.

Before the Gurkhas had time to take up their positions the Japanese launched a furious attack. By early the following morning the Gurkhas had been forced back through the jungle to the open ground of the Box itself. A serious threat to this end of the Box had developed. A hastily organised counter-attack had been unsuccessful. Something had to be done quickly if the Japanese were not to establish themselves permanently.

The squadron of tanks in that part of the defended area came into action and their 75-millimetre guns pumped high explosive shells into the Japanese in the jungle. The enemy advance was held and this gave time for an organised counter-attack by the West Yorkshires to be planned. Munshi Cree and I were on the scene and I told him to counter-attack as soon as he could. Within a very short time Alec Dunlop and B Company were ready to go and in they went supported by the tanks. We both accompanied Company Headquarters, but after a short time, when the attack seemed to be going well, I felt that I should return to my own headquarters to see what other calamities had taken place in my absence.

Fortunately none had, and it was with great relief that I heard from Munshi, on his return about an hour and a half later, that the situation had been restored and the Gurkhas were firmly in position. We had managed to beat off the second attack, the most powerful and resolute so far.

The rest of the 7th February passed uneasily and night came. But it brought no peace. Not long after dark, about 8 p.m., a further attack started. This time from the south-west. It was preceded by shouts in Urdu from members of the "Japanese-inspired Fifth Column" or "JIFFS" for short. These were the followers of Subhas Chandra Bhose, a collaborator of the Japanese. They were often used when the two sides were in close contact in the hopes that they could persuade Indian soldiers to desert. In this they were entirely unsuccessful.

Bursts of rifle and automatic fire broke out about three hundred yards from Box Headquarters. These were accompanied by more shouting and then screams and cries for help. I heard a voice say in the darkness: "Good God, they've got into the hospital."

This was a dreadful thought, as there was very little that could be done until daylight. The defenders were a section of West Yorkshires and twenty walking wounded armed with rifles. It would be sheer folly to send in an infantry attack to drive out the enemy. Only the West Yorkshires were available and they did not know the geography of the hospital, nor would they be able to distinguish between friend and foe in the dark. To call on artillery and mortars was out of the question as our own men would be killed.

I got on the telephone to Munshi to send in a patrol to try and find out what was happening, but the carrier that was sent was driven off by grenades. Clearly the enemy were in greater strength than I had expected. We could only wait for the day to come and take what comfort we might from the fact that the wards, theatres and resuscitation installations were dug down. In the darkness the Japanese might miss them; otherwise, when they discovered that they had entered a hospital, they might take what prisoners they could and leave the remainder.

For some time the shooting and agonised cries went on and then silence reigned, to be broken now and again by one or two shots. Had the Japanese withdrawn? Had all the staff and patients been murdered? A little later the commander of the Field Ambulance arrived at my Command Post. He knew little of what was happening, but thought that only a small enemy party had forced their way into the hospital. He had been in the officers' ward attending to patients when his staff sergeant appeared to tell him that the Japanese had entered the hospital. He had doubted this could be the case, but when the staff sergeant was pointing out where he had seen the enemy, he was shot dead.

The full details of what had happened we learned later from

the pitifully small number of survivors of the thirty-six-hour ordeal, which was only ended when A Company of the West Yorkshires drove the enemy out. The Orderly Medical Officer, Lieutenant Basu of the Indian Army Medical Corps, was lying on a stretcher in the medical inspection room, resting, when a group of men rushed in from the dispensary at about 9 p.m. One man seized his arm, while another threatened him with a bayonet. "*Aiyo*," they ordered. "Come on."

Basu was taken to the Japanese commander, who was in the darkened officers' ward busily scribbling notes. Through an interpreter, the questioning began. "What is your name? What is the name of this place?"

The next question made it clear that the Japanese knew they were in a hospital, and therefore what followed was done quite deliberately and in cold blood. "How many patients are in this hospital?" they asked. "How many of them are British officers? What other personnel are there?"

Basu stalled as best he could. "I am here for three days only," he said. "I don't know much about the organisation here."

"What army units are posted here? How many of them are British and how many Indian?"

"I am a doctor," said Basu. "I am interested in medicine, not military tactics."

At this, one angry Japanese soldier brought the point of his bayonet against the doctor's head as though to finish him off unless he answered the questions more satisfactorily. But the Japanese commander cut in brusquely: "Show us to the telephone."

Basu thought quickly. "It would be dangerous to go near it," he said. "There is a machine-gun post just beside it."

"Then show my men to the operating theatre and the laboratory," ordered the Japanese officer.

Basu could not get out of this. But he took them to the small theatre which was used only for minor operations. As bad luck would have it, while the Japanese hunted round the theatre for anything they could lay their hands on in the way of bandages and cotton wool, they came on three of the hospital officers asleep. They were awakened and made prisoners. With Basu, they were taken back to the commander for further questioning. It was quite clear that the one piece of information the enemy were particularly anxious to discover was where the supply depot was situated. Basu kept up his pretence at ignorance: "I don't know; I tell you I'm only here for a few days. I've been too busy looking after patients to find time to visit the supply depot."

All the officers had their hands tied tightly behind their backs and they were taken to join the British and Indian other ranks, who had been trussed in a similar painful fashion.

The Japanese officer pulled out his sword, waved it over their heads threateningly, sheathed it and withdrew.

The purpose of this melodramatic gesture became clear almost at once. It was intended to cause the maximum alarm to the Indians as an inducement to make them desert to the Japanese to save their lives. The offer came almost immediately.

A man whose appearance suggested he was Indian was brought in.

"There is no need to worry," he assured the Indians. "No need at all to worry. You will be taken to Rangoon to join the Indian Independence League."

"What is your native country?" Basu asked him.

"I come from Maungdaw, in Burma," he answered.

Later that night, the five Indian medical officers were taken to the dispensary and ordered to pack up drugs for the Japanese. They were particularly interested in quinine, morphine, anti-tetanus and anti-gas-gangrene serum. They wanted to know what was in every phial. Those they did not want they threw to one side. Sally, in 9 Brigade Headquarters, could hear the bottles breaking.

When the Japanese had taken all they wanted, they drove the medical officers back to the watercourse where the rest of the captives were seated in acute discomfort because the bindings on their wrists were cutting into the flesh. Those who asked for water were jabbed with a heavy stick by a Japanese sentry.

The following morning a carrier pushed through the bushes towards the medical inspection room and one Japanese soldier dragged the trussed British other ranks out in front so that they should receive the first burst of fire. When the automatic gun of the carrier opened up, the prisoners desperately tried to find shelter, but they were handicapped because their hands were lashed behind their backs. Some of them were killed and others were wounded. The Japanese looked on grinning. They, too, had suffered casualties. They selected six of the Indian soldiers and ordered them to help to carry seven wounded Japanese.

The stretcher-bearers struggled southwards through the jungle the whole of that day. When night came, the Japanese cooked food for themselves but gave none to the Indians. Next morning, still having had nothing to eat, the Indians were given packs to carry and it was not until the early hours of the following morning that they reached their destination. This was the Buthidaung Tunnel, where there were about a thousand Japanese living. The tunnel was used as a store for arms, ammunition, guns and transport. During the day, six Indians came to see them. "You need not worry," they told them. "You will have trouble for a few days, but then you will be sent to Rangoon to work. Our major will fix things and you will not be tortured by the Japanese." One of their visitors, a Madrassi, said: "There are now 400,000 followers of Subhas Bhose. In two months we shall reach Chittagong."

Seven days later, the six Indian soldiers were taken out of the tunnel, ordered to strip off their clothes beside a sixty-foot-deep gorge and then were shot by a Japanese officer and five men. Four were killed and their bodies were kicked down the gorge where already more than twenty bodies lay decomposing. One of the Indian soldiers, a R.I.A.S.C. ambulance driver, was shot through the right arm, left shoulder and right hand by the Japanese officer and managed to fall down the gorge without being kicked. At the bottom he discovered that the sixth member of his party, a labourer, although badly wounded, was still conscious. When the Japanese had gone, he helped the other man to move across the gorge, but after a hundred yards the labourer lost consciousness and he had to leave him. Five days later, after living on water and leaves, the ambulance driver staggered into the regimental aid post of 3/14 Punjab Regiment, one of 9 Brigade's units.

Meanwhile, back at the hospital in the Box, the Japanese raiding party had completed their crime. Wounded prisoners received no attention. Those who cried out were shot or bayoneted. Men lying helplessly in bed were killed. In one shelter were a British lieutenant, a major of the Gurkhas and a Signals sergeant. They were being tended by a captain of the R.A.M.C. Four West Yorkshires, whose defence post was overrun in the attack, joined them. When day came, they lay still so that the Japanese might not notice them. During the morning they heard a shout outside and the R.A.M.C. Captain asked. "What do you want?"

The shout – it sounded like "You go" – was repeated. The Captain shook his head and lay down again. "Who is it?" asked the Lieutenant.

"It's a Jap," said the Captain. At that moment one of the Japanese soldiers appeared and shot him through the right thigh. The Captain shouted: "I am a doctor – Red Cross – I am a medical officer."

The Japanese shot dead the Captain, the Gurkha Major, two British soldiers and a mess servant. The Lieutenant and the three surviving British soldiers lay still. They stayed like that all day, and when darkness came they managed to leave the hospital and find the safety of the nearest West Yorkshire post.

A British private of the R.A.M.C. – one of a party of twenty – survived to describe his experience. He was tied by his neck to another man – as they all were – kicked, cuffed and cracked over the head by rifle butts, and used as a shield on top of a trench by the Japanese when the carrier attacked. Just before dark on February 8 a Japanese officer told the twenty men: "Come and get treatment."

They were taken along a dried-up water course to a clearing with a running stream. Through the whole hot day they had been allowed only two bottles of water between them. And now they stood by the stream. But they were not allowed to drink. The Japanese opened up at them with rifles. Seventeen of them were killed. That night Lieutenant Basu and nine men who had been wounded when a mortar exploded near them lay in a watercourse, some dying, some crying for water. The Japanese shot one man and bayoneted another who cried too loudly. Just before they left, the Japanese stood in front of them, their rifles ready.

"We are Red Cross people," said Basu – he and another doctor both had their stethoscopes slung round their necks. "We are doctors and hospital workers. We have nothing to do with actual warfare."

Most of them wore Red Cross badges on their arms. It made no difference. The Japanese shot them all.

Lieutenant Basu was shot at twice. He was left stunned. At first he was not sure whether he was alive or dead. He felt at his ear, but there was no blood on his fingers. He could still see and his thoughts became clear once more. He realised how vulnerable he was lying there still alive. So he reached out to the body of one of his dead friends and put his hand on the wounds until it was covered with blood, and then he smeared the blood over his face and head and down his shirt front, so that the Japanese would think he, too, was mortally wounded. He slipped groaning into a trench, and there he spent the night.

On the morning of February 9 the West Yorkshires cleared the Japanese out of the hospital. Their task was made all the more difficult because the enemy had camouflaged their machine-gun posts cunningly with stretchers in the wards and theatre. Fortunately, before the attacks began, most of the wounded had been removed from the hospital to a dried-up watercourse lying to the north. As it was, we found in the hospital area the bodies of thirty-one patients and four doctors – doctors whose services were to be desperately needed in the days to come.

XENOPHON

The Siege of Babylon

In 539 BC Cyrus the Great, the founder of the Persian empire, determined to add Babylon to his possessions. The ensuing siege of Babylon is recounted by the Greek historian Xenophon.

CYRUS was marching to Babylon, but on his way he subdued the Phrygians of Greater Phrygia and the Cappadocians, and reduced the Arabians to subjection. These successes enabled him to increase his Persian cavalry till it was not far short of forty thousand men, and he had still horses left over to distribute among his allies at large.

At length he came before Babylon with an immense body of cavalry, archers, and javelin-men, beside slingers innumerable.

When Cyrus reached the city he surrounded it entirely with his forces, and then rode round the walls himself, attended by his friends and the leading officers of the allies. Having surveyed the fortifications, he prepared to lead off his troops, and at that moment a deserter came to inform him that the Assyrians intended to attack as soon as he began to withdraw, for they had inspected his forces from the walls and considered them very weak. This was not surprising, for the circuit of the city was so enormous that it was impossible to surround it without seriously thinning the lines. When Cyrus heard of their intention, he took up his post in the centre of his troops with his own staff round him and sent orders to the infantry for the wings to double back on either side, marching past the stationary centre of the line, until they met in the rear exactly opposite himself. Thus the men in front were immediately encouraged by the doubling of their depth, and those who retired were equally cheered, for they saw that the others would encounter the enemy first. The two wings being united, the power of the whole force was strengthened, those behind being protected by those in front and those in front supported by those behind. When the phalanx was thus folded back on itself, both the front and the rear ranks were formed of picked men, a disposition that seemed calculated to encourage valour and check flight. On the flanks, the cavalry and the light infantry were drawn nearer and nearer to the commander as the line contracted. When the whole phalanx was in close order, they fell back from the walls, slowly, facing the foe, until they were out of range; then they turned, marched a few paces, and then wheeled round again to the left, and halted, facing the walls, but the further they got the less often they paused, until, feeling themselves secure, they quickened their pace and went off in an uninterrupted march until they reached their quarters.

When they were encamped, Cyrus called a council of his officers and said, "My friends and allies, we have surveyed the city on every side, and for my part I fail to see any possibility of taking by assault walls so lofty and so strong: on the other hand, the greater the population the more quickly must they yield to hunger, unless they come out to fight. If none of you have any other scheme to suggest, I propose that we reduce them by blockade."

Then Chrysantas spoke:

"Does not the river flow through the middle of the city, and is it not at least a quarter of a mile in width?"

"To be sure it is," answered Gobryas, "and so deep that the water would cover two men, one standing on the other's shoulders; in fact the city is even better protected by its river than by its walls."

At which Cyrus said, "Well, Chrysantas, we must forego what is beyond our power: but let us measure off at once the work for each of us, set to, and dig a trench as wide and as deep as we can, that we may need as few guards as possible."

Thereupon Cyrus took his measurements all round the city, and, leaving a space on either bank of the river large enough for a lofty tower, he had a gigantic trench dug from end to end of the wall, his men heaping up the earth on their own side. Then he set to work to build his towers by the river. The foundations were of palm-trees, a hundred feet long and more – the palm-tree grows to a greater height than that, and under pressure it will curve upwards like the spine of an ass beneath a load. He laid these foundations in order to give the impression that he meant to besiege the town, and was taking precautions so that the river, even if it found its way into his trench, should not carry off his towers. Then he had other towers built along the mound, so as to have as many guard-posts as possible. Thus his army was employed, but the men within the walls laughed at his preparations, knowing they had supplies to last them more than twenty years. When Cyrus heard that, he divided his army into twelve, each division to keep guard for one month in the year. At this the Babylonians laughed louder still, greatly pleased at the idea of being guarded by Phrygians and Lydians and Arabians and Cappadocians, all of whom, they thought, would be more friendly to themselves than to the Persians.

However by this time the trenches were dug. And Cyrus heard that it was a time of high festival in Babylon when the citizens drink and make merry the whole night long. As soon as the darkness fell, he set his men to work. The mouths of the trenches were opened, and during the night the water poured in, so that the riverbed formed a highway into the heart of the town.

When the great stream had taken to its new channel, Cyrus ordered his Persian officers to bring up their thousands, horse and foot alike, each detachment drawn up two deep, the allies to follow in their old order. They lined up immediately, and Cyrus made his own bodyguard descend into the dry channel first, to see if the bottom was firm enough for marching. When they said it was, he called a council of all his generals and spoke as follows:

"My friends, the river has stepped aside for us; he offers us a passage by his own high-road into Babylon. We must take heart and enter fearlessly, remembering that those against whom we are to march this night are the very men we have conquered before, and that too when they had their allies to help them, when they were awake, alert, and sober, armed to the teeth, and in their battle order. Tonight we go against them when some are asleep and some are drunk, and all are unprepared: and when they learn that we are within the walls, sheer astonishment will make them still more helpless than before. If any of you are troubled by the thought of volleys from the roofs when the army enters the city, I bid you lay these fears aside: if our enemies do climb their roofs we have a god to help us, the god of Fire. Their porches are easily set aflame, for the doors are made of palm-wood and varnished with bitumen, the very food of fire. And we shall come with the pine-torch to kindle it, and with pitch and tow to feed it. They will be forced to flee from their homes or be burnt to death. Come, take your swords in your hand: God helping me, I will lead you on. Do you," he said, turning to Gadatas and Gobryas, "show us the streets, you know them; and once we are inside, lead us straight to the palace."

"So we will," said Gobryas and his men, "and it would not surprise us to find the palace-gates unbarred, for this night the whole city is given over to revelry. Still, we are sure to find a guard, for one is always stationed there."

"Then," said Cyrus, "there is no time for lingering; we must be off at once and take them unprepared."

Thereupon they entered: and of those they met some were struck down and slain, and others fled into their houses, and some raised the hue and cry, but Gobryas and his friends covered the cry with their shouts, as though they were revellers themselves. And thus, making their way by the quickest route, they soon found themselves before the king's palace. Here the detachment under Gobryas and Gadatas found the gates closed, but the men appointed to attack the guards rushed on them as they lay drinking round a blazing fire, and closed with them then and there. As the din grew louder and louder, those within became aware of the tumult, till, the king bidding them see what it meant, some of them opened the gates and ran out. Gadatas and his men, seeing the gates swing wide, darted in, hard on the heels of the others who fled back again, and they chased them at the sword's point into the presence of the king.

They found him on his feet, with his drawn scimitar in his hand. By sheer weight of numbers they overpowered him: and not one of his retinue escaped, they were all cut down, some flying, others snatching up anything to serve as a shield and defending themselves as best they could. Cyrus sent squadrons of cavalry down the different roads with orders to kill all they found in the street, while those who knew Assyrian were to warn the inhabitants to stay indoors under pain of death. While they carried out these orders, Gobryas and Gadatas returned, and first they gave thanks to the gods and did obeisance because they had been suffered to take vengeance on their unrighteous king, and then they fell to kissing the hands and feet of Cyrus, shedding tears of joy and gratitude. And when it was day and those who held the heights knew that the city was taken and the king slain, they were persuaded to surrender the citadel themselves. Cyrus took it over forthwith, and sent in a commandant and a garrison, while he delivered the bodies of the fallen to their kinsfolk for burial, and bade his heralds make proclamation that all the citizens must deliver up their arms: wherever weapons were discovered in any house all the inmates would be put to death. So the arms were surrendered, and Cyrus had them placed in the citadel for use in case of need. When all was done he summoned the Persian priests and told them the city was the captive of his spear and bade them set aside the first-fruits of the booty as an offering to the gods and mark out land for sacred demesnes. Then he distributed the houses and the public buildings to those whom he counted his partners in the exploit; and the distribution was on the principle accepted, the best prizes to the bravest men: and if any thought they had not received their deserts they were invited to come and tell him. At the same time he issued a proclamation to the Babylonians, bidding them till the soil and pay the dues and render willing service to those under whose rule they were placed. As for his partners the Persians, and such of his allies as elected to remain with him, he gave them to understand they were to treat as subjects the captives they received.

After this Cyrus felt that the time was come to assume the style and manner that became a king: and he wished this to be done with the goodwill and concurrence of his friends and in such a way that, without seeming ungracious, he might appear but seldom in public and always with a certain majesty. Therefore he devised the following scheme. At break of day he took his station at some convenient place, and received all who desired speech with him and then dismissed them. The people, when they heard that he gave audience, thronged to him in multitudes, and in the struggle to gain access there was much jostling and scheming and no little fighting. His attendants did their best to divide the suitors, and introduce them in some order, and whenever any of his personal friends appeared, thrusting their way through the crowd, Cyrus would stretch out his hand and draw them to his side and say, "Wait, my friends, until we have finished with this crowd, and then we can talk at our ease." So his friends would wait, but the multitude would pour on, growing greater and greater, until the evening would fall before there had been a moment's leisure for his friends. All that Cyrus could do then was to say, "Perhaps, gentlemen, it is a little late this evening and time that we broke up. Be sure to come early tomorrow. I am very anxious myself to speak with you." With that his friends were only too glad to be dismissed, and made off without more ado. They had done penance enough, fasting and waiting and standing all day long. So they would get to rest at last, but the next morning Cyrus was at the same spot and a much greater concourse of suitors round him than before, already assembled long before his friends arrived. Accordingly Cyrus had a cordon of Persian lancers stationed round him, and gave out that no one except his personal friends and the generals were to be allowed access, and as soon as they were admitted he said:

"My friends, we cannot exclaim against the gods as though they had failed to fulfil our prayers. They have granted all we asked. But if success means that a man must forfeit his own leisure and the good company of all his friends, why, to that kind of happiness I would rather bid farewell. Yesterday," he added, "I make no doubt you observed yourselves that from early dawn till late evening I never ceased listening to petitioners, and today you see this crowd before us, larger still than yesterday's, ready with business for me. If this must be submitted to, I calculate that what you will get of me and I of you will be little enough, and what I shall get of myself will simply be nothing at all. Further," he added, "I foresee another absurd consequence. I, personally, have a feeling towards you which I need not state, but, of that audience yonder, scarcely one of them do I know at all, and yet they are all prepared to thrust themselves in front of you, transact their business and get what they want out of me before any of you have a chance. I should have thought it more suitable myself that men of that class, if they wanted anything from me, should pay some court to you, my friends, in the hopes of an introduction. Perhaps you will ask why I did not so arrange matters from the first instead of always appearing in public. Because in war it is the first business of a commander not to be behindhand in knowing what ought to be done and seeing that it is done, and the general who is seldom seen is apt to let things slip. But today, when war with its insatiable demands is over, I feel as if I had some claim myself to rest and refreshment. I am in some perplexity, however, as to how I can arrange matters so that all goes well, not only with you and me, but also with those whom we are bound to care for. Therefore I seek your advice and counsel, and I would be glad to learn from any of you the happiest solution."

Cyrus paused, and up rose Artabazus the Mede, who had claimed to be his kinsman, and said:

"You did well, Cyrus, to open this matter. Years ago, when you were still a boy, from the very first I longed to be your friend, but I saw you did not need me, and so I shrank from approaching you. Then came a lucky moment when you did have need of me to be your good messenger among the Medes with the order from Cyaxares, and I said to myself that if I did the work well, if I really helped you, I might become your comrade, and have the right to talk with you as often as I wished. Well, the work was done, and done so as to win your praise. After that the Hyrcanians joined us, the first friends we made, when we were hungry and thirsty for allies, and we loved them so much we almost carried them about with us in our arms wherever we went. Then the enemy's camp was taken, and I scarcely think you had the leisure to trouble your head with me – oh, I quite forgave you. The next thing was that Gobryas became your friend, and I had to take my leave, and after him Gadatas, and by that time it was a real task to get hold of you. Then came the alliances with the Sakians, and the Cadousians, and no doubt you had to pay them court; if they danced attendance on you, you must dance attendance on them. So that there I was, back again at my starting-point, and yet all the while, as I saw you busy with horses and chariots and artillery, I consoled myself by thinking, 'when he is done with this he will have a little leisure for me.' And then came the terrible news that the whole world was gathering in arms against us; I could not deny that these were important matters, but still I felt certain if all went well, a time would come at last when you need not grudge me your company, and we should be together to my heart's content, you and I. Now, the day has come; we have conquered in the great battle; we have taken Sardis and Babylon; the world is at our feet, and yesterday, by Mithras! unless I had used my fists a hundred times, I swear I could never have got near you at all. Well, you grasped my hand and gave me greeting, and bade me wait beside you, and there I waited, the cynosure of every eye, the envy of every man, standing there all day long, without a scrap to eat or a drop to drink. So now, if any way can be found by which we who have served you longest can get the most of you, well and good: but, if not, pray send me as your messenger once more, and this time I will tell them they can all leave you, except those who were your friends of old."

This appeal set them all laughing, Cyrus with the rest. Then Chrysantas the Persian stood up and spoke as follows:

"Formerly, Cyrus, it was natural and right that you should appear in public, for the reasons you have given us yourself, and also because we were not the folk you had to pay your court to. We did not need inviting: we were with you for our own sakes. It was necessary to win over the masses by every means, if they were to share our toils and our dangers willingly. But now you have won them, and not them alone; you have it in your power to gain others, and the moment has come when you ought to have a house to yourself. What would your empire profit you if you alone were left without hearth or home? Man has nothing more sacred than his home, nothing sweeter, nothing more truly his. And do you not think," he added, "that we ourselves would be ashamed if we saw you bearing the hardships of the camp while we sat at home by our own firesides? Should we not feel we had done you wrong, and taken advantage of you?"

When Chrysantas had spoken thus, many others followed him, and all to the same effect. And so it came about that Cyrus entered the palace, and those in charge brought the treasures from Sardis thither, and handed them over. And Cyrus when he entered sacrificed to Hestia, the goddess of the Hearth, and to Zeus the Lord, and to any other gods named by the Persian priests.

Otto Skorzeny

The Seizure of Budapest by SS Commandos

Waffen SS Captain Skorzeny leapt to fame for his spectacular rescue of Mussolini from prison in 1943; a year later, Hitler gave Skorzeny another mission impossible: the capture of the Citadel of Budapest, from where Hitler's sometime Hungarian ally Admiral Horthy was trying to negotiate a peace with the Allies.

The "alert" plan, which had now been worked out, provided that I and the detachment under my command should effect a military occupation of the Citadel. I had abandoned the idea of a glider or parachute landing altogether.

It was now time for my troops to come to Budapest. The GOC Corps insisted that there should be no further delay. They left Vienna about the beginning of October, and took up their quarters in the suburbs.

In the first week of October, SS Obergruppenführer Bach-Zelevski also came to Budapest. He had been sent by the FHQ to take charge of all proceedings in the city. Having come from Warsaw, where he had just put down the rising of the Polish Underground, he took care to let us know at our conferences that he was a "strong man". He told us he was determined to be as ruthless as he had been in Warsaw. He had even brought a 65 cm mortar with him, a weapon which had only been brought into play twice before – at the sieges of Sebastopol and Warsaw.

I considered his methods unnecessarily brutal, and said that we would attain our ends better and quicker in other and less objectionable ways. Operation Panzerfaust could succeed without the help of the famous mortar. Many of the officers seemed impressed by Bach-Zelevski's intervention and almost afraid of him, but I disregarded his bad manners, stuck to my point of view and got it accepted.

I could not understand why fifteen or twenty officers should be present at conferences when the alert plan was discussed. It seemed to me that the Hungarian government was bound to hear of them and act accordingly. We received a very alarming report from our Intelligence that General M., commanding the Hungarian Army in the Carpathians, was personally engaged in direct negotiations with the Russians. Of course that information was transmitted to FHQ, but it issued no definite orders as to what counter-measures should be taken. Conference followed conference.

On the night of the 10th October, there was a meeting between Horthy junior and the Yugoslav delegates. The German police were warned in advance, but took no action. The next meeting was to take place on Sunday, the 15th, in the vicinity of the Danube quay. Just before the 15th, FHQ sent General Wenck to Budapest with orders to take command if necessary and issue such orders as he thought fit. The Security Police were determined to take action at the first opportunity and arrest the Protector's son and the Yugoslav delegates. The codeword "Mouse" was chosen for this operation, owing to Niklas's nickname, "Nicky", being mistaken for Micky. The association with "Micky Mouse" was obvious.

The adoption of this plan by the police was based on the supposition that the Protector, to avoid the public exposure of his son, would mend his ways and abandon the plan for a separate peace.

General Winkelman had asked me to have a company of my men ready for that afternoon. He said that he knew that Niklas von Horthy's previous meetings had been guarded by Honved troops. If he was right, I could see that my men were considered in the light of a counterblast. I agreed on condition that I myself should decide how and when they should intervene.

On the Saturday, I received an urgent telegram from Berlin, ordering me, to my sorrow, to send Radl back to Berlin. He was very annoyed, but of course complied.

The 15th October was a bright Sunday. The streets were empty at the time appointed for the rendezvous. My company was in a side street in covered trucks. Captain von Foelkersam kept me in touch with them, as obviously I could not show myself in uniform that day. If I was to appear on the stage, so to speak, I must be inconspicuous. My driver and another man, both Luftwaffe personnel, were taking the air on a seat in the little garden which occupied most of the square. I drove up in my own car shortly before the meeting began. When I entered the square, I noticed a Hungarian military lorry and a private car, which was presumably Horthy's, stationed in front of the building of which we had been told. It took me no time to make up my mind and park my own car right in the path of these vehicles, so that they could not get away in a hurry.

The floor above the offices in this building had been occupied the day before by policemen, who had taken lodgings nearby. Others were to enter it from the street about 10.10 p.m., and make the arrests.

Three Honved officers were sitting in the covered lorry, but could not be seen from the street. Two others were lounging on benches in the gardens. I was standing by my car, pretending to be fiddling with the engine, when the curtain rose on the drama.

The first German policeman had hardly entered the building when there was a burst of machine-pistol fire from the lorry, and the second fell to the ground with a wound in the stomach. The two other Hungarian officers came running out of the gardens, firing their revolvers. I had just time to take cover behind my car when its open door was drilled. Things were getting really hot! Honved soldiers appeared at the windows and on the balconies of houses. The moment the first shots rang out, my driver and his companion rushed up to me, assuming that I had been hit. The driver was shot through the thigh, but could still walk. I gave the agreed signal to my detachment, and we three defended ourselves with our weapons as best we could against the rain of fire from the enemy. It was a most uncomfortable situation, though it only lasted a few minutes.

By then my car was not much more than a sieve. Bullets ricocheting from walls passed unpleasantly near and we could only put our noses out of cover for long enough to have potshots at the enemy and keep them at least 10 to 15 metres away.

Then I heard my men running out of the side street in our direction. Foelkersam had taken the situation in at a glance and posted the first section at the corner of the square, while the others swept through the gardens and began firing at the house-fronts. My first assailants now withdrew to the shelter of a nearby house, which was occupied by Hungarians in some strength. I observed that these men were lining up for an assault and quick thinking inspired us to hurl a number of grenades in the doorway, thereby bringing down the door and some marble slabs, which temporarily blocked the entrance.

With that the fighting ceased. It may have lasted five minutes.

Our policemen now came down from the upper floor, bringing four prisoners with them. The two Hungarians, "Micky Mouse" and his comrade Bornemisza, were bundled into one of our trucks. To conceal their identity, our fellows had tried to roll them up in carpets, with only partial success, I observed, noting the effort required to get the refractory prisoners into the vehicle.

The lorry moved off and my company withdrew. I was anxious to avoid further scuffles, which were only too likely when the enemy recovered from his surprise. Fortunately, our retirement passed without further incident.

Some instinct prompted me to follow the truck. Another car and driver were available for me. Barely a hundred metres from the square, under the Elizabeth bridge, I saw three Honved companies approaching at the double. If they got any nearer, they could easily find themselves involved in a mix-up with my men – an eventuality I was determined to prevent at all costs. Time must be gained somehow, but bluff was my only resource. I told my driver to pull up, and ran towards the officer who appeared to be in command. "Halt your men quick!" I yelled. "There's a hell of a mix-up going on up there! No one knows what's happening! You'd better find out for yourself first!"

The trick came off. The troops halted and the officer seemed undecided what to do. It was lucky for me that he knew some German, as otherwise he might not have understood me. The short pause was vital from my point of view. By now, my own men must have got away in their trucks. "I must get on!" I called to the Hungarian officer, jumped into my car and made for the aerodrome. When I arrived, the two Hungarians were in a plane, and two minutes later they were on their way to Vienna.

My next destination was Corps Headquarters, in a hotel at the top of a hill. Here I met General Wenck. We were all wondering what would happen now. It was known that the Hungarians had been taking military precautions at the Citadel for some days. The garrison had been reinforced and it was said that some of the streets had been mined.

About midday, a call came through from the German Embassy, lodged in a small palace on the Citadel. The Military

Attaché told us that the Citadel was now being officially occupied by Honved troops, and the gates and roads were closed to traffic. He had tried to get away himself, but had been turned back. Shortly afterwards, the telephone wires must have been cut, as we could not get through. The German establishments, of which there were several, were practically isolated.

Just before 2 o'clock, we were told to stand by for a special announcement on the Hungarian wireless. A message from the Regent, Admiral Horthy, came through: "Hungary has concluded a separate peace with Russia!" Now we knew where we were. Our counter-measures must be carried out at once.

Orders for the execution of Operation Panzerfaust were also issued. I thought them premature, and asked that it should be postponed for a few hours, and that the immediate reply to the Hungarian action should be to draw a cordon of German troops round the Citadel. This job was assigned to the 22nd SS Division. The occupation by German troops of the railway stations and other important buildings passed off without incident in the afternoon.

A general was dispatched to the Hungarian GHQ at the front. Unfortunately, he arrived too late. General M. and some of his officers and secretaries had already gone over to the Russians. It surprised us greatly that his action, and the Hungarian wireless announcement, did not have such a serious effect on the Hungarian troops as might have been expected. Generally speaking, they remained where they were and few of the officers followed the example of their commander-in-chief. But it was essential that there should be no delay in preventing the Hungarian War Ministry from following up with an order to capitulate.

At a conference late in the afternoon, it was decided that Operation Panzerfaust should be carried out early in the morning of the 16th. The slight postponement suited me well, as I could put it to good use. I fixed on 6 a.m. for zero-hour, as I considered surprise essential and the early hours were best from that point of view. Foelkersam and I pored for hours over the plan of the Citadel which we had made, and our ideas of the coming action began to assume definite shape.

I projected a concentric assault, which should yet have a focal point in the centre, which I intended to be a detachment approaching from Vienna Street. The factor of surprise would be of greatest effect at that point. I hoped to rush the Vienna gate with little resistance and without too much noise, and suddenly emerge in the square facing the Citadel. A rapid decision should follow automatically. If we could quickly force our way into the presumed centre of the Hungarian resistance, the action would soon be over, with a minimum of casualties on both sides.

We then instructed our units in their specific tasks. We had been allotted one company of Panther and one of Goliath tanks. Incidentally, these little Goliath tanks were a recent addition to German armament. They were radio-controlled, low, handy affairs, with caterpillar tracks and a big explosive charge in the bows. They could prove very useful in breaking down any barricades or gates.

The battalion of the Wiener-Neustadt Kriegsakademie was to attack through the gardens on the southern slope of the Citadel – no small undertaking, as we knew that these gardens had been converted into a complex of trenches, machine-gun emplacements and anti-aircraft gun positions. Its function was to beat down resistance and facilitate the occupation of the castle.

A platoon of the "Mitte" Battalion, reinforced by two Panthers, would attack the western side with the object of forcing one of the entrances at the back, while a platoon of the 600 SS Parachute Battalion made its way into the chain-bridge tunnel passing under the Citadel, cleared out the subterranean passages and reached the ministries of War and Home Affairs above. The rest of the "Mitte" unit, the bulk of the SS parachute battalion, six Panthers and the Goliath company, were to be available for my *coup de main*. The Luftwaffe parachute battalion would be kept in reserve for emergencies.

The orders for the individual assignments were carefully worked out, and about midnight my troops were in position behind the cordon drawn by the 22nd Division.

The streets had worn their usual appearance all day, as the civil population did not seem to have noticed the activities of either the Hungarian or German troops. The coffee houses were full as ever, and did not empty until a very late hour. The news from the stations was equally reassuring; supply trains were coming from Germany and passing through to the front in the ordinary way.

Just after midnight, a high-ranking officer of the Hungarian War Ministry presented himself at Corps Headquarters. He had come by some route unknown to us, and said that he was authorized by his Minister to negotiate. We replied that there could be no negotiations until the Regent's proclamation was withdrawn, and also that it was an unfriendly act to hold the members of the embassy and other German organizations prisoner in the Citadel. At my suggestion, the Hungarians were given until 6 a.m. to decide whether they would remove the mines and barricades in Vienna Street leading to the German Embassy. That time was fixed with a view to my design for a surprise attack on the Citadel with a minimum of bloodshed.

About 3 a.m., I went to my command post at the foot of the Citadel and summoned all my officers. The night was very dark and we had to use our torches when examining our sketch plans. There were a few details to be cleared up, though my officers had worked hard and familiarized themselves thoroughly with the ground. My second-in-command produced some coffee, which was very welcome on such a nerve-racking occasion.

Meanwhile, I had made my final decision on the procedure to be adopted. We must simply march up the hill to the Citadel and do our best to give the impression that nothing unusual was afoot. The men must stay in their trucks. I knew that my order to that effect was taking a big risk, as they would be defenceless for the first few moments if the convoy was attacked, but I had no option if I wanted a quick end to any scuffle. I informed my battalion commanders of my plans and assured them that if it succeeded they could count on speedy help from the Citadel.

I assembled my column and told the officers that as soon as the Vienna Gate had been passed it must split in two and proceed at full speed by two parallel roads to the Citadel square. The company and platoon commanders were given strict instructions as to the use of their arms. They were not to reply to casual shots in their direction, and must do everything in their power to arrive at the rendezvous without firing themselves. The watchword must be: "The Hungarians are not our enemies."

Just before 5.30 a.m., when it was beginning to get light, I took my place in my truck at the head of the column. Behind me, I had two Panthers, followed by a platoon of the Goliath company, and the rest of the unit in their trucks. Automatics were set at safety. Most of the men had slumped in their seats and were enjoying a quiet nap. They had the hardened warriors' gift of snatching a bit of sleep when a really tough job lay ahead.

I took the precaution of sending my second-in-command to

Corps Headquarters, to ascertain whether there had been any change in the situation, but the answer was in the negative, so zero hour was adhered to.

In my truck I had Foelkersam and Ostafel, as well as five NCOs who had been in the Gran Sasso show. I considered them my personal assault group. Each was armed with a machine-pistol, a few hand grenades, and the new *panzerfaust* (bazooka). We were wondering what the Hungarian tanks in the Citadel would do. If necessary our tanks and *panzerfausts* would have to look after them.

At one minute before 6 o'clock, I waved my arm as the signal to switch on. Then I stood up in my truck and pointed upwards several times, whereupon we started off, rather slowly, as it was uphill. I could only hope that none of our vehicles struck a mine, which would have blocked our advance and upset our plan. The Vienna Gate emerged out of the half-light – the way was open! A few Hungarian soldiers stared curiously at us. We were soon at the top. "Gradually accelerate," I whispered to my driver.

On our right was a Honved barracks. "Nasty if we get fired on from the flank," murmured Foelkersam at my side. There were two machine-guns behind sandbags in front of the barracks, but nothing happened. No sound could be heard but the rumble of the Panthers behind.

I chose the side street on the right in which the German Embassy was situated. We could now travel at a good pace without losing the rest of the column. The tanks were doing a good 35 to 40 kilometres to the hour, and at length the Citadel was not more than a thousand metres away and a substantial part of our task had been accomplished.

Now the great detached mass of the War Ministry appeared to the left, and we heard the distant sound of two heavy explosions. Our men must have forced their way through the tunnel. The critical moment was at hand. We were past the War Ministry and in the square in a flash. Three Hungarian tanks faced us, but as we drew level the leading one tilted its gun skywards as a signal that they would not fire.

A barricade of stones had been placed in front of the gate of the Citadel. I told my driver to draw aside and signalled to the leading Panther to charge it. We left our truck and ran behind, while the barricade collapsed under the weight of the 30-ton monster, which continued its irresistible thrust. Levelling its long gun-barrel at the centre of the courtyard, it found itself faced with a battery of six anti-tank guns.

We leaped over the debris of the barricade and burst through the shattered gate. A colonel of the guard got out his revolver to stop us, but Foelkersam knocked it out of his hand. On our right was what appeared to be the main entrance, and we took it at the run, almost colliding with a Honved officer, whom I ordered to lead us straight to the Commandant. He immediately complied, and at his side we rushed up the broad staircase, not failing to notice the elegant red carpet.

On reaching the first floor we turned left into a corridor, and I left one of my men behind to cover us. The officer pointed to a door and we went into a small ante-room where a table had been drawn up to the open window and a man was lying on it firing a machine-gun into the courtyard. Holzer, a short, stocky NCO, clasped the gun in his arms and flung it out of the window. The gunner was so surprised that he fell off the table.

I saw a door on my right, knocked and walked straight in. A Honved Major-General got up and came towards me. "Are you the Commandant?" I asked. "You must surrender the Citadel at once! If you don't, you will be responsible for any bloodshed. You must decide immediately!" As we could hear shots outside, including bursts of machine-gun fire, I added: "You can see that any resistance is hopeless. I have already occupied the Citadel." I was speaking the truth, as I was quite certain that the "Mitte" Battalion, led by the redoubtable Lieutenant Hunke, was just behind me and must have seized all the strategic points.

The Hungarian Major-General was not long in making up his mind: "I surrender the Citadel and will order the ceasefire at once." We shook hands and soon arranged that a Hungarian officer and one of ours should inform the troops fighting in the Citadel gardens of the ceasefire. After ten minutes had passed, no noise of battle could be heard.

Archibald Hurd

The Blocking of Zeebrugge

The Zeebrugge canal was a main base for U-boats during the First World War. When the submarine pens revealed themselves impervious to air attack, the Royal Navy devised a valiant scheme to sink blockships at the sea entrance to the canal.

Situated on the Belgian coast, some twelve miles apart, and facing a little to the west of north, Zeebrugge was in reality but a sea-gate of the inland port of Bruges – the latter being the station to which the enemy destroyers and submarines were sent in parts from the German workshops; where they were assembled; and whence, by canal, they proceeded to sea by way of Zeebrugge and Ostend. Of these two exits, Zeebrugge, the northernmost, was considerably the nearer to Bruges and the more important – Zeebrugge being eight, while Ostend was eleven miles distant from their common base – and to receive an adequate impression of what was subsequently achieved there it is necessary to bear in mind its salient features.

Unlike Ostend, apart from its harbour, it possessed no civic importance, merely consisting of a few streets of houses clustering about its railway station, locks, wharves, and storehouses, its sandy roadstead being guarded from the sea by an immensely powerful crescentic Mole. It was into this roadstead, that the Bruges canal opened between heavy timbered breakwaters, having first passed through a sea-lock, some half a mile higher up. Between the two lighthouses, each about twenty feet above high-water level, that stood upon the ends of these breakwaters, the canal was 200 yards wide, narrowing to a width, in the lock itself, of less than seventy feet.

Leading from the canal entrance to the tip of the Mole, on which stood a third lighthouse, and so out to sea, was a curved channel, about three-quarters of a mile long, kept clear by continual dredging; and this was protected both by a string of armed barges and by a system of nets on its shoreward side. It was in its great sea-wall, however, some eighty yards broad and more than a mile long, that Zeebrugge's chief strength resided; and this had been utilized, since the German occupation, to the utmost extent. Upon the seaward end of it, near the lighthouse, a battery of 6-inch guns had been mounted, other batteries and machine-guns being stationed at various points throughout its length. With a parapet along its outer side, some sixteen feet higher than the level of the rest of the Mole, it not only carried a railway-line but contained a sea-plane shed, and shelters for stores and personnel. It was connected with the shore by a light wood and steel viaduct – a pilework structure, allowing for the passage of the through-current necessary to prevent silting.

Emplaced upon the shore, on either side of this, were further batteries of heavy guns; while, to the north of the canal entrance, and at a point almost opposite to the tip of the Mole, was the Goeben Fort, containing yet other guns covering both the Mole and the harbour. Under the lee of the parapet were dugouts for the defenders, while, under the lee of the Mole itself, was a similar shelter for the enemy's submarines and destroyers. Nor did this exhaust the harbour's defences, since it was further protected not only by minefields, but by natural shoals, always difficult to navigate, and infinitely more so in the absence of beacons.

Even to a greater extent was this last a feature of Ostend, though here the whole problem was somewhat simpler, there being no Mole, and therefore no necessity – though equally no opportunity – for a subsidiary attack. Covered, of course, from the shore by guns of all calibres – and here it should be remembered that there were 225 of these between Nieuport and the Dutch frontier – the single object in this case was to gain the entrance, before the block-ships should be discovered by the enemy, and sunk by his gunners where their presence would do no harm. Since for complete success, however, it was necessary to seal both places, and, if possible, to do so simultaneously, it will readily be seen that, in the words of Sir Eric Geddes – the successor, as First Lord of the Admiralty, to Mr Balfour and Sir Edward Carson – it was, "a particularly intricate operation which had to be worked strictly to timetable." It was also one that, for several months before, required the most arduous and secret toil.

Begun in 1917 while Sir John Jellicoe was still First Sea Lord, the plan ultimately adopted – there had been several previous ones, dropped for military reasons – was devised by Vice-Admiral Roger Keyes, then head of the Plans Division at the Admiralty. From the first it was realized, of course, by all concerned that the element of surprise would be the determining factor; and it was therefore decided that the attempt to block the harbours should take place at night. It was also clear that, under modern conditions of star-shells and searchlights, an extensive use would have to be made of the recent art of throwing out smokescreens; and fortunately, in Commander Brock, Admiral Keyes had at his disposal just the man to supply this need. A Wing-Commander in the Royal Naval Air-Service, in private life Commander Brock was a partner in a well-known firm of firework makers; and his inventive ability had already been fruitful in more than one direction. A first-rate pilot and excellent shot, Commander Brock was a typical English sportsman; and his subsequent death during the operations, for whose success he had been so largely responsible, was a loss of the gravest description both to the Navy and the empire.

The next consideration was the choosing of the block-ships and for these the following vessels were at last selected – the *Sirius* and *Brilliant* to be sunk at Ostend, and the *Thetis*, *Iphigenia*, and *Intrepid* to seal the canal entrance at Zeebrugge. These were all old cruisers, and they were to be filled with cement, which when submerged would turn into concrete, fuses being so placed that they could be sunk by explosion as soon as they had reached the desired position; and it was arranged that motor-launches should accompany them in order to rescue their crews.

So far these general arrangements were applicable to both places; but, as regarded Zeebrugge, it was decided to make a diversion in the shape of a subsidiary attack on the Mole, in which men were to be landed and to do as much damage as possible. Such an attack, it was thought, would help to draw the enemy's attention from the main effort, which was to be the sinking of the block-ships, and, apart from this, would have valuable results, both material and moral. For this secondary operation, three other vessels were especially selected and fitted out – two Liverpool ferry boats, the *Iris* and *Daffodil*, obtained by Captain Grant, not without some difficulty, owing to the natural reluctance of the Liverpool authorities and the impossibility of divulging the object for which they were wanted – and the old cruiser *Vindictive*. This latter vessel had been designed as a "ram" ship more than twenty years before, displacing about 5,000 tons and capable of a speed of some twenty knots. She had no armour-belt, but her bow was covered with plates, two inches thick and extending fourteen feet aft, while her deck was also protected by hardened plates, covered with nickel steel, from a half to two inches thick. Originally undergunned, she had subsequently been provided with ten 6-inch guns and eight 12-pounders.

This was the vessel chosen to convey the bulk of the landing party, and, for many weeks, under the supervision of Commander E. O. B. S. Osborne, the carpenters and engineers were hard at work upon her. An additional high deck, carrying thirteen brows or gangways, was fitted upon her port side; pom-poms and machine-guns were placed in her fighting-top; and she was provided with three howitzers and some Stokes mortars. A special flame-throwing cabin, fitted with speaking tubes, was built beside the bridge, and another on the port quarter.

It was thus to be the task of the *Vindictive* and her consorts to lay themselves alongside the Mole, land storming and demolition parties, and protect these by a barrage as they advanced down the Mole; and, in order to make this attack more effective, yet a third operation was designed. This was to cut off the Mole from the mainland, thus isolating its defenders and preventing the arrival of reinforcements; and, in order to do so, it was decided to blow up the viaduct by means of an old submarine charged with high explosives. Meanwhile the whole attempt was to be supported from out at sea by a continuous bombardment from a squadron of monitors; sea-planes and aeroplanes, weather permitting, were to render further assistance; and flotillas of destroyers were to shepherd the whole force and to hold the flanks against possible attack.

This then was the plan of campaign, one of the most daring ever conceived, and all the more so in face of the difficulty of keeping it concealed from the enemy during the long period of preparation – a difficulty enhanced in that it was not only necessary to inform each man of his particular *rôle*, but of the particular objectives of each attack and the general outline of the whole scheme. That was unavoidable since it was more than likely that, during any one of the component actions, every officer might be killed or wounded and the men themselves become responsible. Nor was it possible, even approximately, to fix a date for the enterprise, since this could only be carried out under particular conditions of wind and weather. Thus the night must be dark and the sea calm; the arrival on the other side must be at high water; and there must above all things be a following wind, since, without this, the smoke screens would be useless. Twice, when all was ready, these conditions seemed to have come, and twice, after a start had been made, the expedition had to return; and it was not until April 22nd, 1918, that the final embarkation took place.

By this time Vice-Admiral Keyes had succeeded Vice-Admiral Bacon in command of the Dover Patrol; and he was therefore in personal charge of the great adventure that he had initiated and planned with such care. Every man under him was not only a volunteer fully aware of what he was about to face, but a picked man, selected and judged by as high a standard, perhaps, as the world could have provided. Flying his own flag on the destroyer *Warwick*, Admiral Keyes had entrusted the *Vindictive* to acting Captain A. F. B. Carpenter, the *Iris* and the *Daffodil* being in the hands respectively of Commander Valentine Gibbs and Lieutenant Harold Campbell. The marines, consisting of three companies of the Royal Marine Light Infantry and a hundred men of the Royal Marine Artillery, had been drawn from the Grand Fleet, the Chatham, Portsmouth, and Devonport Depots, and were commanded by Lieutenant-Colonel Bertram Elliot. The three block-ships that were to be sunk at Zeebrugge, the *Thetis, Intrepid*, and *Iphigenia*, were in charge of Commander Ralph S. Sneyd, Lieutenant Stuart Bonham-Carter, and Lieutenant E. W. Billyard-Leake; while the old submarine *C3* that was to blow up the viaduct was commanded by Lieutenant R. D. Sandford. In control of the motor-launches, allotted to the attack on Zeebrugge, was Admiral Keyes' flag-captain, Captain R. Collins, those at Ostend being directed by Commander Hamilton Benn, M. P. – the operations at the latter place being in charge of Commodore Hubert Lynes. Also acting in support, was a large body of coastal motor-boats under Lieutenant A. E. P. Wellman, and a flotilla of destroyers under Captain Wilfred Tomkinson, the general surveying of the whole field of attack – including the fixing of targets and firing-points – being in the skilful hands of Commander H. P. Douglas and Lieutenant-Commander F. E. B. Haselfoot.

Included among the monitors were the *Erebus* and *Terror*, each mounting 15-inch guns, to operate at Zeebrugge; and the *Prince Eugene, General Crauford*, and *Lord Clive*, carrying 12-inch guns, and the *Marshal Soult*, carrying 15-inch guns, to assist at Ostend. To the old *Vindictive* Admiral Keyes had presented a horseshoe that had been nailed for luck to her centre funnel; and, to the whole fleet, on its way across, he signalled the message, "St. George for England." Few who received that message expected to return unscathed, and in the block-ships none; but it is safe to say that, in the words of Nelson, they would not have been elsewhere that night for thousands.

Such then were the forces that, on this still dark night, safely arrived at their first rendezvous and then parted on their perilous ways, some to Zeebrugge and some to Ostend. It was at a point about fifteen miles from the Belgian coast that the two parties separated; and, since it is impossible to follow them both at once, let us confine ourselves at first to the former. Theirs was the more complicated, though, as it afterwards proved, the more swiftly achieved task, the first to arrive on the scene of action, almost at the stroke of midnight, being the old cruiser *Vindictive* with her two stout little attendants. These she had been towing as far as the rendezvous; but, at this point, she had cast them off, and they were now following her, under their own steam, to assist in berthing her and to land their own parties. Ahead of them the small craft had been laying their smoke-screens, the north-east wind rolling these shorewards, while already the monitors could be heard at work bombarding the coast defences with their big guns. Accustomed as he was to such visitations, this had not aroused in the enemy any particular alarm; and it was not until the *Vindictive* and the two ferry-boats were within 400 yards of the Mole that the off-shore wind caused the smoke-screen to lift somewhat and left them exposed to the enemy. By this time the marines and bluejackets, ready to spring ashore, were mustered on the lower and main decks; while Colonel Elliot, Major Cordner, and Captain Chater, who were to lead the marines, and Captain Halahan, who was in charge of the bluejackets, were waiting on the high false deck.

It was a crucial moment, for there could be no mistaking now what was the *Vindictive*'s intention. The enemy's star-shells, soaring into the sky, broke into a baleful and crimson light;

while his searchlights, that had been wavering through the darkness, instantly sprang together and fastened upon the three vessels. This, as Captain Carpenter afterwards confessed, induced "an extraordinarily naked feeling," and then, from every gun that could be brought to bear, both from the Mole and the coast, there burst upon her such a fire as, given another few minutes, must inevitably have sunk her. Beneath it Colonel Elliot, Major Cordner, and Captain Halahan, all fell slain; while Captain Carpenter himself had the narrowest escape from destruction. His cap – he had left his best one at home – was two or three times over pierced by bullets, as was the case of his binoculars, slung by straps over his back; while, during the further course of the action, both his searchlight and smoke-goggles were smashed.

The surprise had so far succeeded, however, that, within less than five minutes, the *Vindictive*'s bow was against the side of the Mole, and all but her upper works consequently protected from the severest of the enemy's fire. Safe – or comparatively so – as regarded her water-line, she was nevertheless still a point-blank target; her funnels were riddled over and over again, the one carrying the horse-shoe suffering least; the signal-room was smashed and the bridge blown to pieces, just as Commander Carpenter entered the flame-throwing cabin; and this in its turn, drawing the enemy's fire, was soon twisted and splintered in all directions. It was now raining; explosion followed explosion till the whole air quaked as if in torment; and meanwhile a new and unforeseen danger had just made itself apparent. Till the harbour was approached, the sea had been calm, but now a ground-swell was causing a "scend" against the Mole, adding tenfold not only to the difficulties of landing, but of maintaining the *Vindictive* at her berth. In this emergency, it was the little *Daffodil* that rose to and saved the situation. Her primary duty, although she carried a landing party, had been to push the *Vindictive* in until the latter had been secured; but, as matters were, she had to hold her against the Mole throughout the whole hour and a quarter of her stay there. Even so, the improvised gangways that had been thrust out from the false deck were now some four feet up in the air and now crashing down from the top of the parapet; and it was across these brows, splintering under their feet, and in the face of a fire that baffled description, that the marines and bluejackets had to scramble ashore with their Lewis guns, hand-grenades, and bayonets.

Under such conditions, once a man fell, there was but little hope of his regaining his feet; and it was only a lucky chance that saved one of the officers from being thus trodden to death. This was Lieutenant H. T. C. Walker, who, with an arm blown away, had stumbled and fallen on the upper deck, the eager storming parties, sweeping over him until he was happily discovered and dragged free. Let it be said at once that Lieutenant Walker bore no malice, and waved them good luck with his remaining arm. The command of the marines had now devolved upon Major Weller; and, of the 300 or so who followed him ashore, more than half were soon to be casualties. But the landing was made good; the awkward drop from the parapet was successfully negotiated thanks to the special scaling-ladders; the barrage was put down; and they were soon at hand-to-hand grips with such of the German defenders as stayed to face them. Many of these were in the dug-out under the parapets, but, seeing that to remain there was only to be bayoneted, they made a rush for some of their own destroyers that were hugging the lee of the Mole. But few reached these, however, thanks to the vigour of the marines and the fire of the machine-guns from the *Vindictive*'s top, while one of the destroyers was damaged by hand-grenades and by shells lobbed over the Mole from the *Vindictive*'s mortars.

Meanwhile the *Vindictive* was still the object of a fire that was rapidly dismantling all of her that was visible. A shell in her fighting-top killed every man at the guns there except Sergeant Finch of the Royal Marine Artillery, who was badly wounded, but who extricated himself from a pile of corpses, and worked his gun for a while single-handed. Another shell, bursting forward, put the whole of a howitzer crew out of action, and yet a third, finding the same place, destroyed the crew that followed.

Fierce as was the ordeal through which the *Vindictive* was passing, however, that of the *Iris* was even more so. Unprotected, as was her fellow the *Daffodil*, boring against the side of the larger *Vindictive*, the *Iris*, with her landing-party, was trying to make good her berth lower down the Mole, ahead of Captain Carpenter. Unfortunately the grapnels with which she had been provided proved to be ineffective owing to the "scend," and, with the little boat tossing up and down, and under the fiercest fire, two of the officers, Lieutenant-Commander Bradford and Lieutenant Hawkins, climbed ashore to try and make them fast.

Both were killed before they succeeded, toppling into the water between the Mole and the ship, while, a little later, a couple of shells burst aboard with disastrous results. One of these, piercing the deck, exploded among a party of marines, waiting for the gangways to be thrust out, killing forty-nine and wounding seven; while another, wrecking the ward-room, killed four officers and twenty-six men. Her Captain, Commander Gibbs, had both his legs blown away, and died in a few hours, the *Iris* having been forced meanwhile to change her position, and take up another astern of the *Vindictive*.

Before this happened, however, every man aboard her, as aboard the *Vindictive*, *Daffodil*, and upon the Mole, had been thrilled to the bone by the gigantic explosion that had blown up the viaduct lower down. With a deafening roar and a gush of flame leaping up hundreds of yards into the night, Lieutenant Sandford had told them the good tidings of his success with the old submarine. Creeping towards the viaduct, with his little crew on deck, he had made straight for an aperture between the steel-covered piles, and to the blank amazement and apparent paralysis of the Germans crowded upon the viaduct, had rammed in the submarine up to her conning-tower before lighting the fuse that was to start the explosion.

Before himself doing this, he had put off a boat, his men needing no orders to tumble into her, followed by their commander, as soon as the fuse was fired, with the one idea of getting away as far as possible. As luck would have it, the boat's propeller fouled, and they had to rely for safety upon two oars only, pulling, as Lieutenant Sandford afterwards described it, as hard as men ever pulled before. Raked by machine-gun fire and with shells plunging all round them, most of them, including Lieutenant Sandford, were wounded; but they were finally borne to safety by an attendant picket-boat under his brother, Lieutenant-Commander F. Sandford.

That had taken place about fifteen minutes after the *Vindictive* and her consorts had reached their berths, and a few minutes before the block-ships, with *Thetis* leading, had rounded the light-house at the tip of the Mole. In order to assist these to find their bearings, an employee of Commander Brock, who had never before been to sea, had for some time been firing rockets from the after cabin of the *Vindictive*; and presently they came in sight, exposed as the *Vindictive* had been, by the partial blowing back of their smoke screen. Steaming straight ahead for their objectives, they were therefore opposed by the intensest fire; and the spirit in which they proceeded is well illustrated by what had just taken place on board the *Intrepid*. It had been previously arranged that, for the final stage of their journey, the crews of the block-ships should be reduced to a minimum; but, when the moment came to disembark the extra men, those on the *Intrepid*, so anxious were they to remain, actually hid themselves away. Many of them did in fact succeed in remaining, and sailed with their comrades into the canal.

The first to draw the enemy's fire, the *Thetis*, had the misfortune, having cleared the armed barges, to foul the nets – bursting through the gate and carrying this with her, but with her propellers gathering in the meshes and rendering her helpless. Heavily shelled, she was soon in a sinking condition, and Commander Sneyd was obliged to blow her charges, but not before he had given the line, with the most deliberate coolness, to the two following block-ships – Lieutenant Littleton, in a motor-launch, then rescuing the crew.

Following the *Thetis* came the *Intrepid*, with all her guns in full action, and Lieutenant Bonham-Carter pushed her right into the canal up to a point actually behind some of the German batteries. Here he ran her nose into the western bank, ordered his crew away, and blew her up, the engineer remaining down below in order to be able to report results. These being satisfactory, and every one having left, Lieutenant Bonham-Carter committed himself to a Carley float – a kind of lifebuoy that, on contact with the water, automatically ignited a calcium flare. Illuminated by this, the *Intrepid*'s commander found himself the target of a machine-gun on the bank, and, but for the smoke still pouring from the *Intrepid*, he would probably have been killed before the launch could rescue him.

Meanwhile the *Iphigenia*, close behind, had been equally successful under more difficult conditions. With the *Intrepid*'s smoke blowing back upon her, she had found it exceedingly hard to keep her course, and had rammed a dredger with a barge moored to it, pushing the latter before her when she broke free. Lieutenant Billyard-Leake, however, was able to reach his objective – the eastern bank of the canal entrance – and here he sank her in good position, with her engines still working to keep her in place. Both vessels were thus left lying well across the canal, as aeroplane photographs afterwards confirmed; and thanks to the persistent courage of Lieutenant Percy Dean, the crews of both block-ships were safely removed.

With the accompanying motor-launch unhappily sunk as she was going in, Lieutenant Dean, under fire from all sides, often at a range of but a few feet, embarked in *Motor-Launch 282* no less than 101 officers and men. He then started for home, but, learning that there was an officer still in the water, at once returned and rescued him, three men being shot at his side as he handled his little vessel. Making a second start, just as he cleared the canal entrance, his steering-gear broke down; and he had to manœuvre by means of his engines, hugging the side of the Mole to keep out of range of the guns. Reaching the harbour mouth he then, by a stroke of luck, found himself alongside the destroyer *Warwick*, who was thus able to take on board and complete the rescue of the block-ships' crews.

It was now nearly one o'clock on the morning of the 23rd; the main objects of the attack had been secured; and Captain Carpenter, watching the course of events, decided that it was time to recall his landing-parties. It had been arranged to do so with the *Vindictive*'s siren, but this, like so much of her gear, was no longer serviceable; and it was necessary to have recourse to the *Daffodil*'s little hooter, so feebly opposed to the roar of the guns. Throughout the whole operation, humble as her part had been, the *Daffodil* had been performing yeoman's service, and, but for the fine seamanship of Lieutenant Harold Campbell, and the efforts of her engine-room staff, it would have been quite impossible to re-embark the marines and bluejackets from the Mole. In the normal way her boilers developed some 80-lbs steam-pressure per inch; but, for the work of holding the *Vindictive* against the side of the Mole, it was necessary throughout to maintain double this pressure. All picked men, under Artificer-Engineer Sutton, the stokers held to their task in the ablest fashion; and, in ignorance of what was happening all about them, and to the muffled accompaniment of bursting shells, they worked themselves out, stripped to their vests and trousers, to the last point of exhaustion.

Nor did their colleagues on board the *Vindictive* fall in any degree short of the same high standard, as becomes clear from the account afterwards given by one of her stokers, Alfred Dingle. "My pigeon," he said, "was in the boiler-room of the *Vindictive*, which left with the other craft at two o'clock on Tuesday afternoon. We were in charge of Chief Artificer-

Engineer Campbell, who was formerly a merchant-service engineer and must have been specially selected for the job. He is a splendid fellow. At the start he told us what we were in for, and that before we had finished we should have to feed the fires like mad. 'This ship was built at Chatham twenty years ago,' he said, 'and her speed is 19 knots, but if you don't get 21 knots out of her when it is wanted, well – it's up to you to do it anyway.' We cheered, and he told us, when we got the order, to get at it for all we were worth, and take no notice of anybody. We were all strong fellows, the whole thirteen of us . . . The *Vindictive* was got to Zeebrugge; it was just before midnight when we got alongside the Mole. We had gas-masks on then, and were stoking furiously all the time, with the artificer-engineer backing us up, and joking and keeping us in the best of spirits. Nobody could have been down-hearted while he was there. There is no need to say it was awful; you know something from the accounts in the papers, although no written accounts could make you understand what it was really like . . . Well, there we were, bump, bump, bump against the Mole for I don't know how long, and all the time the shells shrieking and crashing, rockets going up, and a din that was too awful for words, added to which were the cries and shrieks of wounded officers and men . . . Several times Captain Carpenter came below and told us how things were going on. That was splendid of him, I think. He was full of enthusiasm, and cheered us up wonderfully. He was the same with the seamen and men on deck . . . I can't help admiring the marines. They were a splendid lot of chaps, most of them seasoned men, whilst the bluejackets (who were just as good) were generally quite young men. The marines were bursting to get at the fight and were chafing under the delay all the time . . . While we were alongside I was stoking and took off my gas-mask, as it was so much in the way. It was a silly thing to do, but I couldn't get on with the work with it on. Suddenly I smelt gas. I don't know whether it came from an ordinary shell, but I know it was not from the smoke screen, and you ought to have seen me nip round for the helmet. I forgot where I put it for the moment, and there was I running round with my hand clapped on my mouth till I found it. In the boiler-room our exciting time was after the worst was over on shore. All of a sudden the telegraph rang down, 'Full speed ahead,' and then there was a commotion. The artificer-engineer shouted 'Now for it; don't forget what you have to do – 21 knots, if she never does it again.' In a minute or two the engines were going full pelt. Somebody came down and said we were still hitched on to the Mole, but Campbell said he didn't care if we towed the Mole back with us; nothing was going to stop him. As a matter of fact, we pulled away great chunks of the masonry with the grappling irons, and brought some of it back with us. Eventually we got clear of the Mole, and there was terrific firing up above. Mr Campbell was urging us on all the time, and we were shoving in the coal like madmen. We were all singing. One of the chaps started with 'I want to go home,' and this eventually developed into a verse, and I don't think we stopped singing it for three and a half hours – pretty nearly all the time we were coming back. In the other parts of the ship there wasn't much singing, for all the killed and wounded men we could get hold of had been brought on board, and were being attended to by the doctors and sick bay men. I don't know if we did the 21 knots, but we got jolly near it, and everybody worked like a Trojan, and was quite exhausted when it was all over. When we were off Dover the Engineer-Commander came down into the boiler-room and asked Artificer-Engineer Campbell, 'What have you got to say about your men?' He replied, 'I'm not going to say anything for them or anything against them; but if I was going to hell to-morrow night I would have the same men with me.' "

Not until the Mole had been cleared of every man that could possibly be removed did the *Vindictive* break away, turning in a half-circle and belching flames from every pore of her broken funnels. That was perhaps her worst moment, for now she was exposed to every angry and awakened battery; her lower decks were already a shambles; and many of her navigating staff were killed or helpless. But her luck held; the enemy's shells fell short; and soon she was comparatively safe in the undispersed smoke-trails, with the glorious consciousness that she had indeed earned the Admiral's "Well done, *Vindictive*."

Robert Mason

Chickenhawk

A helicopter pilot, Mason flew more than 1,000 assault missions for the US Army in Vietnam between August 1965 and July 1966.

The number of wounded we were carrying was growing fast. That week Leese and I flew more than a hundred wounded to the hospital tent. Other slicks carried a similar number.

When there was room and time, we carried the dead. They had low priority because they were no longer in a hurry. Sometimes they were thrown on board in body bags, but usually not. Without the bags, blood drained on the deck and filled the Huey with a sweet smell, a horribly recognizable smell. It was nothing compared to the smell of men not found for several days. We had never carried so many dead before. We were supposed to be winning now. The NVA were trapped and being pulverized, but the pile of dead beside the hospital tent was growing. Fresh recruits for graves registration arrived faster than they could be processed.

Back at our camp, I was feeling jittery after seeing too much death. I heard that two pilots had got caught on the ground.

Nate and Kaiser had gone to rescue them. Nate was almost in tears as he talked to us in the Big Top. "The stupid assholes. They had been relieved to return for fuel. But you know Paster and Richards: typical gunship pilots. Somehow they think their flex guns make them invulnerable. Anyway, on the flight back they were alone and spotted some VC or NVA or somebody on the ground and decided to attack. Nobody knows how long they were flying around there, because they called after they got hit. When Kaiser and I got there about ten minutes later, the Huey was just sitting there in a clearing looking fine. There were two gunships with us, and they circled around first and took no fire.

Kaiser and I went behind the grounded ship. When we landed. I saw a red mass of meat hanging off a tree branch. It turned out to be Paster, hanging by his feet with his skin ripped off. There was nobody else around. The guns kept circling around and a Dust Off landed behind us. I got out, Kaiser stayed with the ship. The medic jumped out and ran with me." Nate kept patting his breast pockets, looking for his pipe. He never found it. "Paster's skin hung down in sheets and covered his head. The bastards had even cut off his cock. They must have just started on Richards, because we found him lying half naked about a hundred feet away in the elephant grass. His head was almost off." Nate stopped for a second, looking pale. "I almost threw up. Richards and I went to flight school together. The medics cut Paster down and stuffed him into a body bag." He shook his head, holding back tears. "Remember how Richards always bragged about how he knew he'd survive in the jungle if he got shot down? Shit, he even went to jungle school in Panama. If anybody'd be able to get away, it'd be Richards."

Nate's story hit hard. I remembered Richards and his jungle-school patch. Big deal, jungle expert. You got a hundred feet on your one big chance to evade the enemy. All that training down the drain. The thought of his wasting all that training brought tears to my eyes.

The pace remained hectic. The next day several assaults were made to smaller LZs near X-Ray to broaden our front against the NVA. Farris was assigned the command ship in a company-size flight, a mix of ships from the Snakes and the Preachers. We were going to a small, threeship LZ. He picked me to be his pilot.

Everyone was tense. Radio conversations were terse. The grunts in the back looked grim. Even Farris looked worried. The NVA were being surrounded, and we knew they had to fight.

Farris and I would be in the first group of three to land. The company, each ship carrying eight grunts, trailed out behind us.

As the flight leader. Farris had the option to fly from any position in his flight. He chose the second ship. A theory from the developmental days of the air-assault concept said that the flight commander supposedly got a better idea about what was happening from the middle or even the end of the formation. Really big commanders flew high above us, for the best view of all.

I think this was my first time as a command-ship pilot, and I was all for survival. I would've been very happy flying the brigade commander up there at 5000 feet, or Westmoreland to his apartment in Saigon. It's amazing how many places I considered being besides there.

In assaults, we usually started drawing fire at 100 feet, sometimes at 500. This time we didn't.

At 500 feet, on a glide path to the clearing, smoke from the just completed prestrike by our artillery and gunships drifted straight up in the still air. There had to be one time when the prep actually worked and everybody was killed in the LZ. I hoped this might be it.

Fighting my feeling of dread. I went through the automatic routine of checking the smoke drift for wind direction. None. We approached from the east, three ships lined up in a trail, to land in the skinny LZ. But it was too quiet!

At 100 feet above the trees, closing on the near end of the LZ, the door gunners in Yellow One started firing. They shot into the trees at the edge of the clearing, into bushes, anywhere they suspected the enemy was hiding. There was no return fire. The two gunships on each side of our flight opened up with their flex guns. Smoke poured out of them as they crackled. My ears rang with the loud but muffled popping as my door gunners joined in with the rest. I ached to have my own trigger. With so many bullets tearing into the LZ, it was hard to believe anyone on the ground could survive.

The gunships had to stop firing as we flared close to the ground because we could be hit by ricocheting bullets. Still no return fire. Maybe they *were* all dead! Could this be the wrong spot?

My adrenaline was high, and I was keenly aware of every movement of the ship. I waited for the lurch of dismounting troopers as the skids neared the ground. They were growling and yelling behind me, psyched for battle. I could hear them yelling above all the noise. I still can.

My landing was synchronized with the lead ship, and as our skids hit the ground, so did the boots of the growling troops.

At the same instant, the uniformed regulars from the North decided to spring their trap. From at least three different directions, they opened up on our three ships and the off-loading grunts with machine-gun crossfire. The LZ was suddenly alive with their screaming bullets. I tensed on the controls, involuntarily leaning forward, ready to take off. I had to fight the logical reaction to leave immediately. I was light on the skids, the troops were out. Let's go! Farris yelled on the radio for Yellow One to go. They didn't move.

The grunts weren't even making it to the trees. They had leapt out, screaming murderously, but now they dropped all around us, dying and dead. The lead ship's rotors still turned, but the men inside did not answer. I saw the sand spurt up in front of me as bullets tore into the ground. My stomach tightened to stop them. Our door gunners were firing over the prone grunts at phantoms in the trees.

A strange quietness happened in my head. The scene around me seemed far away. With the noise of the guns, the cries of the gunners about everybody being dead, and Farris calling for Yellow One to go, I thought about bullets coming through the Plexiglas, through my bones and guts and through the ship and never stopping. A voice echoed in the silence. It was Farris yelling "Go! Go! Go!"

I reacted so fast that our Huey snapped off the ground. My adrenaline seemed to power the ship as I nosed over hard to get moving fast. I veered to the right of the deadly quiet lead ship, still sitting there. The door gunners fired continuously out both sides. The tracers coming at me now seemed as thick as raindrops. How could they miss? As a boy I made a game of dodging raindrops in the summer showers. I always got hit eventually. But not this time. I slipped over the treetops and stayed low for cover, accelerating. I veered left and right fast, dodging, confounding, like Leese had taught me, and when I was far enough away, I swooped up and away from the nightmare. My mind came back, and so did the sound.

"What happened to Yellow Three?" a voice said. It was still on the ground.

The radios had gone wild. I finally noticed Farris's voice saying, "Negative, White One. Veer left. Circle back." Farris had White One lead the rest of the company into an orbit a couple of miles away. Yellow One and Yellow Three were still in the LZ.

I looked down at the two ships sitting quietly on the ground. Their rotors were turning lazily as their turbines idled. The machines didn't care, only the delicate protoplasm inside them cared. Bodies littered the clearing, but some of the thirty grunts we had brought in were still alive. They had made it to cover at the edge of the clearing.

Farris had his hands full. He had twelve more ships to get in and unloaded. Then the pilot of Yellow Three called. He was still alive, but he thought his partner was dead. His crew chief and gunner looked dead, too. He could still fly.

Two gunships immediately dove down to escort him out, machine guns blazing. It was a wonderful sight to see from a distance.

Only Yellow One remained on the ground. She sat, radios quiet, still running. There was room behind her to bring in the rest of the assault.

A grunt who found himself still alive got to a radio. He said that he and a few others could keep some cover fire going for the second wave.

Minutes later, the second group of three ships was on its way in, and Farris told me to return to the staging area. I flew back a couple of miles to a big field, where I landed and picked up another load of wild-eyed boys.

They also growled and yelled. This was more than just the result of training. They were motivated. We all thought that this was the big push that might end it all. By the time I made a second landing to the LZ, the enemy machine guns were silent. This load would at least live past the landing.

Somebody finally shut down Yellow One's turbine when we left. Nobody in the crew could. They were all patiently waiting to be put into body bags for the trip home.

Why I didn't get hit I'll never know. I must have read the signs right. Right? They started calling me "Lucky" after that mission.

That afternoon, while the sunset glowed orange behind Pleiku in the distance, Leese and I and some others walked over to the hospital tent.

We came to see the bodies. A small crowd of living stood watching the growing crowd of dead. Organization prevailed. Bodies on this pile. Loose parts here. Presumably the spare arms and legs and heads would be reunited with their owners when they were pushed into the bags. But graves registration had run out of body bags, and the corpses were stacked without them.

New arrivals, wounded as well as dead, were brought over from the helicopters. A medic stood in the doorway of the operating tent diverting some of the stretchers away. Some cases were too far gone. Bellies blown open. Medics injected morphine into them. But morphine couldn't change the facts. I stared at one of the doomed men, fifty feet away. He saw me, and I knew that he knew. His frightened eyes widened, straining to live. He died. After a few minutes somebody came by and closed his eyes.

A new gunner, a black kid who had until recently been a grunt, had come over with me and Leese. We stayed back, but he had gone closer to the pile of bodies just to look. He started wailing and crying and pulling at the corpses and had to be dragged away. He had seen his brother at the bottom of the heap.

Two days later there was a lull in the fighting, at least as far as our company was concerned. We were given the day off. You could hear a collective sigh of relief. Compared to Happy Valley, this was *action*, and living through a year of it seemed unlikely.

What do you do on your first day off after weeks of action when you're feeling tired, depressed, and doomed on a hot, wet day at Camp Holloway, Vietnam? You get in a deuce-and-a-half and go into Pleiku and drink your brains out. That's what you do.

I rode in with Leese and Riker. Kaiser and Nate. Connors and Banjo, and Resler. I remember drinking beer all afternoon – effective because I usually didn't drink – first at one bar and then another. They all blended into one. Though I had started with Resler as my companion. I somehow found myself with Kaiser at a table in the Vietnamese officers' club that night.

"We see Americans as being apelike, big and clumsy with hairy arms," a Vietnamese lieutenant was saying to Kaiser. "Also, you all smell bad, like greasy meat."

Kaiser had got into a conversation with a racist of the opposite race. I watched the two men hate each other while I drank the genuine American bourbon that the Vietnamese lieutenant had so kindly bought us.

"Of course, you won't be offended if I continue?" asked the lieutenant.

"Naw," said Kaiser, squinting. "It's okay. I don't give a fuck what a slope thinks anyway." He belted down another shot.

The two men continued to trade heartfelt insults, the gist of which revealed normally submersed beliefs. Kaiser disclosed the widespread resentment among the Americans that the

ARVN units apparently would not or could not fight their own battles. The lieutenant demonstrated that the ARVN resented being rescued by such oafish, unjustifiably wealthy gorillas who were taking over everything in their country, including their women.

After an hour of drinking and insults, Kaiser ended our stay by telling the old "pull the plug" joke to the lieutenant. That was the cynical solution to differentiating between friend and foe in Vietnam and ending the war. The joke had us putting all the "friendlies" on boats in the ocean where they would wait while the remaining people, the enemy, were killed. Then, as the punch line went, we would pull the plug, sink the ships.

Kaiser seemed almost surprised that the lieutenant didn't laugh. Instead the insulted man got up and left. Soon, looks from the other Vietnamese officers told us we weren't welcome. We left to continue our party at another bar.

Somehow we missed the truck back to camp. We went back to the officers' club and borrowed one of their Jeeps. We didn't tell them we were borrowing it; it was made in America, after all. When the Jeep was found the next day, parked in the Camp Holloway motor pool, it caused a stink. No one knew who it was who took it, but Farris looked awfully suspicious when it was reported at the morning briefing. He had noticed that we had come back by ourselves.

"Pretty slick, guys," he said after the briefing.

"Hey, Captain Farris, not us!" Kaiser looked sincere. "It was an inside job."

"Inside job?"

"That's right, sir. Those little guys will do anything to discredit us Americans. You should have heard the stupid things they said about us last night. 'Hairy apes. Greasy meat. Stupid.' No, sir, it doesn't surprise me a bit, knowing how they really feel about us."

"Right." Farris sighed. "Well, Mr. Kaiser, from now on, whenever you get a chance to party it up in town, I'll be keeping you company."

"Captain?" Kaiser looked at Farris, distressed.

"What if you guys had been stopped at the gate? Two warrant officers. You need a captain along to keep you out of trouble. Besides, I don't want to ride around in a deuce-and-a-half when I know you guys can get a Jeep."

Frank Richards

Christmas in the Trenches 1914

Richards, a signalman with the Royal Welsh Fusiliers, served on the Western Front for the entirety of the First World War. He consistently refused any promotion, but won both the Military Medal and the Distinguished Conduct Medal. His account of the famous Christmas truce of 1914 is from his memoir, Old Soldiers Never Die.

On Christmas morning we stuck up a board with "A Merry Christmas" on it. The enemy had stuck up a similar one. Platoons would sometimes go out for twenty-four hours' rest – it was a day at least out of the trench and relieved the monotony a bit – and my platoon had gone out in this way the night before, but a few of us stayed behind, to see what would happen. Two of our men then threw their equipment off and jumped on the parapet with their hands above their heads. Two of the Germans did the same and commenced to walk up the river bank, our two men going to meet them. They met and shook hands and then we all got out of the trench. Buffalo Bill rushed into the trench and endeavoured to prevent it, but he was too late: the whole of the Company were now out, and so were the Germans. He had to accept the situation, so soon he and the other company officers climbed out too. We and the Germans met in the middle of no-man's-land. Their officers was also now out. Our officers exchanged greetings with them. One of the German officers said that he wished he had a camera to take a snapshot, but they were not allowed to carry cameras. Neither were our officers.

We mucked in all day with one another. They were Saxons and some of them could speak English. By the look of them their trenches were in as bad a state as our own. One of their men, speaking in English, mentioned that he had worked in Brighton for some years and that he was fed up to the neck with this damned war and would be glad when it was all over. We told him that he wasn't the only one that was fed up with it. We did not allow them in our trench and they did not allow us in theirs. The German Company-Commander asked Buffalo Bill if he would accept a couple of barrels of beer and assured him that they would not make his men drunk. They had plenty of it in the brewery. He accepted the offer with thanks and a couple of their men rolled the barrels over and we took them into our trench. The German officer sent one of his men back to the trench, who appeared shortly after carrying a tray with bottles and glasses on it. Officers of both sides clinked glasses and drunk one another's health. Buffalo Bill had presented them with a plum pudding just before. The officers came to an understanding that the unofficial truce would end at midnight. At dusk we went back to our respective trenches.

We had a decent Christmas dinner. Each man had a tin of Maconochie's and a decent portion of plum pudding. A tin of Maconochie's consisted of meat, potatoes, beans and other vegetables and could be eaten cold, but we generally used to fry them up in the tin on a fire. I don't remember any man ever suffering from tin or lead poisoning through doing them in this way. The best firms that supplied them were Maconochie's and Moir Wilson's and we could always depend on having a tasty dinner when we opened one of their tins. But another firm that supplied them at this time must have made enormous profits out of the British Government. Before ever we opened the first tins that were supplied by them we smelt a rat. The name of the firm made us suspicious. When we opened them our suspicions were well founded. There was nothing inside but a rotten piece of meat and some boiled rice. The head of that firm should have been put against the wall and shot for the way they sharked us troops. The two barrels of beer were drunk, and the German officer was right: if it was possible for a man to have drunk the two barrels himself he would have bursted before he had got drunk. French beer was rotten stuff.

Just before midnight we all made it up not to commence firing before they did. At night there was always plenty of firing by both sides if there were no working parties or patrols out. Mr Richardson, a young officer who had just joined the Battalion and was now a platoon officer in my company wrote a poem during the night about the Briton and the Bosche meeting in no-man's-land on Christmas Day, which he read out to us. A few days later it was published in *The Times* or *Morning Post*, I believe. During the whole of Boxing Day we never fired a shot, and they the same, each side seemed to be waiting for the other to set the ball a-rolling. One of their men shouted across in English and inquired how we had enjoyed the beer. We shouted back and told him it was very weak but that we were very grateful for it. We were conversing off and on during the whole of the day. We were relieved that evening at dusk by a battalion of another brigade. We were mighty surprised as we had no whisper of any relief during the day. We told the men who relieved us how we had spent the last couple of days with the enemy, and they told us that by what they had been told the whole of the British troops in the line, with one or two exceptions, had mucked in with the enemy. They had only been out of action themselves forty-eight hours after being twenty-eight days in the front-line trenches. They also told us that the French people had heard how we had spent Christmas Day and were saying all manner of nasty things about the British Army.

Going through Armentières that night some of the French women were standing in the doors spitting and shouting at us: "You no bon, you English soldiers, you boko kamerade Allemenge." We cursed them back until we were blue in the nose, and the Old Soldier, who had a wonderful command of bad language in many tongues, excelled himself. We went back to Erquinghem on the outskirts of Armentières and billeted in some sheds. Not far from the sheds was a large building which had been converted into a bath-house for the troops. We had our first bath one day in the latter end of November, and on the twenty-seventh of December we had our second. Women were employed in the bathhouse to iron the seams of our trousers, and each man handed in his shirt, underpants and socks and received what were supposed to be clean ones in exchange; but in the seams of the shirts were the eggs, and after a man had his clean shirt on for a few hours the heat of his body would hatch them and he would be just as lousy as ever he had been. I was very glad when I had that second bath, because I needed a pair of pants. A week before whilst out in the village one night I had had a scrounge through a house and found a magnificent pair of ladies' bloomers. I thought it would be a good idea to discard my pants, which were skin-tight, and wear these instead, but I soon discovered that I had made a grave mistake. The crawlers, having more room to manœuvre in, swarmed into those bloomers by platoons, and in a few days time I expect I was the lousiest man in the company. When I was stripping for the bath Duffy and the Old Soldier noticed the bloomers, and they both said that I looked sweet enough to be kissed.

Richard Hillary

Burned

Hillary, a Spitfire pilot with 603 Squadron, was shot down on 3 September 1940.

I was falling. Falling slowly through a dark pit. I was dead. My body, headless, circled in front of me. I saw it with my mind, my mind that was the redness in front of the eye, the dull scream in the ear, the grinning of the mouth, the skin crawling on the skull. It was death and resurrection. Terror, moving with me, touched my cheek with hers and I felt the flesh wince. Faster, faster. . . . I was hot now, hot, again one with my body, on fire and screaming soundlessly. Dear God, no! No! Not that, not again. The sickly smell of death was in my nostrils and a confused roar of sound. Then all was quiet. I was back.

Someone was holding my arms.

"Quiet now. There's a good boy. You're going to be all right. You've been very ill and you mustn't talk."

I tried to reach up my hand but could not.

"Is that you, nurse? What have they done to me?"

"Well, they've put something on your face and hands to stop them hurting and you won't be able to see for a little while. But you mustn't talk: you're not strong enough yet."

Gradually I realized what had happened. My face and hands had been scrubbed and then sprayed with tannic acid. The acid had formed into a hard black cement. My eyes alone had received different treatment: they were coated with a thick layer of gentian violet. My arms were propped up in front of me, the fingers extended like witches' claws, and my body was hung loosely on straps just clear of the bed.

I can recollect no moments of acute agony in the four days which I spent in that hospital; only a great sea of pain in which I floated almost with comfort. Every three hours I was injected with morphia, so while imagining myself quite coherent, I was for the most part in a semi-stupor. The memory of it has remained a confused blur.

Two days without eating, and then periodic doses of liquid food taken through a tube. An appalling thirst, and hundreds of bottles of ginger beer. Being blind, and not really feeling strong enough to care. Imagining myself back in my plane, unable to get out, and waking to find myself shouting and bathed in sweat. My parents coming down to see me and their wonderful self-control.

They arrived in the late afternoon of my second day in bed, having with admirable restraint done nothing the first day. On the morning of the crash my mother had been on her way to the Red Cross, when she felt a premonition that she must go home. She told the taxi-driver to turn about and arrived at the flat to hear the telephone ringing. It was our Squadron Adjutant, trying to reach my father. Embarrassed by finding himself talking to my mother, he started in on a glamorized history of my exploits in the air and was bewildered by my mother cutting him short to ask where I was. He managed somehow after about five minutes of incoherent stuttering to get over his news.

They arrived in the afternoon and were met by Matron. Outside my ward a twittery nurse explained that they must not expect to find me looking quite normal, and they were ushered in. The room was in darkness; I just a dim shape in one corner. Then the blinds were shot up, all the lights switched on, and there I was. As my mother remarked later, the performance lacked only the rolling of drums and a spotlight. For the sake of decorum my face had been covered with white gauze, with a slit in the middle through which protruded my lips.

We spoke little, my only coherent remark being that I had no wish to go on living if I were to look like Alice. Alice was a large country girl who had once been our maid. As a child she had been burned and disfigured by a Primus stove. I was not aware that she had made any impression on me, but now I was unable to get her out of my mind. It was not so much her looks as her smell I had continually in my nostrils and which I couldn't disassociate from the disfigurement.

They sat quietly and listened to me rambling for an hour. Then it was time for my dressings and they took their leave.

The smell of ether. Matron once doing my dressing with

three orderlies holding my arms; a nurse weeping quietly at the head of the bed, and no remembered sign of a doctor. A visit from the lifeboat crew that had picked me up, and a terrible longing to make sense when talking to them. Their inarticulate sympathy and assurance of quick recovery. Their discovery that an ancestor of mine had founded the lifeboats, and my pompous and unsolicited promise of a subscription. The expectation of an American ambulance to drive me up to the Masonic Hospital (for Margate was used only as a clearing station). Believing that I was already in it and on my way, and waking to the disappointment that I had not been moved. A dream that I was fighting to open my eyes and could not: waking in a sweat to realize it was a dream and then finding it to be true. A sensation of time slowing down, of words and actions, all in slow motion. Sweat, pain, smells, cheering messages from the Squadron, and an overriding apathy.

Finally I was moved. The ambulance appeared with a cargo of two somewhat nervous A.T.S. women who were to drive me to London, and, with my nurse in attendance, and wrapped in an old grandmother's shawl, I was carried aboard and we were off. For the first few miles I felt quite well, dictated letters to my nurse, drank bottle after bottle of ginger beer, and gossiped with the drivers. They described the countryside for me, told me they were new to the job, expressed satisfaction at having me for a consignment, asked me if I felt fine. Yes, I said, I felt fine; asked my nurse if the drivers were pretty, heard her answer yes, heard them simpering, and we were all very matey. But after about half an hour my arms began to throb from the rhythmical jolting of the road. I stopped dictating, drank no more ginger beer, and didn't care whether they were pretty or not. Then they lost their way. Wasn't it awful and shouldn't they stop and ask? No, they certainly shouldn't: they could call out the names of the streets and I would tell them where to go. By the time we arrived at Ravenscourt Park I was pretty much all-in. I was carried into the hospital and once again felt the warm September sun burning my face. I was put in a private ward and had the impression of a hundred excited ants buzzing around me. My nurse said good-bye and started to sob. For no earthly reason I found myself in tears. It had been a lousy hospital, I had never seen the nurse anyway, and I was now in very good hands; but I suppose I was in a fairly exhausted state. So there we all were, snivelling about the place and getting nowhere. Then the charge nurse came up and took my arm and asked me what my name was.

"Dick," I said.

"Ah," she said brightly. "We must call you Richard the Lion Heart."

I made an attempt at a polite laugh but all that came out was a dismal groan and I fainted away. The house surgeon took the opportunity to give me an anaesthetic and removed all the tannic acid from my left hand.

At this time tannic acid was the recognized treatment for burns. The theory was that in forming a hard cement it protected the skin from the air, and encouraged it to heal up underneath. As the tannic started to crack, it was to be chipped off gradually with a scalpel, but after a few months of experience, it was discovered that nearly all pilots with third-degree burns so treated developed secondary infection and septicaemia. This caused its use to be discontinued and gave us the dubious satisfaction of knowing that we were suffering in the cause of science. Both my hands were suppurating, and the fingers were already contracting under the tannic and curling down into the palms. The risk of shock was considered too great for them to do both hands. I must have been under the anaesthetic for about fifteen minutes and in that time I saw Peter Pease killed.

He was after another machine, a tall figure leaning slightly forward with a smile at the corner of his mouth. Suddenly from nowhere a Messerschmitt was on his tail about 150 yards away. For two seconds nothing happened. I had a terrible feeling of futility. Then at the top of my voice I shouted, "Peter, for God's sake look out behind!"

I saw the Messerschmitt open up and a burst of fire hit Peter's machine. His expression did not change, and for a moment his machine hung motionless. Then it turned slowly on its back and dived to the ground. I came-to, screaming his name, with two nurses and the doctor holding me down on the bed.

"All right now. Take it easy, you're not dead yet. That must have been a very bad dream."

I said nothing. There wasn't anything to say. Two days later I had a letter from Colin. My nurse read it to me. It was very short, hoping that I was getting better and telling me that Peter was dead.

Slowly I came back to life. My morphia injections were less frequent and my mind began to clear. Though I began to feel and think again coherently I still could not see. Two V.A.D.s fainted while helping with my dressings, the first during the day and the other at night. The second time I could not sleep and was calling out for someone to stop the beetles running down my face, when I heard my nurse say fiercely, "Get outside quick: don't make a fool of yourself here!" and the sound or footsteps moving towards the door. I remember cursing the unfortunate girl and telling her to put her head between her knees. I was told later that for my first three weeks I did little but curse and blaspheme, but I remember nothing of it. The nurses were wonderfully patient and never complained. Then one day I found that I could see. My nurse was bending over me doing my dressings, and she seemed to me very beautiful. She was. I watched her for a long time, grateful that my first glimpse of the world should be of anything so perfect. Finally I said:

"Sue, you never told me that your eyes were so blue."

For a moment she stared at me. Then, "Oh, Dick, how wonderful," she said. "I told you it wouldn't be long"; and she dashed out to bring in all the nurses on the block.

I felt absurdly elated and studied their faces eagerly, gradually connecting them with the voices that I knew.

"This is Anne," said Sue. "She is your special V.A.D. and helps me with all your dressings. She was the only one of us you'd allow near you for about a week. You said you liked her voice." Before me stood an attractive fair-haired girl of about twenty-three. She smiled and her teeth were as enchanting as her voice. I began to feel that hospital had its compensations. The nurses called me Dick and I knew them all by their Christian names. Quite how irregular this was I did not discover until I moved to another hospital where I was considerably less ill and not so outrageously spoiled. At first my dressings had to be changed every two hours in the day-time. As this took over an hour to do, it meant that Sue and Anne had practically no time off. But they seemed not to care. It was largely due to them that both my hands were not amputated.

Sue, who had been nursing since seventeen, had been allocated as my special nurse because of her previous experience of burns, and because, as Matron said, "She's our best girl and very human." Anne had been married to a naval officer killed in the *Courageous*, and had taken up nursing after his death.

At this time there was a very definite prejudice among the regular nurses against V.A.D.s. They were regarded as painted society girls, attracted to nursing by the prospect of sitting on the officers' beds and holding their hands. The V.A.D.s were rapidly disabused of this idea, and, if they were lucky, were finally graduated from washing bedpans to polishing bed-tables. I never heard that any of them grumbled, and they gradually won a reluctant recognition. This prejudice was considerably less noticeable in the Masonic than in most hospitals: Sue, certainly, looked on Anne as a companionable and very useful lieutenant to whom she could safely entrust my dressings and general upkeep in her absence. I think I was a little in love with both of them.

The Masonic is perhaps the best hospital in England, though at the time I was unaware how lucky I was. When war broke out the Masons handed over a part of it to the services; but owing to its vulnerable position very few action casualties were kept there long. Pilots were pretty quickly moved out to the main Air Force Hospital, which I was not in the least eager to visit. Thanks to the kind-hearted duplicity of my house surgeon, I never had to; for every time they rang up and asked for me he would say that I was too ill to be moved. The Masonic's great charm lay in that it in no way resembled a hospital; if anything it was like the inside of a ship. The nursing staff were very carefully chosen, and during the regular blitzing of the district, which took place every night, they were magnificent.

The Germans were presumably attempting to hit Hammersmith Bridge, but their efforts were somewhat erratic and we were treated night after night to an orchestra of the scream and crump of falling bombs. They always seemed to choose a moment when my eyes were being irrigated, when my poor nurse was poised above me with a glass undine in her hand. At night we were moved into the corridor, away from the outside wall, but such was the snoring of my fellow sufferers that I persuaded Bertha to allow me back in my own room after Matron had made her rounds.

Bertha was my night nurse. I never discovered her real name, but to me she was Bertha from the instant that I saw her. She was large and gaunt with an Eton crop and a heart of gold. She was engaged to a merchant seaman who was on his way to Australia. She made it quite clear that she had no intention of letting me get round her as I did the day staff, and ended by spoiling me even more. At night when I couldn't sleep we would hold long and heated arguments on the subject of sex.

She expressed horror at my ideas on love and on her preference for a cup of tea. I gave her a present of four pounds of it when I was discharged. One night the Germans were particularly persistent, and I had the unpleasant sensation of hearing a stick of bombs gradually approaching the hospital, the first some way off, the next closer, and the third shaking the building. Bertha threw herself across my bed; but the fourth bomb never fell. She got up quickly, looking embarrassed, and arranged her cap.

"Nice fool I'd look if you got hit in your own room when you're supposed to be out in the corridor," she said, and stumped out of the room.

An R.A.S.C. officer who had been admitted to the hospital with the painful but unromantic complaint of piles protested at the amount of favouritism shown to me merely because I was in the R.A.F. A patriotic captain who was in the same ward turned on him and said: "At least he was shot down defending his country and didn't come in here with a pimple on his bottom. The Government will buy him a new Spitfire, but I'm damned if it will buy you a new arse."

One day my doctor came in and said that I could get up. Soon after I was able to totter about the passages and could be given a proper bath. I was still unable to use my hands and everything had to be done for me. One evening during a blitz, my nurse, having led me along to the lavatory, placed a prodigiously long cigarette-holder in my mouth and lighted the cigarettee in the end of it. Then she went off to get some coffee. I was puffing away contentedly when the lighted cigarette fell into my pyjama trousers and started smouldering. There was little danger that I would go up in flames, but I thought it advisable to draw attention to the fact that all was not well. I therefore shouted "Oi!" Nobody heard me. "Help!" I shouted somewhat louder. Still nothing happened, so I delivered myself of my imitation of Tarzan's elephant call of which I was quite proud. It happened that in the ward opposite there was an old gentleman who had been operated on for a hernia. The combination of the scream of falling bombs and my animal cries could mean only one thing. Someone had been seriously injured, and he made haste to dive over the side of the bed. In doing so he caused himself considerable discomfort: convinced of the ruin of his operation and the imminence of his death, he added his cries to mine. His fears finally calmed, he could see nothing humorous in the matter and insisted on being moved to another ward. From then on I was literally never left alone for a minute.

For the first few weeks, only my parents were allowed to visit me and they came every day. My mother would sit and read to me by the hour. Quite how much she suffered I could only guess, for she gave no sign. One remark of hers I shall never forget. She said: "You should be glad this has to happen to you. Too many people told you how attractive you were and you believed them. You were well on the way to becoming something of a cad. Now you'll find out who your real friends are." I did.

John Nicol

Interrogation

John Nicol, a navigator in a RAF Tornado, was brought down by enemy fire in the first Gulf War and subsequently captured. And interrogated.

They left me standing against a wall in a classic stress position, designed to weaken, to break down resistance. In this position, my forehead was flat against the wall, my feet about twenty inches away from it; I was stretched right up onto my toes, arms handcuffed behind my back. My forehead was supporting my entire bodyweight against the cold surface, a surface flat as purgatory. Every time I tried to move from that position, somebody punched me, whacked me back into it. They left me like that. I tried to move my head, somebody smacked it hard against the wall, a staggering blow. My head must be thicker than I realised, or had I passed out? I tried to move my arms, manacled behind my back now for bloody hours. The handcuffs were still of the ratchet type, tightening automatically if I tried to move. Because of the beating, they were racked up tight to the last notch again, biting into my wrists, a cold, insistent metallic cutting agony. The flesh on the wrists themselves I could feel ballooning up once more over the edges of the cuffs, a two-inch step of swollen flesh, the fingers entirely numb, fluid from the sores flowing out around the steel. My shoulder muscles seized suddenly with the tearing torture of cramp, unbearable, it must be relieved. The arches of my feet were clenched hard with the strain of being on tiptoes, slow rivers of fire burned through the muscle, sinews quivering. Every fibre was shaking now with the impossible effort of maintaining the posture; I had to move, but I knew they were standing, watching, waiting for a twitch, the whip poised, the baton raised. I moved. They beat me to the floor . . .

They dragged me up, walked me around for a little bit to disorientate me, which wasn't difficult, as I was in darkness all the time. Now somebody stood me with my back to another wall in another part of the building – only this time my head was not against the wall, I was standing slightly away from it. I could not see, but I could sense somebody to one side. Suddenly, wham! they smashed my head back hard against the wall. Two seconds later, two hours later, impossible to judge, smash! my head crashed off the wall again. Nobody was asking any questions. Crack! I was thinking, "Ask me something, ask me something, at least let me say something, or hit me so I fall unconscious . . ." Crack! This went on – it could have been thirty times, it could have been a thousand times, I have no idea. Nobody asked me anything after this, nothing at all. Now, dazed, stunned like a chicken before its throat is cut, I was worried, I had lost track of how they were going to interrogate me; this was not going by the book any more. They left me. They put me in a cell with John, for the first time, but we knew they would be listening.

He said, "I've told them something."

"Don't worry, it's no problem."

They took me back out of that room, along a corridor, and attached me to the frame of what was obviously a bare iron bedstead. At the other end of the room I could hear someone being questioned. "Zaun," he said. "My name is Zaun." I wondered who Zaun was. From his accent, he was obviously American. That meant we were not alone. He had to be a flyer like ourselves.

They pushed me down onto the bed, but the agony in my wrists and hands, still ratcheted behind my back, made me catapult straight back upright again.

"What is it?" asked the guard.

"My arms. My arms." He uncuffed me, loosened off the ratchets, and cuffed my wrists back together, but this time in front. This was absolute heaven, once the worst of the pain wore off. Utterly exhausted, I curled up on the bedsprings and fell into a dead sleep. It was daylight when I awoke. The guard undid my blindfold. I looked around. I was in what we came to call "the dormitory". Several other beds, some with mattresses, were dotted around.

"It's OK," said the guard, "I'm a friend. How are you?" Thinking this was another part of the interrogation, or some sort of trick, I shook my head.

"No," he said. "Look, I'm trying to help you . . ." He brought out some cigarettes from an inside pocket.

"Do you need to see a doctor, or anything?" he asked.

At this point, I desperately, desperately wanted to say something to this guy, whom I christened "Ahmed". Though horribly scarred, his was the first friendly face in what felt already like a long age, a potential soft patch in a particularly unpleasant experience. His was the first friendly voice. But still, at the back of my mind, was the idea that he might be a more subtle form of quiz-master. He was very good though. He brought me some cold lentil soup, and I drank a tiny amount; but despite not having touched anything for twenty-four hours, I was just not interested in food; there were other things to think about.

"Where do you come from?" he asked.

"I cannot answer that question."

The heavy brigade suddenly came in; Ahmed had my blindfold back on just in time. They began interrogating me there and then. "Stand up!" Punch, kick. I blocked them with the usual response. They became furious. "You will be sorry, Nichol, you will be sorry." A few more blows. They left.

Once they had gone, Ahmed chained me back to the bed by one wrist and took off the blindfold again. My interrogators left me there all that day. And all that day, the air raids came in, the jets screamed overhead, the Triple-A mountings on the roof hammered away, bombs crumped and rattled nearby. During one of these raids, Ahmed was in the room. A bomb went off right next to the building we were in. He looked out of the window, and then said, casually, "Someone has just been killed down there," and he pointed. I fell asleep again in the afternoon.

Later, still chained to the bed, a heavy kick in the ribs woke me: "What's your name? Rank? Number?"

"8204846."

"Where did you come from?"

"I cannot answer that question."

"You *will* be sorry, you know that, don't you, Nichol?"

"I cannot answer that question."

"We will come back for you soon. You will be sorry."

I know I'm going to be sorry. I'm already bloody sorry.

Thè psychological terror, the psychological torture, is just as great as the physical torture. You are shit-scared. You desperately want him to come and get you, as soon as possible, to get it over with, so that you can break, so that you can tell him something. But you haven't suffered enough yet, they haven't done enough to you yet to tell them anything. You haven't suffered enough to let anyone down, you haven't suffered enough to let yourself down; but you *do* want it to be over, to reach an end.

In the evening they came for me. They unshackled me, put the blindfold on, hauled me upright, dragged me down the stairs, round the streets, back into the interrogation centre; the familiar journey, almost routine by now. The chair: they threw me down into it. One guy was holding my arm on one side, one on the other. I knew in my heart of hearts that this was the time, I knew that it was going to get really rough now.

I was sitting with the solid fist of my own fear in my stomach.

"What squadron are you from?"

"I cannot answer that . . ."

Bang! somebody punched me in the face. Blood came pouring out of my face onto my lap, dripping. I could feel it warm on my thighs. On my lower half, I was wearing a flying-suit, a chemical-warfare suit, long-johns underneath all that, but I could still feel the blood dripping warm onto the upper part of my legs. Someone was hitting me in the face, over and over again. Question. Then somebody standing just to one side hit me hard across the skull with a solid piece of wood. Thwack! My head rang to the blow like some kind of bell. There were brilliant aching lights flashing behind the blindfold. You really do see stars. I was in the middle of the Milky Way. Question.

"I cannot answer . . ." A kick in the stomach – how he got to my stomach I don't know, they were still holding me down on the chair. I fell over to one side in the chair, my gorge rising; they dragged me up by the hair. Question.

"I cannot . . ." Whack! Someone punched me again, someone hit me with the wood, dazzling bright lights and the sudden downward spiral into blackness. Now I was disorientated, my brain was really starting to shut down, but still I thought, "It's going to take more than this, it's going to take more than this. I'm not breaking down without good cause." Somebody dragged my boot off, tearing it away with a furious wrench. "What on earth? What are they going to do to me now?" Whack! A plastic pipe filled with something hard hit me across the shins. A biting agony across the shins, on and on, biting. Question.

And now, somebody grabs the hair at the nape of my neck,

and begins stuffing tissue-paper down the back of my T-shirt. That is appalling. This is terrifying now. I am sitting in a darkened room in the middle of enemy territory, and somebody has just stuffed tissue-paper down the back of my neck. "What are they doing that for?" I know straightaway what they are doing that for, I can imagine only too well. "Shit, they are going to set me on fire!" Now I really want him to ask me another question, I *am* sorry, I want to say something, I want to tell him something, anything. But he doesn't ask me a question. He just sets fire to the paper.

I throw my head violently from side to side, to try to escape from the burning, to try to shake the tissue-paper clear of my neck. They are still whacking my shins. Quite soon, mercifully soon, somebody behind me slaps out the flames.

"What squadron are you from?"

"Fifteen."

I had had enough.

Anonymous Wehrmacht Soldiers

Last Post

The letters below are from the last German post from Stalingrad before the surrender on 31 January 1943. They, along with hundreds of others, never reached their destinations, being impounded – on the Führer's orders – to find what they revealed about Wehrmacht morale. Effectively there wasn't any: only 2.1 per cent of the letters approved of the war; 3.4 per cent were vengefully opposed, and the rest were indifferent, doubtful or sceptical.

Once again I have held your picture in my hand. As I gaze at it my mind was filled with the memory of what we shared together on that glorious summer evening in the last year of peace, as we approached our house through the valley of flowers. The first time we found each other it was only the voice of our hearts that spoke; later came the voice of love and happiness. We talked of ourselves and the future that stretched out before us like a gaily coloured carpet.

That carpet is no more. The summer evening is no more; nor is the valley of flowers. And we are no longer together. Instead of the gaily coloured carpet there is an endless field of whiteness; there is no longer any summer, only winter, and there is no longer any future – not for me, at all events, and thus not for you either. All this time I have had a strange sensation which I could not explain, but today I know that it was fear for you. Over those many thousands of miles I was conscious that you felt the same about me. When you get this letter, listen very hard as you read it; perhaps you will hear my voice. We are told that we are fighting this battle for Germany, but only very few of us here believe that our senseless sacrifice can be of any avail to the homeland.

. . . So now you know that I am not coming back. Break it gently to Mother and Father. It has given me a terrible shock and the worst possible doubts about everything. Once I was strong and believed; now I am small and unbelieving. Much of what is going on here I shall never know about; but even the little bit I am in on is too much to stomach. Nobody can tell me that my comrades died with words like "Germany" or "Heil Hitler!" on their lips. It cannot be denied that men are dying; but the last word a man speaks goes out to his mother or the person he loves most, or else it is merely a cry for help. I have already seen hundreds fall and die, and many, like myself, were in the Hitler Youth. But all those who could still do so shouted for help or called out the name of someone who could not really do anything for them.

The Führer has solemnly promised to get us out of here. This has been read out to us, and we all firmly believed it. I still believe it today, because I simply must believe in something. If it isn't true, what is there left for me to believe in? I would have no more use for the spring and the summer or any of the things that make life happy. Let me go on believing, dear Grete; all my life – or eight years of it, at least – I have believed in the Führer and taken him at his word. It's terrible the way people out here are doubting, and so humiliating to hear things one cannot contradict because the facts support them.

If what we were promised is not true, then Germany will be lost, for no other promises can kept after that. Oh, these doubts, these terrible doubts. If only they were already dispelled!

. . . I have now written you twenty-six letters from this accursed city, and you have sent me seventeen replies. Now I shall write just once again, and then no more. There, I have said it at last. I have long wondered how to word this fateful sentence in such a way as to tell you everything without its hurting too much.

I am bidding you farewell because the die has been cast since this morning. I shall entirely disregard the military side of things in this letter; that is purely a concern of the Russians. The only question now is how long we shall hold out: it may be a few days or a few hours. You and I have our life together to look back upon. We have respected and loved one another and waited two years. In a way it's a good thing this interval has elapsed, for though it has intensified our desire to be together again it has also greatly helped to estrange us. The passage of time is also bound to heal the wounds caused by my not returning.

In January you will be twenty-eight, which is still very young for such a pretty woman. I am glad I have been able to pay you this compliment so often. You will miss me a lot, but that is no reason why you should shut yourself off from other human beings. Allow a few months to pass, but no more than that. Gertrud and Claus need a father. Remember that you have to live for the children, and don't make too much song and dance about their father. Children forget quickly, particularly at that age. Take a good look at the man of your choice and pay special heed to his eyes and handshake, just as you did in our own case, and you will not be disappointed. Most of all, bring the children up to be upright men and women who can hold their heads high and look everyone straight in the face. I am writing these lines with a heavy heart – not that you would believe me if I said I found it easy – but don't worry, I am not afraid of what is to come. Always tell yourself – and the children, too, when they are older – that their father was never a coward and that they must never be cowards either.

. . . I was going to write you a long letter, but my thoughts keep disintegrating like those houses under gunfire. I have still ten hours left before this letter must be handed in. Ten hours are a long time when you are waiting; but they are short when you are in love. I am not at all nervous. In fact it has taken the East to make a really healthy man of me. I have long since stopped catching colds and chills; that's the one good thing the war has done. It has bestowed one other thing on me, though – the realization that I love you.

It's strange that one does not start to value things until one is about to lose them. There is a bridge from my heart to yours, spanning all the vastness of distance. Across that bridge I have been used to writing to you about our daily round and the world we live in out here. I wanted to tell you the truth when I came home, and then we would never have spoken of war again. Now you will learn the truth, the last truth, earlier than I intended. And now I can write no more.

There will always be bridges as long as there are shores; all we need is the courage to tread them. One of them now leads to you, the other into eternity – which for me is ultimately the same thing.

Tomorrow morning I shall set foot on the last bridge. That's a literary way of describing death, but you know I always liked to write things differently because of the pleasure words and their sounds gave me. Lend me your hand, so that the way is not too hard.

. . . What a calamity it is that the war had to come! All those beautiful villages laid waste and none of the fields tilled. And the most dreadful thing of all is how many people have died. Now they all lie buried in an enemy land. What a calamity, indeed! Be glad, all the same, that the war is being fought in a distant country and not in our beloved German homeland. That's a place it must never reach, or else the misery will be even worse. You must be really grateful for that and go down on your knees to thank your God. "On the banks of the Volga we stand on guard . . ." For all of you and for our homeland. If we were not here, the Russians would break through and wreck everything. They are very destructive and there are millions of them. They don't seem to care about the cold, but we feel it terribly.

I am lying in a hole in the snow and can only creep away to a cellar for a few hours at nightfall. You have no idea how much good that does me. We are at hand, so you have no need to be afraid. But our numbers get less and less, and if it goes on like this there will soon be no more of us. Germany has plenty of soldiers, though, and they are all fighting for the homeland. All of us want peace to come soon. The main thing is that we win. All keep your fingers crossed!

. . . I am finding this letter hard enough to write, but that is nothing like as hard as you are going to take it! The news it bears is not good news, I am afraid. Nor has it been improved by the ten days I waited. Our situation is now so bad that there is talk of our soon being entirely cut off from the outer world. A short while back we were assured that this post would go off quite safely, and if only I knew there would still be another opportunity to write I should wait a little longer. But that is just what I don't know, and for better or worse I must get this off my chest.

The war is over for me. I am in a field hospital in Gumrak waiting to be evacuated by air. Much as I long to get away, the deadline keeps being put off. My home-coming will be a great joy to us both, but the state in which I come will give you no cause for joy. It makes me quite desperate to think of lying before you as a cripple. But you must know sooner or later that both my legs have been shot off. I am going to be quite honest with you. My right leg is completely smashed and amputated below the knee; the left one has been taken off at the thigh. The medical officer thinks that with artificial limbs I should be able to run around like any normal person. The M.O. is a good man and means well. I hope he turns out to be right. Now you know it in advance. Dear Elise, if only I knew what you are thinking. I think of nothing else and have all day long to do it. And you are very much in my thoughts. Time and again I have wished I was dead, but that is a grave sin and does not bear mentioning.

. . . If there is a God, you told me in your last letter, He will bring me back to you safe and soon. And, you went on, God will always give His protection to a man like myself – a man who loves flowers and animals, has never done wrong to anybody, and is devoted to his wife and child.

I thank you for those words: I always carry the letter next to my heart. But, my dearest, if one weighs your words, and if you make God's existence dependent on them, you are faced with a terribly grave decision. I am a religious man, and you were always a believer. Now all that will have to change if we both draw the logical conclusions from our attitudes to date, for something has intervened which destroys everything we believed in. I am looking for the right words in which to say it. Or have you already guessed what I mean? There seemed to me to be such an odd tone about your last letter of 8th December. It's now the middle of January.

For a long time to come, perhaps for ever, this is to be my last letter. A comrade who has to go to the airfield is taking it along with him, as the last machine to leave the pocket is taking off tomorrow morning. The situation has become quite untenable. The Russians are only two miles from the last spot from which aircraft can operate, and when that's gone not even a mouse will get out, to say nothing of me. Admittedly several hundered thousand others won't escape either, but it's precious little consolation to share one's own destruction with other men.

Franciszek Radziszowski

Is Killing a Pleasure?

Radziszowski was an officer of the reserve army in the Polish Army during the German invasion of 1939.

During the last few months of 1943 several articles have appeared in the Press dealing with personal experiences of snipers and guerrilla fighters. Among them was one entitled "I like killing Germans", by Lieutenant Ludmila Pavlichenko, a Russian woman sniper, who claims to have shot over 300 Germans and enjoyed it. Now I am definitely sure that if you take ten guerrillas and asked them whether they like killing their enemies, nine would reply negatively. Because, in spite of whatever is said about human nature and its lust for killing and destroying, a normal person cannot enjoy taking life. One does it when in danger or in emotion, but even then it's far from being a pleasure. This opinion is based not only on conversations with those who spent many months in guerrilla fighting and skirmishes in Poland, but also on my own experiences.

My first "kill" happened on 4 September 1939. Being an officer of the reserve and mobilised, I was then in command of a small signal unit, attached to one of our armies near Rzeszow in southern Poland. Soon after an air raid warning, word came that enemy parachutists had been dropped not far from our positions. They were all rounded up within two hours, all except one, whose whereabouts nobody could ascertain. In half an hour he killed all four men operating the radio station and kept firing from time to time at anybody who showed as much as a finger. By a lucky chance, one of my men saw the German fire from a barn but was killed by his next shot. Lying in a ditch I was able to crawl the distance unseen, creep around and enter the barn from the other end. In the first moment, I couldn't see him, but his fire gave him away. Lying comfortably in a pile of hay, he fired through the spaces between logs without showing so much as an inch of his muzzle outside. He didn't see or hear me, so I fired two shots at him. He turned round, into a half-sitting position, taking aim at me, so I fired once more. He fell on his back, yelled something and then rolled to the ground. I went away without looking at him. I couldn't. He was the first man I killed.

During the next ten days, we were driven very hard by German armoured columns and infantry and had to retreat eastward. Several times we were cut off and had to shoot our way out. Man for man, every one of us had to fight and kill – for his own life.

I had no thoughts on that subject at the time. I just picked one of them, kept him in the sights and waited till he came near enough for me to see his face, and then fired somewhere into his middle. Approaching with a gun in his hand and in a bent forward position, he usually fell on his face. And all I could feel at such moments was being glad that it wasn't I lying there in the dirt. Even now, after nearly four years. I feel no emotion when I think about the Germans I shot in the fields of southern Poland. Neither do I feel proud.

But these, and others, when I was in the defence of Lwow, were kills at a distance. Not a great distance, but all I had to do was to pull the trigger. It's quite different when you feel the flesh of your enemy under your fingers or knife! This happened later.

After nearly two weeks of incessant bombing and shelling of Lwow, having no more ammunition and no hope, we decided to have our last go at the Germans. Eight of us, with grenades, knives and revolvers, went out in an attempt to get the crew of a German mortar battery. But we ran into ambush before we managed to do anything. Knocked against a tree by the explosion of several grenades thrown at us simultaneously, I awoke to find myself badly shaken and in company of three other of my men, all wounded, and under German guardianship. The rest were killed. I was taken to a nearby village of Skolimow and next morning put before a German court-martial and sentenced to be shot. The sentence was, however, postponed for eight days, during which I was cross-examined, and it was decided that I should be shot on the Monday morning.

There were 18 of us locked in a barn, around which sentries were posted. We could see them through spaces between logs.

We noticed, during the week, that as the nights grew old, sentries became rather quiet, perhaps fell asleep. We decided to escape.

In the straw on which we lay, we found a bayonet, left there perhaps by some soldier in the earlier days of war. With this, a stick and our hands, we dug beneath the wall, as quietly as possible, until we had a hole big enough for a man to slip through. Trying to steady my legs which were trembling, I crawled in and put my head out.

It wasn't a dark night. A piece of moon was peeping through the leaves of a willow. About three yards away, sitting with a rifle across his legs and a big revolver in his hand, was the sentry – apparently sleeping. I pulled myself out, but broke the edges of the hole and the sound of falling grit woke him. Before he realised what was happening. I was on top of him and at his throat.

It was an awful job. I don't know why, or how, but my mind worked clearly all the time. I could feel the muscles in his throat moving under my fingers and saw deadly fear in the whites of his eyes. On my face, I felt a few bursts of his breath mixed with saliva and I knew that I must kill him, choke him to death, if I was to be free and alive. So I put all my strength into the grip, but he still struggled. Then all at once he gave a shiver and relaxed. It was the bayonet which the man after me had put into him. I released my grip and he fell to the ground, blood pouring from his mouth. We took to the woods.

Hearing no search after us, we waited all day till the next night came and then, cautiously, avoiding roads and villages, headed north. In a few days we reached Tarnawatka, a village some 60 kilometres north, where from a peasant we learned that there was a group of Polish guerrillas under Colonel Tomaszewski operating in the forests. He himself was going to join them and was willing to act as a guide.

For three weeks, in small groups, we played hide and seek with the Germans, ambushed their patrols, built obstructions for their cars, shot the occupants, made raids on ammunition stores and slept in forests, barns or ditches.

And again, it was a question of patience, steady nerves and hands. There was time to think about it, too. Many a time, lying with a gun near a path down which a German patrol was bound to come, I thought, within half an hour or more, I shall probably kill a man, the same kind of being as I am, but whose ideas on living are different. A man, who with thousands of others like him, invaded our country, killed so many of our people, and would kill me too, given a chance, and that's why I'm going to kill him, though I should much prefer another solution. When the time came, I fired and killed – but I didn't like it.

But in one instance, I was very glad and rather happy to have shot a German. Not because if I hadn't I would have been shot myself – that realisation came later – but because it was a hard job. And though it may sound trivial to use that expression in the serious business of killing, it turned into a sort of sporting match between us two – and I won.

In Lublin Voyevodship we got the news that a German supply column would pass south. This being a much wooded part of the country, we had no difficulty in preparing an ambush on the road. Several trees were partly sawn, so as to fall on the road after a slight pull of the ropes or a hand grenade explosion. We took our positions, let the motorcyclists through for our men down the road to take care of and waited for the convoy about half a mile behind.

I was given a look-out job at the end from which the Germans were coming, to raise alarm if help should be coming for them and to stop those who tried to run back.

A small convoy appeared. Only two of the trees fell on the road but that was enough to stop them. The motorcyclists, I was told later, were the first to be killed, and the rest in confusion fired in every direction, ran into the woods straight into our guns. Two ran back, one fell but the other kept on towards me. And though it was only 50 to 60 yards – missed. He saw me and before I could fire again, jumped into a ditch. From there he threw two grenades. I replied with shots whenever he showed an inch above the ground, but missed each time, being unable to take good aim. Four of his bullets came uncomfortably near, throwing earth into my face.

I realised that the fight with the convoy was over, heard my group moving away, but couldn't withdraw myself; neither could he. We had to shoot it out.

I called out to him in German, but all I got back was a dirty name and another grenade which fell short. I knew I couldn't wait there long, for some of them would be bound to come down the road soon. I tried to shoot off, and upon him, a telegraph pole, but had to give it up as requiring too many rounds.

There was some dry grass in my hole, a few dry sticks and

plenty of leaves. The breeze was in the right direction. I made a small bunch of sticks, leaves and grass, set fire to it and threw it into the leaves ahead of me. This brought more shots from the German, but before he realised what was happening, hidden by smoke, I was out of my hole and behind a big pile of cut wood that gave me a good aiming position. I got him with my next shot and then went over to see him. He was about 23, ginger-haired, good-looking. The bullet had entered his right shoulder, passed through his chest and torn a big piece out of his left side. He had a surprised expression.

It took me two days to rejoin my group who had given me up for dead.

In another encounter, a bullet grazed the top of my head. Nothing serious but enough to lose consciousness and get into German hands again. Why they did not shoot me on the spot I never learned. I was sent by car to Krasnystaw, and there included in a batch of prisoners which was marched off to Lublin.

We arrived on a dark night and had to wait several hours in the market place. It happened that I sat on a grating over a sewer. It was dark, sentries were posted at the entrances to the market place, so without being seen, except by prisoners, two of us lifted the cover, went in and the others replaced it.

After about an hour, the prisoners were marched off. We waited for a while, then with some difficulty lifted the cover and got out. We were unable to stand owing to the cramp which paralysed our muscles, so we crawled and rolled, without being seen, into the nearest yard. There we lay massaging our legs, and before morning, disappeared into the woods again.

I had my family in Warsaw, the other man a sister in Gydnia, so we decided to go and see them, having heard nothing from them since war started. In Warsaw, during the next few weeks, hunted by Gestapo and seeing how they treated my people, I learned to hate them as only a Pole can hate a German.

There was one I am sorry to have killed. It was early spring, 1940. I was trying to get out of Poland to join the Polish Army in France. With a friend, I decided to cross the frontier to Hungary. To save our legs from crossing the Carpathian Mountains, we smuggled ourselves on a freight car. Next morning, the sun came up, and tired of sitting in a cramped position, we pulled aside the canvas cover to stretch and get some warmth. The German guard saw us. He didn't fire, just walked along the planks to our truck and started threatening us with handcuffs. We pushed him over the side and he fell on the buffers. There he hung, yelling, so I had to kick his hands off. He fell under the wheels and was crushed. I can still remember the look on his face. I hate it.

Now, from England, I am engaged in a peaceful job of bringing food and guns across the Atlantic. But often at sea, when I recall the bloody times in Poland and my experiences there, I am again sure that a normal man or woman will never enjoy killing another. And if what Ludmila Pavlichenko said of herself was true, it is not true of all of us.

Sources & Acknowledgements

The editor has made every effort to secure permission for the use of extracts appearing in this anthology. If any omission have been made, it will of course be corrected in future editions and appropriate recompense made. In the case of any queries please contact the editor c/o the publishers.

O'Brien, Tim, "FNG" is an extract from *If I Die in a Combat Zone*, Calder & Boyars, 1973. Copyright © 1973, Tim O'Brien.

Douglas, Keith, "First Action" is an extract from *Alamein to Zem Zem*, Oxford University Press, 1979. Copyright © Marie J. Douglas, 1966, 1979.

Lawrence, T.E., "Guerrilla Attack on a Turkish Outpost" is an extract from *Revolt in the Desert*, George H. Doran Co., 1927.

Grattan, William, "The Storming of Ciudad Rodrigo" is an excerpt from *Adventures with the Connaught Rangers, 1809–1814*, Charles Oman (ed.), Edward Arnold, 1902.

Churchill, Winston, "The 21st Lancers at Omdurman" is an extract from *A Roving Commission*, Scribners 1930. Copyright © 1930, 1939, Winston S. Churchill.

Lewis, Lloyd, "Bloody Shiloh" is an extract from *Sherman: Fighting Prophet*, Harcourt, Brace & Co., 1932.

Southey, Robert, "Nelson at the Nile" is an extract from *The Life of Nelson*, John Murray, 1913.

Gibson, Guy, "Enemy Coast Ahead" is an extract from *Enemy Coast Ahead*, Michael Joseph, 1946. Copyright (1946), the estate of Guy Gibson.

Rudel, Hans, "Stukas Dive-Bomb the Soviet Fleet", quoted in *Bombs Away!*, Stanley M. Ulanoff (ed.), Doubleday & Company Inc, 1971.

Berent, Mark E., "Night Mission over Nam" is an extract from USAF *Air Force/Space Digest*, 1971.

Neame, Philip, "2 Para at Goose Green" is an extract from *Above All, Courage*, Max Arthur, Sidgwick & Jackson, 1985. Copyright 1985, Max Arthur.

Fuchida, Mitsuo, "Tora! Tora! Tora!", quoted in *Bombs Away!*, Stanley M Ulanoff (ed.), Doubleday & Company Inc, 1971.

Livy, "Hannibal at Cannae", quoted in Ernest Hemingway, *Men at War*, Crown Publishers, Inc., 1942.

Creasey, Edward, "The Battle that Saved the West" is an extract from *The Fifteen Decisive Battles of the World*, Richard Bentley & Son., 1880.

Parker, John, "Not Men But Devils" is an extract from *Inside the Foreign Legion*, Judy Piatkus (Publishers) Ltd., 1998. Copyright © 1998 John Parker.

Williamson, Gordon, "The Short Life and Many Kills of Michael Wittmann" is an extract from *Aces of the Reich*, Arms & Armour, 1989. Copyright © 1989 Gordon Williamson.

Sasson, Siegfried, "Trench Raid at Mametz Wood", is an extract from *Diaries, 1915–18*, Faber, 1983. Copyright © 1983 the estate of Siegfried Sassoon.

Farwell, Byron, "Rorke's Drift" is an extract from *Queen Victoria's Little Wars*, Allen Lane, 1973. Copyright © 1973 Byron Farwell.

McCudden, James, "Death of an Ace" is an extract from *Five Years in the Royal Flying Corps*, Aeroplane & General Publishing, 1940.

Munro, Ross, "The Canadians at Dieppe" is an extract from *Gauntlet to Overlord*, Macmillan, 1945.

Xenophon. "The March of the Ten Thousand to the Sea", quoted in Ernest Hemingway, *Men at War*, Crown Publishers, Inc., 1942.

Saunders, Hilary St George, "The Drop" is an extract from *Red Beret*, Michel Joseph, 1950. Copyright © 1950 Hilary St George Saunders.

Deane-Drummond, Anthony, "Return Ticket" is an extract from *Return Ticket*, Collins, 1967.

Poolman, Kenneth, "Zepp Sunday" is an extract from *Zeppelins Over England*, White Lion, 1971.

Roosevelt, Theodore, "The Rough Riders at Santiago" is an extract from *The Rough Riders*, Kegan, Paul, Trench, Trubner & Co., 1899.

Virgil, "The Trojan Horse" is an extract from the *Aeneid*, quoted in Ernest Hemingway, *Men at War*, Crown Publishers, Inc., 1942.

Waldron, T.J. & James Gleason, "The Frogmen" is an extract from *The Frogmen*, The Elmfield Press, 1974. Copyright © 1950 Evans Bros Ltd.

Moss, W. Stanley, "The Kidnapping of General Kreipe" is an extract from *Ill Met By Moonlight*, Harrap, 1950. Copyright © 1950 W. Stanley Moss.

Marbot, Baron de, "Lisette" is an extract from *Memoirs*, trans A.J. Butler, Longmans Green & Co., 1893.

Lejeune, Louis-Francois, "1812: The Retreat from Moscow" is an extract from *The Memoirs of Baron Lejeune*, trans & ed. Mrs Arthur Bell, Longmans, Green & Co., 1897.

Prescott, William H., "The Death of Montezuma" is an extract from *The Conquest of Mexico*, University of Chicago Press, 1966.

Hoss, Rudolph, "I, The Commandant of Auschwitz", is an extract from *Commandant of Auschwitz: The Autobiography of Rudolph Hoss*, trans Constantine FitzGibbon, Pan, 1961. Copyright © Comite International des Camps.

Russell, Lord, "The Bataan Death March" is an extract from *The Knights of Bushido*, Cassell, 1958. Copyright © 1958 Lord Russell of Liverpool.

Frank, Anne, "Diary of a Jewish Girl in Hiding" is an extract from *The Diary of a Young Girl*, Vallentine, Mitchell, 1952.

Black Elk, "Wounded Knee" is an extract from *Black Elk Speaks*, John G Neihardt (ed.) University of Nebraska Press, 1972. Copyright © 1961 University of Nebraska Press. Reprinted by permission.

Tomalin, Nicholas, "Killing Cong" (originally "Zapping Charlie Kong") is from the *Sunday Times*, 5 June 1966. Reprinted by permission of News International.

Fuhrman, Claus, "Götterdämmerung: The Fall of Berlin", quoted in *The Faber Book of Reportage*, John Carey (ed.), Faber, 1989.

Robertson, Terence, "The Hunting of *U-99*" is an except from *The Golden Horseshoe*, Evans Bros, 1955.

Cochrane, Thomas, "The Cruise of the *Speedy*" is an extract from *Adventures Afloat*, Thomas Nelson & Sons, 1907.

Gallager, O.D., "The Loss of the *Repulse*" is from the *Daily Express*, 12 December 1941.

Mitsuru, Yoshida, "Requiem for *Yamato*" is an extract from *Requiem for Battleship* Yamato, BCA, 1999. Copyright © 1985 University of Washington Press. Reprinted by permission of Constable & Robinson.

William of Poitiers, "Hastings" is an extract from "The Deeds of William, Duke of the Normans and King of the English" in *English Historical Documents 1042–1189*, trans D.C. Douglas and G.W. Greenaway, Eyre and Spottiswoode, 1953.

Caesar, Julius, "The Defeat of the Nervii" is an extract from *The Conquest of Gaul*, trans S.A. Handford, Penguin, 1951. Copyright © 1951 S.A. Handford.

Froissart, John, "English Archers at Crécy" is an extract from *Chronicle*, trans Thomas Johnes, 1803–4.

Bowlby, Alex, "Officer Selection" is an extract from *The Recollections of Rifleman Bowlby*, Corgi, 1971.

Hall, Roger, "Finest Hour" is an extract from *Clouds of Fear*, Bailey Brothers and Swinfen Ltd, 1975. Copyright © 1975 R.M.D. Hall.

Tactitus, Cornelius, "Boudicca's Revolt" is an extract *from The Annals of Imperial Rome*, trans Michael Grant, Penguin, 1956. Trans copyright ©1956, 1971 Michael Grant.

Evans, Geoffrey, "Admin Box" is an extract from *The Desert and the Jungle*, Corgi, 1961.

Xenophon, "The Siege of Babylon" is an extract from *The Education of Cyrus or Cyropaedia*, trans Henry Graham Dakyns, Dent, 1992.

Skorzeny, Otto, "The Seizure of Budapest by SS Commandos" is an extract from *Skorzeny's Special Missions*, Robert Hale, 1957.

Hurd, Sir Archibald, "The Blocking of Zeebrugge" is an extract from *Sons of the Admiralty*, Constable, 1919.

Mason, Robert, "Chickenhawk" is an extract from *Chickenhawk*, Corgi, 1984. Copyright © 1983 Robert Mason.

Richards, Frank, "Christmas in the Trenches 1914" is an extract from *Old Soldiers Never Die*, Faber & Faber, 1933.

Hillary, Richard, "Burned" is an extract from *The Last Enemy*, Richard Hillary, Macmillan & Co., 1943.

Nicol, John, "Interrogation" is an extract from *Tornado Down*, John Peters and John Nicol, Michael Joseph, 1992. Copyright © 1992 John Peters and John Nicol. Reproduced by permission of Penguin UK.

Anonymous Wehrmacht Soldiers, "Last Post" is an extract from *Last Letters from Stalingrad*, Methuen, 1956.

Radziszowski, Franciszek, "Is Killing a Pleasure?" is from *New Statesman*, 1 January 1944.